Time Travelers

**Time Travelers:
Victorian Encounters with Time and History,
with a Foreword by Mary Beard**

Edited by
Adelene Buckland and Sadiah Qureshi

The University of Chicago Press :: Chicago and London

The University of Chicago Press, Chicago 60637

The University of Chicago Press, Ltd., London
© 2020 by The University of Chicago
All rights reserved. No part of this book may be used or reproduced in any manner whatsoever without written permission, except in the case of brief quotations in critical articles and reviews. For more information, contact the University of Chicago Press, 1427 E. 60th St., Chicago, IL 60637.
Published 2020
Printed in the United States of America

29 28 27 26 25 24 23 22 21 20   1 2 3 4 5

ISBN-13: 978-0-226-67665-4 (cloth)
ISBN-13: 978-0-226-67679-1 (paper)
ISBN-13: 978-0-226-67682-1 (e-book)
DOI: https://doi.org/10.7208/chicago/9780226676821.001.0001

Library of Congress Cataloging-in-Publication Data

Names: Buckland, Adelene, editor. | Qureshi, Sadiah, editor. | Beard, Mary, 1955– writer of foreword.
Title: Time travelers : Victorian encounters with time and history / edited by Adelene Buckland and Sadiah Qureshi ; with a foreword by Mary Beard.
Description: Chicago : University of Chicago Press, 2020. | Includes bibliographical references and index.
Identifiers: LCCN 2019045162 | ISBN 9780226676654 (cloth) | ISBN 9780226676791 (paperback) | ISBN 9780226676821 (ebook)
Subjects: LCSH: History—Philosophy—History—19th century. | Archaeology and history.
Classification: LCC D16.8 .T486 2020 | DDC 907.2/041—dc23
LC record available at https://lccn.loc.gov/2019045162

♾ This paper meets the requirements of ANSI/NISO Z39.48-1992 (Permanence of Paper).

For Alice Eleanor
and
For Sufyan Qureshi

Contents

	Foreword	ix
	Mary Beard	
	Introduction	xiii
	Adelene Buckland	

Part One: Narratives
1	Looking to Our Ancestors	3
	Sadiah Qureshi	
2	Looking Around the World	24
	Peter Mandler	
3	The World Beneath Our Feet	42
	Adelene Buckland	

Part Two: Origins
4	Ad Fontes	67
	Simon Goldhill	
5	In the Beginning	86
	Helen Brookman	
6	Under False Pretenses	107
	Astrid Swenson	
7	Through the Proscenium Arch	126
	Rachel Bryant Davies	

Part Three: Time in Transit

8	On Pilgrimage	155
	Michael Ledger-Lomas	
9	Across the Divide	176
	David Gange	
10	At Sea	196
	Clare Pettitt	

Part Four: Unfinished Business

11	Looking Forward	223
	Jocelyn Paul Betts	
12	How We Got Here	242
	Daniel C. S. Wilson	

Acknowledgments	261
List of Contributors	263
List of Illustrations	265
Index	267

Foreword

Mary Beard

One of the most memorable Victorian images of London is Gustav Doré's engraving, published in 1872, of "a New Zealander" sitting on the banks of the river Thames, at some unspecified time in the future, sketching the ruins of the city. He has chosen the spot where London Bridge had once stood, the recognizable remnants of St. Paul's Cathedral still dominating the skyline. It was (and still is) a pointed reminder of the fragility of cultural and political hierarchies. We are asked to imagine that London's grandeur and power have gone. The curious visitor from what was once an imperial colony is here busy converting the ruins of the metropolis into his own *cultural* capital. It is not so different from what British aesthetes, antiquarians, and assorted mi'lords had done with the ruins of ancient Rome itself (they had, as it were, *sketched* them into submission); but the tables are now turned. And the scene knowingly chimes with the predictions made by some ancient Romans 2,000 years earlier about the fate of their own empire. When, in one of the most notorious cases of imperial brutality, Rome's armies destroyed the city of Carthage in 146 BCE, an eyewitness caught the Roman commander weeping at the sight of his awful handiwork—and asked him what caused the tears. It was,

FOREWORD

FIGURE FW.1. Gustave Doré, "The New Zealander," in *London: A Pilgrimage*, by Gustave Doré and Blancahrd Jerrold (London: 1972), orig. pub. in *Harper's Weekly Supplement*, May 31, 1873, 473. Metropolitan Museum of Art, Harris Brisbane Dick Fund, 1928.

he made clear, because he knew that Rome itself would one day suffer the same fate. Here, exactly that has happened to the British Empire.

The "New Zealander" was, in fact, a cliché in nineteenth-century media. The direct inspiration for the image goes back to a review of von Ranke's *History of the Popes* in the *Edinburgh Review* of October 1840, by Thomas Babington Macaulay—who ended his critique of the book by speculating that, for all its faults, the Catholic church might still be thriving "when some traveller from New Zealand shall, in the midst of a vast solitude, take his stand on a broken arch of London Bridge to sketch the ruins of St. Paul's." Macaulay may well have been surprised by how popular variations on this theme became. Within ten years Dante Gabriel Rossetti had written a poem, *The Burden of Nineveh*, in which future archaeologists from the Southern hemisphere (Australia this time, not New Zealand) had dug up the Assyrian sculptures from the ruins of the British Museum and, assuming that they had found the cult objects of the "natives," had taken them back home. Already in January 1865, several years before Doré's engraving, *Punch* satirized the poor New Zealander as a dreadful journalistic stereotype that had outlived its usefulness: "The retirement of this veteran is indispensible. He can no longer be suffered to impede the traffic over London Bridge. Much wanted at the present time in his own country. May return when London is in ruins."

It is partly because "this veteran" was such a cliché of Victorian views of both past and future that we chose the Doré engraving as the logo of the Cambridge Victorian Studies Group, whose work is showcased in this volume. The group was established in 2006, thanks to generous funding from the Leverhulme Trust, for a five-year project entitled "Past vs Present in Victorian Britain: Abandoning the Past in an Age of Progress." Over its lifetime, this involved five principal coordinators, eight postdoctoral research fellows, and three doctoral students, almost all of whom are represented here—not to mention excellent administrators and other kinds of essential backup, largely provided by the Faculty of Classics in Cambridge.

Doré's image spoke to our concerns, about the edgy boundary between past and present, nostalgia and progress. The five coordinators (Simon Goldhill, Peter Mandler, Clare Pettitt, Jim Secord, and myself) originally came together out of a shared sense of curiosity. All our work had at least one foot in the Victorian period and in its engagement with the past (from fossils and novels to the dissection of biblical or classical texts and the growth of archaeological societies). But we had never had the opportunity, even in an interdisciplinary university, of bringing our interests together. We had not found the space to reflect on the links (or not) between different forms of the rediscovery of the past—or the relationship, as the title of the project hints, between radical notions of progress in the British nineteenth century and radically new ways of analyzing the many pasts that the period faced, discovered, invented, or discarded. Increasingly we became concerned (in a way that prefigures more recent questions about "decolonizing" the curriculum) with how "our" curriculum had been established in the nineteenth century and with the challenge and fun of trying to shine an interdisciplinary light onto the period in which the now-conventional academic disciplines were still being defined. And, of course, as the New Zealander insisted, we had repeatedly to face the ways that so many of the systems of knowledge we explored were embedded in the politics of empire and theories of ethnicity.

It is hard to assess impartially the success of a project of which one has been a part. I am proud to point to the publications that came out of the Cambridge Victorian Studies Group (as well as this volume, you can find many listed in its bibliography, on subjects as diverse as displayed peoples, Victorian epic poetry, school textbooks or "heritage management"). I am proud to reflect on how successful the postdocs and PhD students have been (almost all of whom have gone on from the project to long-term academic jobs). But what I shall always remember most are the talks, discussions, and the opportunities that we had to discover a very practical, day-to-day side of "interdisciplinarity." The Leverhulme Trust is a wonderful organization, committed to open-ended, risky, and not rigidly target-driven inquiry. My most vivid memories will always be those regular lunchtime meetings in which we read together a variety of nineteenth-century texts. I cherish those occasions when the classicists in the room would (rightly) point out that the author in question was actually quoting the Roman poet Horace, and would then for a moment sit back rather smugly—before one of the "proper" Victorianists would add that he was also referring to a famous nineteenth-century divorce case of which (I confess) I had never heard. And so the sharing went on.

This was interdisciplinary scholarship in its most raw, real, and enjoyable form. I hope readers and the Leverhulme Trust (and, let's pretend, that New Zealander too) enjoy the results we offer here.

Introduction

Adelene Buckland

The Victorians, perhaps more than any Britons before them, were diggers and sifters of the past. At the beginning of the nineteenth century—in England, if not in Scotland—classics, mathematics, and the Bible were the staples of an educational curriculum specifically geared to train a gentlemanly elite; by its end, new subjects were open to increasing numbers of new (sometimes female) students, from history and literature to geology and archaeology. Neither was this historical interest confined solely to formal education. Digging for the railways and clambering up cliff faces, the Victorians discovered many of the hideous primeval monsters we now call the dinosaurs; selling shows and lectures to an increasingly interested public, they built many of Britain's most important national museums and galleries; in their libraries and in their studies, they deciphered ancient texts at hectic rate, and they traveled far and wide to foreign lands searching for traces of biblical and mythological cities. New pasts emerged in the century, in debates over human origins or in discoveries of new texts, facts and artifacts, and seemingly well-known pasts were shaken up by new tools and methods of scrutiny. If the study of the past had been, in the eighteenth century, the province of a handful of elites, in the nineteenth century new technologies and economic development meant that the past, in all its brilliant detail, was for the first time the property of the many, not the few.

The nineteenth century was not, of course, the first century in which historical understanding had been of deep significance to British culture and identity. But the intensity and range of that century's preoccupations with the past was unprecedented and helped give rise to many recognizable and still-significant intellectual disciplines (geology, archaeology, and evolutionary biology, as well as professional historical and literary studies, for instance). These disciplines were shaped, too, by the growing accessibility of multiple histories to ever-broader social groups: working- and middle-class men, women, and children participated in the construction of new histories in diverse ways. Nonetheless, not everybody was included equally. As Dipesh Chakrabarty has famously put it, "Historicism—and even the modern, European idea of history—one might say, came to non-European peoples in the nineteenth century as somebody's way of saying 'not yet' to somebody else," of consigning other cultures to the "waiting-room" of history, deemed not yet fit to govern, not yet fit to participate in the global economy on equal terms with British and other European countries.[1] It follows that new modes of historical discovery produced and reflected ideas about national identity that often depended for their conceptual unity on what Johannes Fabian has termed "the denial of coevalness" to non-Europeans.[2] We might add that other social groups (including, but not at all limited to women) could also be denied full participation in history-making, or full representation in written histories, on the grounds that they too were not fully modern or were not agents of historical or evolutionary change. Embedded within the very structures of the historical disciplines that took shape in the nineteenth century are a wide range of cultural, racial, and economic prejudices. The ways in which the shape and structure of modern historical disciplines still resemble those that emerged in the nineteenth century have often naturalized and reproduced those prejudices, from the colonial politics of much writing about the Anthropocene to the deeply entrenched (and often illusory) divisions between the arts and sciences.

Time Travelers attempts to unpack these profuse and contradictory Victorian pasts in order to offer a vivid new picture of the Victorian world and its historical obsessions. Until now, historians have tended to claim that, as the Victorians created powerful new technologies that propelled them into a frightening and unknown future, they clung for safety to what they already knew—building shiny new railway terminals in the style of medieval Gothic or throwing new energy into ancient rituals, from Christmas to the Morris dance.[3] But *Time Travelers* shows instead that the past was no safe ground in this period. Voyaging into a bewildering pageant of pasts, Victorians were constantly beset by toil, trouble, excitement, and drama. Just as it is today, the past could be controversial territory into which to venture. On the one hand, archaeologists developed self-consciously "scientific" methods for excavating the past, even as they set their spades to uncovering biblical sites and cities, so that scientific study was motivated by theological catfights and a troubled but resurgent faith. On the other, uncovering worlds for which there was no traditional authority, geologists won readers and converts by quoting epic poets and classical historians in the pages of books that sold as quickly as novels. And just as the Victorians took pleasure in dressing up on stage or in order to travel to historic sites incognito, they also had little compunction in dressing down those pasts they thought too boring, too rude, or too uncertain to be useful in the present.

Importantly, too, the past was a terrain opening up a wide variety of positions and perspectives. Poets, artists, educationalists, travelers, historians, archaeologists, men and women of science, preservationists, architects, novelists, critics, museum-builders, and scholars—far from ignoring one another as they might today—often shared methods, tools, and texts as they struggled to comprehend a myriad of lost and ancient worlds and to invent themselves in relation to these pasts. To a large degree—and particularly in the invention and reinvention of a wide variety of historical disciplines, from evolutionary biology to psychology—people in the nineteenth century created the perceptual frameworks through which we continue to understand the past. There were other modes and techniques, too, that made order and structure out of this profusion. One idea that has lingered long about the Victorians as particular nineteenth-century subjects is that they were nostalgic worshippers of a staid and rosy-tinted image of the past. At the same time, as many historians of the Global South have demonstrated, such disciplines were often built upon ignoring, eliminating, and appropriating other possible forms of historical imagining and experience.[4] In fact, while the repeated derogation of the Victorians as mere nostalgia-mongers might be seen as a critique of their egoistic historiographies, we might instead comprehend it as a critical reflex produced by our inheritance of those histories. It is at once a failure to recognize *both* the profusion and perplexity of the Victorian historical imagination (its uncertainties, risks, hesitations, and doubts), *and* a failure to confront the full extent of colonialist intellectual legacies as they were produced in defiance of those doubts. To put it another way: the Victorians appear as nostalgic only when we uncritically assume the positions of the Eurocentric and colonial forms of history they bequeathed to us: history as nationalism and nostalgia may be understood as coevally produced myths delimiting the sheer range of potential histories in the service of Victorian articulations of power. It follows that we need to be both more cautious and more radical in our responses to those myths, and to pay attention to the profusion, egoism, prolixity, and bewilderment that produced them. We ought to pay attention to the uncertain *process* of forming histories as much as to the myths they produced. And destabilizing the disciplinary forms in which historical knowledge appears marks one possible way of attempting to do this.

In this book, then, we seek to recapture the excitement, controversy—even the dangers—of nineteenth-century encounters with the past, as well as the vigorous creative and intellectual energies they propelled and were propelled by. In each essay, we begin to answer a set of interrelated questions. How, for instance, did new, or at least reformulated, Victorian historical disciplines (archaeology, anthropology, evolutionary biology, Classics, theology, and English literature, for instance), grow out of earlier disciplinary practices, methods, and questions? How did they borrow from one another, or share practices, methods, and questions, as they emerged? What drove this newly voracious appetite for historical inquiry in the nineteenth century? And what kinds of experiences and perspectives were privileged, or otherwise critical, for driving (or feeding) this appetite? What kinds of narratives emerged about the past—and how securely did the Victorians believe in the stories they told? Are historians and critics in the twenty-first century still telling those same stories? What can we learn from recovering the voices, forces, and details that got lost in the process of disciplinarization?

One of our central arguments—that the Victorians helped to invent a vibrant multiplicity of pasts in ever-greater profusion—also presents one of our greatest challenges. Clearly any attempt to offer an overarching account of all the many kinds of Victorian pasts—and of the Victorians' many methods of engaging with and creating them—is doomed to failure. We make no claim to comprehensiveness. Instead, in order to escape conventional views on the Victorians and their pasts, we attempt here to offer a kaleidoscopic rather than an encyclopedic view: each author in the collection explores a characteristically nineteenth-century perspective on the times that went before them. Standing at different vantage points—surveying the vastness of deep time from the position of the educated scholar or the child with a new toy, of the pilgrim or the seafarer, for instance—and setting those different vantage points in juxtaposition, we hope to reveal new and exciting patterns that begin to make some sense of the dizzying proliferation of pasts the Victorians present to us. What, we ask, was it like to be the geologist Charles Lyell, imagining ancient human beings roaming the earth as it emerged from the deep waters inch by inch over millennia? How were pasts imagined from amid the ruins of ancient empires different from those imagined in the hubbub of a New World metropolis? Which pasts would be discarded, and which cherished as foundational?

Since we contend that these decisions gave rise to the invention of the modern disciplines as means of organizing and making sense of the chaos of the past, every essay in this collection runs across and between the disciplines we have inherited from the Victorians to recapture the energetic production (and co-production) of texts, methods, questions, tools, and subjects. We consider Victorians engaging with the past across social groups, institutions, audiences, and literary forms that were themselves changing fast. Our use of perspectival positions as an organizing theme for the book is intended to help us move beyond our disciplinary training and to attempt—however difficult the attempt might be—to see what the past looked or felt like from a series of *particular* positions in the Victorian period, no matter how many different kinds of knowledge or different methods of attaining it those particular positions embodied.

We begin with the narratives the Victorians told about the oldest and deepest of pasts. To that end we look backward, with the Victorians, to the origins of humankind, outward around the world as encounters with past civilizations in other continents transformed the very notion of what it meant to be "civilized," and inward to debates about humankind as they were modified by encounters with an often frustratingly elusive geological past. In each case, and in radically different ways, the encounter with deep time prompted the Victorians to ask soul-searching questions about what it meant to be human: one species or many?; the agents or the subjects of extinction? Were humans a special species for whom the world was created, or bestial beings trapped in bodies that gave them only a limited perspective on a world that had long predated their existence?

That the Victorians posed these questions is nothing new. But the perspectival approach we adopt in this volume helps us to uncover richer Victorian answers to those questions than scholarship has previously allowed. As Sadiah Qureshi forcefully reminds us in her essay on theories of the origins of humanity, for instance, there was "a substantial *proliferation*, not homogenization, of theories of human difference" in the nineteenth century, "because older classifications did not disappear but *competed* with newer ways of cleaving humanity into natural kinds." And if the "arc of disillusionment" is still useful as a model by which to interpret public perceptions of empire in the nineteenth century, nonetheless Peter Mandler reveals in his essay for this collection that this did not apply in all times and in all places. And his point that there were increasingly diverse ideas about the ways in which peoples should be imagined, engaged with, or obliterated speaks to the volume as a whole. There were flourishing modes of engaging with the past increasingly enabled by new technologies, philosophies, and print cultures—new ways of looking at, handling, smelling, digging up, dissecting, archiving, collecting, displaying, exploring, imagining, critiquing, analyzing, and ignoring it.[5] In addition, these increasingly diverse forms of encounter also produced, and were produced by, a proliferation of values, narratives, and ideologies by which the Victorians handled those pasts and imagined their relationships with them. Postcolonial theory has been fundamental in efforts to make sense of the formulations of Victorian encounters with colonized peoples, and as ubiquitous as Victorian evolutionary or progressive narratives of human cultures may seem, for instance, it remains crucial to pay close attention to the unevenness of those narratives and the other stories and judgments Victorians also told and made.[6] Doing so reveals that these stories and judgments are far more various and conflicted, often self-consciously so, than we often imagine. As the Victorians discovered other cultures around the world—often encountering lost cities in the midst of far-flung jungles, evidence of powerful and distant civilizations—they did not always seek to conquer them. Sometimes, they simply lost themselves in the cultures of peoples whose achievements seemed to exceed their own.

In addition, as I show in the third essay in this part, immersion in other times and places brought on serious self-reflection not only about the *history* of humanity, but also about the implications of human belatedness in the story of the earth's past for the comprehension of that past in the first place. The imagination was often (and still is often) posited as a key attribute of the enlightenment claim to human exceptionalism, and the geological imagination—the ability to imagine millions of years that predated human existence—offered potentially powerful proof of the powers of the human mind. Nonetheless, I show that supernatural or magical beings were often deployed by geological writers as tools to aid the human contemplation of geological immensity, in ways that raised questions about the limits of human knowledge of the past.

We continue with part 2, entitled "Origins." We put it in second rather than originary position partly because—as scholars in a variety of disciplines have long attested—the Victorians more frequently revealed the frustratingly phantasmal and inaccessible nature of beginnings than succeeded in inventing or finding new ones. The structure of *Time Travelers* attempts to perform at least something of the disruption to simple narrative patterning that we argue was fundamental to the Victorians' multiple encounters with the past. But we consider origins second, too, because here we move from the exploration of global and universal pasts to the study of particular texts, particular artefacts, and particular locations in detail, emphasizing the multiplicity as well as the imaginary status of Victorian "originality." As such, we examine the Victorians' newly immersive approach to archives and their reinvigorated study of old sources, their translations of ancient and medieval texts, and the problems they faced as they found multiple pasts (archaeological, architectural, geological, textual, human, animal) all buried in a single site. In both the structuring of this part and its attention to detail, then, our perspectival method continues to yield an account of the *proliferation* of Victorian pasts and of Victorian modes of dealing with them. Simon Goldhill's essay "Ad Fontes," concerned with the Victorian impulse to return to original sources and the difficulty of interpreting those "original" biblical and classical texts that turned out so often to exist in multiple, corrected, redacted, or secondary forms, acts as a bridge between the first and second parts. Taking up my discussion of how Victorians articulated a need for imagination in the interpretation of rocks and fossils, Goldhill reminds us that this exploration of deep time offered a quite concrete and deliberate overlap (particularly through the figure of Niebuhr) with growing concerns about the nature of "original" scriptural and classical sources.

Goldhill suggests that the desire to return to the original prompted more and more fictionalized routes by which to *reach* those sources, or increasingly fictionalized versions *of* them. Helen Brookman's essay on the invention of Caedmon, whose nine-line hymn is his only surviving work, as the "father" of English poetry, helps us understand the fabrications required to create stable "origins" in a radically different mode. The irreverence and "giddiness" of many of Goldhill's examples is replaced here by an account of the Victorians' scholarly, deeply and genuinely felt set of tactics for creating these "origins." What for Goldhill is self-conscious "fiction" in the discovery or invention of origins, for Brookman is "creation." The difference is critical. For, while the creation of "Caedmon" and the pursuit of many different kinds of origins and originality in his image required invention and imagination, this was always at the same time a serious and authenticating project. The obviously divine resonance of the word *creation* helps capture a profound ambivalence at the heart of many Victorian encounters with the past: creation here means both imagining, inventing, or fabricating *and* bringing into (real, verifiable) existence. *Making up* the past could be a legitimate and truthful strategy for *making* it, for reproducing the feel of archaic worlds for which the Victorians had only traces of evidence, for getting to know a place and time that was irredeemably lost.

All four essays in part 2 blend interest in both the textual and material traces of the past in ways that get at the heart of one of the key methods of this collection: the emphasis on exploring unusual archival sources to illuminate new areas of Victorian historical experience. Goldhill uses the incompleteness of the strata to consider the imaginative modes deployed in the study of texts; Brookman's analysis of efforts to inscribe Caedmon's memory into the English canon shows that they gave rise to multiple origin stories and invented evidence. And while Astrid Swenson's essay on fakes and forgeries is the first essay to focus mostly on material evidence, it draws on Goldhill and Brookman's discussions of the power of imaginary evocations of the past and the curious mixture of ephemerality and intransigence in its material—even stone—traces. Swenson's fabrications and fictions are more overt and deliberate than the examples in the two previous essays, but they bring out the sheer intellectual energy and exuberance of efforts at fakery, as well as the complicity of many Victorian consumers in their production. If it has often been suggested that the Victorians were anxious about fakery, obsessive in their determinations to expose fraud and discover authenticity, Swenson reveals that it very often did not matter where a thing came from or whether or not it was original. Fakes were often *required* to replace lost or imaginary originals and to evoke the required feelings about the past that were the ultimate object of historians' or consumers' desires.

This, Swenson reveals, was political. Forgery and fakery could enable social climbing, inspire uncertainty about the "origins" not only of artefacts but also their purveyors, and take advantage of increasingly fluid class boundaries. Rachel Bryant-Davies develops this idea by tackling subjects still not regarded with the seriousness they deserve—as equally significant as their classical "originals"—burlesque theater, and children's toys. She reveals them instead to have been fundamental forms of classical encounter in early nineteenth-century Britain. Playing with the past was felt to be democratic, liberating, provocatively racy. What is important here is not the "original" source text and its "adaptation" or "reception" in the modern period, but the fusion of modern and ancient materials, mixed and cut, ignored and reinvented with abandon. New audiences—children, radicals, showmen, and women, among others—not only remade the classics but ignored, plundered, imitated, faked, abused, and poked fun at them as they wished—flagrantly, and with great commercial success. Such a playful mode of engaging with the classical past was always inevitably entangled with "higher" forms of encounter, and Bryant-Davies reveals a much more licentious classical culture in the nineteenth century than we have yet properly acknowledged. Most pertinently, her essay suggests that these texts and objects—whether in plays, parodies, or burlesques ridden with anachronisms—could produce a more immediate, visceral, visual, and therefore *authentic* sense of the classical than the scholarship of university-educated men.

If this is the case, there are many implications for the study of Victorian philosophical, scientific, and literary encounters. Those Victorian realisms literary scholars hold so dear might be newly up for grabs: "authenticity" was less a closely coveted guarantee of truthfulness and reality and more a subject position to be adopted by those whose aims and agendas it suited (and to be disregarded by those for whom fakery, forgery, imagination, creation, or fiction were more useful). Authenticity was valuable to those Victorians for whom it was helpful, and it was only one of a wide variety of possible modes of encounter with the past. It may even have been only as reliable as less seemingly authentic modes at evoking a *sense* of what it *felt* like to belong to the past. "Authentic" texts and objects might also, in some cases, have been hopeless at yielding the information required to understand any given past. They may have been less penetrative tools for exploring historic worlds than carefully reconstructed fakes. At the very least, part 2 reveals that the Victorians were far more open to these many possibilities and far more conscious of the wide range of relations between truth and fiction and their relative usefulness than we are in the habit of giving them credit for.

If our first two parts explore the many-splendored pasts of the Victorians from fixed perspectives as points of view, in part 3, "Time in Transit," we think about "perspectives" as embodying different kinds of approaches. And we move to the global position with which we began, by taking the transitions we have explored throughout the book, from one place to another, or one viewpoint to another, as our subject. Importantly, the writers in part 3 disagree, at least partly, with some of the arguments in part 2. While so far the playfulness, willful fictionality, and burlesque nature of Victorian encounters with "original" sources from the past have been emphasized, Michael Ledger-Lomas develops Brookman's emphasis in relation to deep faith, considering similar kinds of searches for points of origin or places to rest. The past had real, material consequences for lived experience in nineteenth-century Britain; in particular, for Ledger-Lomas, that most sacred and ancient form of journeying, the pilgrimage, offers a metaphor for this book's attempts to disrupt the neat teleologies we have previously attributed to Victorian accounts of the past. As Ledger-Lomas reveals, the pilgrimage itself—the journey—often became more important to pilgrims than the site to which they traveled, in ways that make single teleological lines impossible to trace. The individual pilgrimage was often filled with detour and disappointment; the history of Victorian religious experience—as evidenced by these accounts of the conflicted aims and values of Victorian pilgrims—was deeply divided; and the histories of religious experiences that these journeys implied or invoked were often disrupted by the material experience (and diversions) of the journeys themselves. In their very materiality, Victorian pilgrimages often provoked or embodied a loss of faith in a single, unifying story of Christianity, often with painful consequences.

In David Gange's account of *fin-de-siècle* attitudes toward death, this loss of faith in a single, unifying story of Christianity could nonetheless be bolstered through comparative religion. Arguing that historians have tended to focus on the mid-century, with its evangelical pieties and its mass mourning rituals, in their evocations of Victorian mourning, Gange shifts focus to late-century explorations of Egyptian hieroglyphic texts such as the *Book of the Dead*, which were increasingly brought into line with Christian myths, motifs, and stories to articulate one grand religion of both ancient and global reach. Nonetheless, while this attempt to correlate religious belief across centuries and cultures offered an attempt to make one story from many, it also represents, for Gange, the *end* of the Victorian "cult of death" long before its traditional end date of the First World War. Instead, late Victorians increasingly turned to exotic pasts to shore up Christian faith because they were already experiencing "the end of a consensus." The readings of the *Book of the Dead* Gange offers here may then represent a last-gasp effort to recapture a loss of certainty in faith and in the afterlife.

While pilgrimages and belatedly understood ancient cultures may have provoked a wide range of existential questions for Victorian pilgrims, travelers, archaeologists, and their readers, these journeys at least were on solid ground. In the age of the steamship, it is important that we recognize the experiential nature of traveling in its expanding range of Victorian forms, and the ways in which this may have perpetuated the disruption of a continuous sense of time. In her essay for this collection, Clare Pettitt speculates that, as growing numbers of Victorian voyagers lost their sense of time while traveling across the ocean on steamships, as cables were laid on the sea bed for telegraphs, and as ancient sea creatures were discovered and displayed in collections countrywide, even the ocean, that most powerful metaphor for eternity, became littered with the spoils of time. Did the Victorians find security in all these startling pasts, or were they simply left at sea?

Pettitt's essay offers a bridge to part 4, which—as might be expected, considering its placement in the book—considers endings and futures as much as origins, beginnings, and long-dead pasts by considering unfinished business. In doing so, it considers what it meant to be living inside a present so enriched, expanded, and multiplied by the proliferation of competing and contested histories. In ending with endings, however, we also seek to revisit some of the questions of our first essays—the questions of narrative raised by Qureshi and Mandler—and to put them in (literally) new perspectives. Those earlier essays asked, Are the stories we use to frame our experiences of the ever-changing relationship between past and present shaped by those experiences, or do they shape them? To what extent are those narratives adequate for capturing the complexity and multiplicity of historical experience? To what extent did particular Victorians engage with the inadequacies and the possibilities of those narratives, and how consciously or unconsciously did they do so? In returning to these preoccupations, Jocelyn Betts asks, more specifically, How do we narrate the present moment in relation to its pasts and possible futures? Betts shows how historical theories could lead Victorians to thunderously proclaim the apocalypse, or to nervously anticipate their own demise: the history of the human race could often suggest a future in which destruction seemed to many imminent, even inevitable. Resonating with Mandler and Buckland, Betts troubles our image of the Victorians as a proud and boastful people, imagining themselves at the pinnacle of history as the civilized beings all other races and cultures could only aspire to emulate. Instead, the uncertainty and contingency of the historical moment, and of the futures that history might imply, destabilized the possibility of characterizing or understanding even the present. As Daniel C. S. Wilson writes in the same part of the book, this was increasingly felt to be "a society not yet ripe for evaluation." Historical self-awareness produced as much doubt and introspection as confidence and triumphalism.

Both Betts and Wilson also note that, while history has tended to focus on sages and prophets, on the Carlyles and Wellses who responded to what Wilson calls "the public desire for guidance," there were many countervailing philosophies that emphasized contingency and transition, the ways in which history is not determined by single causes or governed by fatalistic processes. It is no coincidence that both essays examine largely forgotten men in these stories at key points in the history of disciplinary formation. Sidelining prophecy, many Victorian writers and thinkers became concerned with how to feel "at home" in this newly peculiar, sharply historical present. That they are now largely forgotten is not a measure of their intellectual or historic significance, but a by-product of the forces of disciplinarization in which they lost out to fields built on the valorization of "sages and prophets," and in which different modes of historical encounter came to take precedence. So, while William Henry Smith's combination of essay, philosophy, novel, poetry, and history in his 1857 novel *Thorndale* was pertinent for readers struggling to make sense of the past in the 1850s, by the 1870s it no longer felt relevant. And the holistic, integrated historical economics of William Cunningham and William Ashley in the late nineteenth century were forgotten when economics was given a newly ahistorical definition, and a de-economized "history" emerged alongside it. Our own continued indebtedness to these disciplinary formations, we contend, has blinded us to the much more interrogative and less certain Victorians who were left behind by them. We have often seen nineteenth-century writers and thinkers as naively committed both to the possibility of a "realist" and empirical discourse that could adequately describe the world and to the grand narrative as a mode of making sense of its history. Despite rich scholarship to the contrary, there is still a widespread general perception of a Victorian unwavering belief in human and technological progress and of a nineteenth-century faith in Britain's position at the apex of civilization (at least until late-century fears about degeneration began to set in). And though we have long registered the frustration many nineteenth-century people felt in the search for universal truths, for stable and secure origins, and for various forms of authenticity, still we have generally rested easy in the assumption that they sought those certainties—particularly in the past—and that they took refuge in the possibility that, come what may, those certainties were out there somewhere. The (re)invention of Morris dancing, or of Christmas, tells well-trammeled stories of Victorian sentimentalism and nostalgia, the anchoring of present uncertainties in the golden sands of the past. But while this may have been a significant feature of Victorian historical understanding, it is only one feature, and one we have been conditioned to see by our inheritance of the very disciplinary formations the Victorians produced in order to make sense of the sheer multiplicity of pasts that confronted them. Borrowing tools the Victorians evolved over a century to explain the past, we have forgotten the struggles, dilemmas, contests, competing voices, and hybrid forms that went into producing them. Exploring little-known sources and unusual archives, and taking a variety of perspectival positions on the past, no matter what kinds of disciplinary knowledge or skills those positions require us to come to grips with, we seek to begin the process of recovering those struggles and forms, and to remember and interpret the richness and diversity of Victorian historical experience so long obscured by disciplinary specialization.

Time Travelers therefore offers original perspectives that explore, rather than fully describe, the richness and diversity of the Victorians' engagement with the past. It remains to say that these perspectives emerge from a long, shared conversation between each of the authors of our volume, who spent five years between 2006 and 2011 as members of the Cambridge Victorian Studies Group, funded by the Leverhulme Trust. Furthermore, the discussions we held throughout the duration of that project have continued long beyond that shared institutional space, as we have become colleagues and collaborators across a variety of other institutions and projects. This book constitutes the most collaborative of our shared outputs, and, as such, it deploys the best techniques we found for speaking across disciplines with one another and for crossing, undermining, or transcending the Victorian disciplinary formations we have inherited to recover the much more eclectic and surprising encounters the Victorians had with the past. While we continue to depend upon the specialist knowledge and skills we have each acquired in our own disciplines, few of us any longer seek to speak only to members of our own respective academic tribes. For five years we had the luxury of sharing methods, practices, languages, ideas, and objects of study. This book is our attempt to bring together those different skills and knowledges to stimulate new debate about the Victorians and the many pasts they inhabited and produced.

Notes

1. Dipesh Chakrabarty, *Provincializing Europe*, (Princeton, NJ: Princeton University Press, 2000), 8.
2. Johannes Fabian, *Time and the Other* (New York: Columbia University Press, 2014), 35.

3. See Mark Girouard, *The Return to Camelot: Chivalry and the English Gentleman* (New Haven, CT: Yale University Press, 1981) on the rehabilitation of a concept of "chivalry" in the Victorian period in relation to an invented medieval past; for a more wide-ranging interrogation of Victorian medievalisms, see Florence S. Boos, ed., *History and Community: Essays in Victorian Medievalism* (New York: Garland, 1992). For classic accounts of the invention of a variety of national pasts, traditions, and rituals in the Victorian period, see Eric Hobsbawm, "Mass-Producing Traditions: Europe, 1870–1914"; and David Cannadine "The Context, Performance, and Meaning of Ritual: The British Monarchy and the Invention of Tradition, c. 1820–1977," both in *The Invention of Tradition*, ed. Eric Hobsbawm and Terence Ranger (Cambridge: Cambridge University Press, 1983), 263–307, and 101–65, respectively; see also David Lowenthal, *The Past Is a Foreign Country* (Cambridge: Cambridge University Press, 1986). On the popular interest in "merrie England" as a place of benevolent and stable social hierarchy, see Peter Mandler, "'In the Olden Time': Romantic History and English National Identity, 1820–50," in Laurence Brockliss and David Eastwood, eds., *A Union of Multiple Identities: The British Isles, c.1750–c.1850* (Manchester: Manchester University Press, 1996), 78–92; and Peter Mandler, *The Fall and Rise of the Stately Home* (New Haven, CT: Yale University Press, 1997), which places this interest in the first three-quarters of the century. On the persistence of these forms of historical engagement into the late nineteenth and early twentieth centuries, see Paul Readman, "The Place of the Past in the English Culture, c. 1890–1914," *Past & Present* 186, no. 1 (2005): 147–99. Andrew Sanders, *In the Olden Time: Victorians and the British Past* (New Haven, CT: Yale University Press, 2013) has more recently argued that the past was often presented as violent only with relish, in order to make the present seem safer and more progressive.

4. See Chakrabarty, *Provincializing Europe* (2000). See also, for example, Nihad M. Farooq, *Undisciplined* (New York: New York University Press, 2016); Susan D. Dion, *Braiding Histories* (Vancouver: University of British Columbia Press, 2009); Eckhardt Fuchs and Benedikt Stuchey, *Across Cultural Borders* (Langham, MD: Rowman & Littlefield, 2002); and Sujit Suvasundaram, *Islanded* (Chicago: University of Chicago Press, 2013).

5. Of course, there has been much scholarship on each of these activities, but they have largely been viewed through a monodisciplinary lens, with a focus on only one or two of these types of activity: on collecting, for instance, on which there has been a wealth of study, attention has gathered around art-historical collecting (see Dianne Sachko Macleod, "Art Collecting and Victorian Middle Class Taste," *Art History* 10, no. 3 [1987]: 328–50), or the collecting and display of foreign artifacts (see the brilliant Saloni Mathur, *India by Design: Colonial History and Cultural Display* [Berkeley: University of California Press, 2007]), or on collecting and display by naturalists and men and women of science (e.g., Robert E. Kohler, *All Creatures: Naturalists, Collectors, and Biodiversity, 1850–1950* [Princeton, NJ: Princeton University Press, 2006]; Anne Secord, "Corresponding Interests: Artisans and Gentlemen in Nineteenth Century Natural History," *British Journal for the History of Science* 27, no. 4 [1994]: 383–408; and Samuel Alberti's extensive work in essays

such as "Placing Nature: Natural History Collections and Their Owners in Nineteenth-Century Provincial England," *British Journal for the History of Science* 35, no. 3 [2002]: 291–311). This volume seeks to build on that rich and extensive scholarship to give an overall sense of the profusion of nineteenth-century pasts and their relationships to one another.

6. See Peter Bowler, *The Invention of Progress: The Victorians and the Past* (Oxford: Basil Blackwell, 1989) for a classic account by a historian of science of the ubiquity of "progress" as a Victorian model for giving shape to the past. See Sadiah Qureshi, *Peoples on Parade: Exhibitions, Empire, and Anthropology in Nineteenth Century Britain* (Chicago: University of Chicago Press, 2011) for an argument about the need to replace "self/other" formulations with more subtle and wide-ranging accounts of the encounter between British and foreign peoples in the period; see also Anne McClintock's "The Angel of Progress: Pitfalls of the Term," *Social Text* 31/32 (1992): 84–98, for a compelling account of the ways in which the self/other binary risks reducing the very hybridity and nuanced attention to geopolitical it was designed to enable.

Part One: Narratives

1

Looking to Our Ancestors

Sadiah Qureshi

New understandings of being human emerged in the nineteenth century. Until the 1860s, humans were believed to have originated in Asia and diffused across the globe after the post-diluvial grounding of Noah's Ark. Human history was often calibrated using biblical chronologies and conflated with the remains of literate peoples. In the early to mid-nineteenth century, the prospect of another, unimaginably deep past arose. From pulpits and Sunday schools to learned societies and exhibition venues, the lay and the learned asked, Had humans been placed on earth by a benevolent God, or had they descended from the apes? Were humans one species or many, and how might anyone tell the difference? Had people wandered the earth since geological antiquity, or were they recent interlopers? Had some peoples become extinct, and were there others doomed to die? In disciplinary terms, these discussions helped found and sharpen divisions between history, geology, archaeology, anthropology, and paleontology.

Modern disciplinary divides make it difficult to recapture the experience of debating human descent, antiquity, and evolution; however, transcending the legacies of discipline formation remains the best way of understanding how human pasts were reimagined. Histories of race and evolution have substantial literatures devoted to the reception of evolutionary theories and descent.[1] Recent interest in human antiquity has highlighted how our pasts were focal points for broader concerns about national identity, colonial policy, and the role of human beings in geohistory.[2] Nonetheless, we know far less about how discussions on antiquity, evolution, and descent were encountered in broader public circles. Building on a growing body of work, "Looking to Our Ancestors" suggests that disputes about antiquity must be ranked alongside those on evolution and descent.[3] By tracing discussions of human prehistory in public spaces, we see how debates on ancient humans involved making choices about which pasts could be known, which should be rejected, and which should be incorporated into human history. In particular, we see how nineteenth-century engagements with human pasts allowed for the proliferation, rather than homogenization, of views about what it meant to be human and what the differences between us signify. The discovery of fossilized humans, for instance, established the reality of prehistoric human existence; however, ancestral extinctions were quickly argued to presage the most devastating effects of modern settler colonialism through violent dispossession.

Petrified People

In June 1840, No. 18 Leicester-Square hosted the fossil of an "Antediluvian Child" (fig. 1). The proprietors proclaimed that even the comparative anatomist Georges "Cuvier, the most celebrated of his age, has denied the existence of human fossil remains: and others pretend that, previous to the Deluge, the human race had no existence." Found "near Brussels," the fossil was exhibited as "demonstrative evidence" of the "truth of the Holy Bible," proving that people *had* roamed the earth before the Great Flood. Visitors had a fortnight to examine the specimen before the proprietors left to exhibit the child in Paris before "Members of the Academy of Science."[4] Interest was generated as far afield as Australia.[5] It may seem obvious that a human fossil would draw significant attention; however, the "Ante-diluvian Child" exhibit was extraordinary in several respects. First, it was exceptionally expensive. The child was advertised on the same newspaper page as George Catlin's North American Indian Gallery at the Egyptian Hall, a concert at the Royal Surrey Zoological Gardens, and daguerreotypes of the European continent, all of which cost a shilling. The exhibition was also remarkably early in the debates on extinction and human antiquity.[6]

EXHIBITION
OF A
HUMAN FOSSIL
18, Leicester Square,
AN ANTE-DILUVIAN CHILD,
Found in a supercritacy soil
At Diehgen, near Brussels, (Belgium.)

There does not perhaps exist in the world a discovery so rare and so wonderful in the matter of Geological science, Until now there never was found but the bones of animals to have turned into a fossil state. Cuvier, the most celebrated of his age, has denied the existence of human fossil remains; and others pretend that previous to the Deluge the human race had no existence.

Here, then. is at once a demonstrative evidence of the falsity of their opinion, and a clear indication of the truth of the Holy Bible.

The Proprietors of this fossil being obliged to return to Paris, where they are called by the Members of the Academy of Science, cannot possibly stay longer than a fortnight in London, as their departure is fixed for the 20th of June, 1840.

Admission £1. each Person.
The Exhibition is open every day from 11 till 4 o'Clock, Sundays excepted.

W. R. Newman, Printer, Widegate-street, Bishopsgate.

FIGURE 1.1. Handbill, "Human Fossil," Geological Society of London, LDGSL/547: image number 10–03. Reproduced by permission of the Geological Society of London.

In the later eighteenth and early nineteenth centuries, a new scientific consensus proposed that extinction was an endemic, possibly even foreordained, natural process.[7] The permanent loss of species, such as the Mauritian dodo, were known, but attributed to human actions. Proposing that extinction was an endemic natural process raised religious and intellectual conundrums. For theists, extinction appeared to undermine the perfection one might expect of Divine creation and contradicted the belief that God created all possible forms of life, and that these would continue to exist to reveal Divine plenitude. Deists preferred explanations of species loss rooted in secular understandings of migration and transmutation: it seemed entirely plausible that apparently extinct species might be found in uncharted territories or that older forms might have transmuted, or evolved, into their present forms. Thus, in the early nineteenth century, extinction, migration, and transmutation "were treated as alternatives, as it were on a par with one another"; none was "obviously more plausible than the others," and every one "entailed grave difficulties and further problems."[8] Important early studies of extinction were focused on megafauna, such as the megatherium, a giant sloth first discovered in Argentina in 1788, and the mastodon. Cuvier compared fossilized mastodon bones to living elephants in articles published between 1796 and 1806.[9] Such detailed research established Cuvier as the premier authority on extinction and the reality of endemic natural loss by the 1820s.[10]

Extinction's new epistemological status raised urgent questions regarding our ancestors. Had humans roamed among the mammoths? Had older species given rise to new species? Were there extinct archaic human forms? Since the 1820s, geologists had suspected that the Earth's human presence was established much earlier than accepted, but the issue remained unresolved and contentious. As late as 1825, authoritative figures such as Cuvier insisted that human *fossils* did not exist. Unfathomably old human remains were known, but such "petrified" bones were considered insufficient evidence to recalibrate human history.[11] Instead, humanity's age was indirectly addressed by deciphering whether humans and extinct animals had lived together.[12] Early British cave finds included bones in Kent's Hole (1824) and, more dramatically, a skeleton in Paviland Cave in South Wales (1823). Some finds were dismissed as inauthentic; others were at classified as old, but not ancient, because they were not found in the same deposits as older animal bones. Ambiguity bred caution. William Buckland discovered the Red Lady of Paviland, as she was later named, and believed she was a Roman buried much later than the animal remains. Biblical chronology underpinned nineteenth-century histories of nations and peoples.[13] Buckland reconciled the Bible's account of human novelty with Earth's antiquity by proposing a universal deluge in the recent past.[14] Given his geological expertise, classifying the Red Lady as Roman sustained broader skepticism. In 1827, Paul Tournal found human bones and pottery shards in Languedoc in southern France. His published account insisted that the remains were mixed with extinct animal bones.[15] In 1829, the well-known excavator of French cave sites Jules de Christol found human and extinct animal bones mixed together in caves near Sommières in France. He insisted that if the animal remains were fossils, so too were the human bones. Both Tournal and de Christol published preliminary reports that challenged Cuvier's position. He contested the findings, and uncertainty remained.

In the 1830s the case for human antiquity strengthened substantially.[16] In 1833 Philippe-Charles Schmerling sent a paper to the *Société Géologique* on two human skulls and tools found among animal bones in caves he had excavated in Liège in Belgium.[17] The larger skull was found under a thick bed of rock speckled with extinct animal bones. Schmerling concluded that the human bones must be ancient. Flints and animal bones adapted for use as tools or ornaments corroborated his inference. Schmerling's medical education, scholarly publications, use of lithographs, insistence that he had personally excavated undisturbed cave sites and a paper submitted to a major learned society all made the Belgian fossils famous and difficult, but not impossible, to dismiss. Geologists such as Lyell and Buckland still inferred later burial as the best explanation (as with the Red Lady). Shortly afterward, Boucher de Perthes made similar finds in Abbeville in 1838. He confidently claimed that his finds proved the existence of ancient humans, but he was accused of drawing bold conclusions from insufficient evidence.[18] Ultimately, the "Ante-diluvian Child" exhibition profited from the search for indisputably human fossils. Despite Cuvier's death eight years earlier, his disbelief created intrigue, while the child's Belgian origin obliquely associated it with the strongest evidence for human antiquity ever found. The promotional patter worked.

Invitations to see the antediluvian child were taken up by journalists, naturalists, and geologists. The *Morning Post*'s article was generously peppered with scare quotations referring to the child. The "nodule of flint" was said to slightly resemble an "infant, wanting both arms," but one "'destitute of every trace of human organisation." The writer felt that "we have never seen such a comical caricature of a backbone as this siliceous 'child'" and so cannot "flatter our geological readers that there is any ground of serious interest or novel speculation attaching to the specimen." The paper acknowledged that the origin of such flint masses had caused great controversy at the recent British Association of Science meeting in Newcastle. Moreover, while some formations of an "*assez bizarre*" nature had been found at Diehgen, and more were known from England and Guadeloupe, the child was unlikely to "add to our knowledge on the subject, for it is unaccompanied by any description of its previous (geological) relations." Having procured the specimen for forty francs but now valuing it at fifty thousand francs, the owner hoped to make an ample fortune. In a telling final flourish, the article observed, "*Nihil ex Nihilo fit.*"[19] The *Patriot* felt that the fossil resembled the "head and trunk of an infant, completely formed.... The head is perfect—the nape of the neck, the articulations of the vertebrae, the bones of the throat, the chest, shoulders, and parts of the arms equally so, and the ribs are distinctly visible. The right arm is broken short off by the shoulder; the left, which is unmutilated, adheres to the side, and is sunk into it. The lower extremities are indistinct, being thrown up into a circular mass below the abdomen." The journalist confirmed that there was no other geological "problem whose solution offers greater interest than that which depends on the existence or absence of the human antediluvian fossil." [20]

Advertisements explicitly welcomed London's scientific communities. The *Patriot* enthusiastically noted that Members of the Geological Society were invited to inspect the specimen and hoped that their inspection could finally set the question of human antiquity "at rest." The hefty admission fee may have been carefully calibrated to appeal to affluent gentlemen, since it was far beyond the means of most workers and commensurate with the membership cost of elite metropolitan societies and gentlemen's clubs. William Buckland, a popular lecturer at the University of Oxford and twice president of the Geological Society, left unimpressed: "I saw the above, with Mr. Pentland and Sir Francis Chantrey, 12th June, 1840. It is nothing but concretion of chert in the Brussels sand, something like the head and trunk of a child without legs." Yet Buckland kept the copy of the promotional handbill on which he penned this observation. His son, the naturalist Francis Buckland, incorporated the chert's tale into a passage on human mummies in the *Curiosities of Natural History* (1860).[21] These details are intriguing. William's dismissal of the child's humanity suggests strong disappointment. Perhaps he dared not raise his hopes of a genuine find; nonetheless, he parted with (one might say gambled) a substantial sum to be sure. He was not alone. The Geological Society of London's archives hold a promotional handbill for the exhibition and a Conté and chalk drawing of the fossil (fig. 2).[22] Their accession in the society's collection suggests that Buckland's peers were curious enough to create permanent archival mementos. Likewise, Francis inserted the anecdote of his father's visit into a passage on exhibited human remains, thereby transforming the child into a curiosity of natural history. The specimen may have disappointed William and convinced few journalists. Nonetheless, the

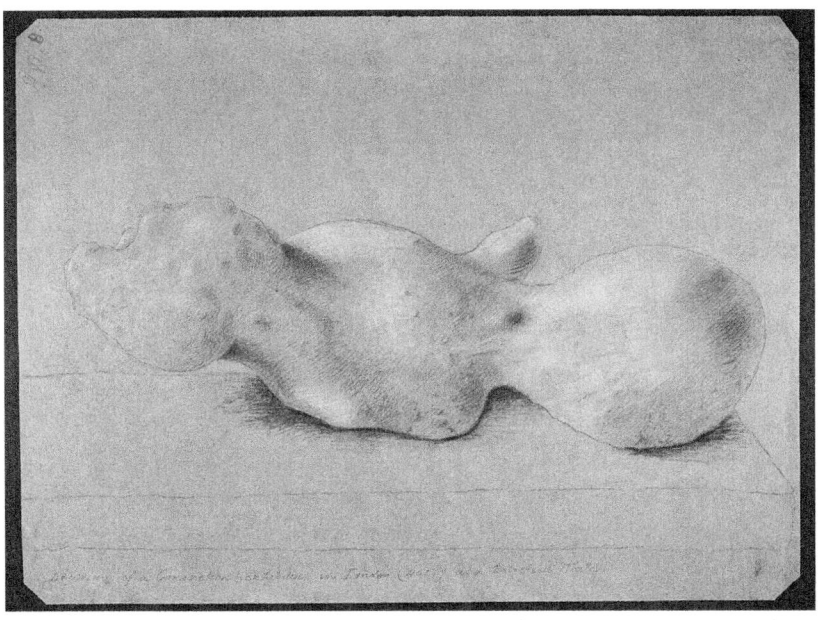

FIGURE 1.2. Drawing of the "Human Fossil," Geological Society of London, LDGSL/547, image number 10-04. Reproduced by permission of the Geological Society of London.

child's reception illuminates why the search for our ancient ancestors was significant for men of science and impresarios. Once the child was removed to Paris, visitors were left to choose from more lasting displays of human antiquity.

Daniel Wilson was a curator for the Society of Antiquaries of Scotland Museum and is remembered for introducing the term *prehistory* to English speakers. The society was founded in 1780. From the outset, their collections featured items such as whale jawbones; exotic birds and monstrosities (given to the Royal Society in 1823); ethnographic collections, including donations from the Captain Cook's voyages; Native American clothing; a Canadian ice boat; and numerous ancient weapons, including bronze axes and flint arrowheads. In 1781, the society leased a house in Old Town—the first of many homes for the collections, including Castle Hill, George Street, and, between 1826 and 1844, the Royal Institution Building. In 1843, the museum returned to George Street, but the society's collections became national property in 1851 in exchange for permanent accommodation for the collection and the society's meetings.[23] In the late 1840s, Wilson reclassified the collections to exhibit new theories of human prehistory.[24]

Wilson's *Synopsis of the Museum of the Society of Antiquaries of*

Scotland (1851) claimed that the ancient relics provided "rare and curious illustrations of National Manners and of the state of [a] Society altogether obsolete."[25] Wilson grouped the vitrines according to their age. First came the "Stone Period" and "Bronze Period": both were classed as "CELTIC, from the earliest traces of the occupation of Europe by a human population."[26] He created cases with mixed artifacts from the "Stone and Bronze Periods." Although the Iron period began before Caesar invaded Britain, there was no section devoted to ferrous items because of their rapid oxidization in unprotected environments. Wilson used taxa he claimed were "generally received among modern Archaeologists." He felt that the "Roman and Greek," "Anglo-Saxon," and "Mediaeval" divisions were intelligible enough to be self-explanatory and added three final sections entitled "Egyptian," "Indian and Mexican," and "Miscellaneous." The human remains, including several skulls, were all classified as Celtic, and many were listed as having been found with animal bones and bronze weapons, or in stone coffins. The museum was open to the public, and in 1841 it hosted more than four thousand visitors, including the queen and Prince Albert. By 1850, numbers swelled to seventeen thousand visitors per year.

Dividing prehistory into three ages originated in Scandinavia. The Three Age system was formally developed by C. J. Thomsen, who curated the antiquarian collections at National Museum of Denmark (est. 1807). Throughout the 1820s and 1830s, he refined long-standing observations that stone, bronze, and iron artifacts produced a repeated sequence in the archaeological record. He argued that each material was fashioned into similar items, such as weapons or jewelry, and that these items appeared together repeatedly and in the same archaeological contexts. Essentially, they produced "*assemblages of finds.*"[27] Thomsen promoted his system and tied it to ancient Scandinavian histories.[28] Thomsen's technique created a means of dating antiquities in relationship to existing chronologies even as they were being challenged. The Three Age system was adopted relatively quickly in Scandinavia and aggressively promoted by figures such as the Danish archaeologist J. J. A. Worsaae, who traveled to London, Edinburgh, and Dublin between 1846 and 1847. Wilson probably knew about the Scandinavian developments through his relationship with the Norwegian historian Peter Andreas Munch.[29] Under Wilson's charge, Scottish antiquarian collections gained new importance in studies of ancient humans, although he reclassified the collections before human antiquity was widely accepted. Thus, he based his displays on the *relative* age of the objects rather than conclusive knowledge of their absolute age. Nonetheless, the ground was yielding increasingly convincing evidence of human antiquity.

Within twenty years of the antediluvian child's unearthing, fossilized human remains and prehistoric tools helped establish a new consensus in Britain and Europe that ancient humans had existed and become extinct in "deep time."[30] In 1856 fragments of a skull, thigh bones, and other remains were discovered in a cavern above the banks of the Neander River, near Düsseldorf, Germany. They were only later reinterpreted as an unknown kind of human with an "extraordinary form of the skull . . . hitherto not known to exist, even in the most barbarous races."[31] Just two years after the discovery of the "Neanderthal," an unknown and undisturbed cave was discovered in Brixham by John Philp. News quickly reached the amateur geologist William Pengelly, who stopped further work and arranged for members of the Geological Society of London to excavate the cave. Months of careful digging produced hundreds of bones and, crucially, seven flint tools. In 1859, having personally visited Brixham, Charles Lyell presented a paper at the annual meeting of the British Association for the Advancement of Science, where he announced that ancient humans and mammals had coexisted. Some sceptics remained, but Lyell's paper accurately represented a new accord among virtually all British geologists and a significant cohort of Continental specialists.[32] The new consensus on human antiquity transformed understandings of ancient Europeans and contributed to scientific debates on race, methodology, and empire.

Interpreting Human Antiquity

Europeans had to contend with a bleak new vision of their past. Originally, the Celts were understood to have arrived in Europe only a few centuries before Julius Caesar led the Roman invasions of 55 and 54 BCE. After the 1860s, the Celts were believed to have led lowly existences for millennia. The earliest, most important attempt to present the new consensus for the lay public was Lyell's *Geological Evidences of the Antiquity of Man* (1863), which presented evidence of ancient humans alongside discussions of Darwin's evolutionary theory, glaciation, and the first appearance of ancient Europeans. Darwin and other naturalists were unhappy with Lyell's presentation of their work.[33] Nonetheless, the book's reception finally shifted the focus from whether humans and mammoths were contemporaries to deciphering their absolute age in the geological record. Two years later, John Lubbock's *Pre-Historic*

Times (1865) summarized articles he had published in the previous four years for lay readers. Lubbock divided human prehistory into four great epochs: the "Palaeolithic" Stone Age, in which Europe was home to people, mammoths, cave bears, and woolly rhinoceroses; the "Neolithic," or "polished Stone Age," in which there was no knowledge of any metal except gold; the Bronze Age, in which bronze was fashioned into arms and other sharp implements; and, finally, the Iron Age, in which bronze was still used for ornaments but the ferrous metal was preferred material for arms and knives. Lubbock's original division of the Stone Age into the Palaeolithic and Neolithic extended, consolidated, and refined the Three Age system in Britain.

In institutional terms, debates were often conducted under the aegis of the Ethnological Society of London (est. 1843) precisely because the disciplinary divisions between archaeology, prehistory, and anthropology were significantly more fluid than they are now.[34] Under Lubbock's presidency, prehistory and archaeological finds were central concerns of the Ethnological Society.[35] For instance, the *Transactions of the Ethnological Society of London* praised the new geological work: "In the well-known works of Sir Chas. Lyell and Sir John Lubbock, the Geological evidence of Man's antiquity is set forth with such fullness and perspicuity, as to carry conviction to the mind of every sincere search after the truth." Given the unease some naturalists felt about Lyell's expositions, such enthusiasm for both Lyell's and Lubbock's work in the pages of a journal such as the *Transactions* is notable. It helps us rethink the significance of the 1860s for reconfiguring the human past. This decade has often been associated with the overthrowing of older elastic and stadial theories of human variation, especially race; yet, both Lyell and Lubbock argued that humans were progressively civilized, with some varieties having achieved greater advances than others.[36] Ultimately, extending human history into deep time enabled the persistence of progressive developmentalism—later social evolutionism—and thus entrenched stadial visions of civilization. Therefore, the 1860s witnessed a substantial *proliferation*, not homogenization, of theories explicating human difference because older classifications *competed* with newer attempts to cleave humans into distinct species.[37] In a relatively short period, the conjectural prehistories and stadial notions of development of the enlightenment came to underpin new classifications of progressive human development.

Lubbock and Lyell drew upon, and made their work relevant for, scholars interested in human descent and race.[38] Throughout the eighteenth and early nineteenth centuries, "race" was an elastic category defined by a broad variety of cultural, environmental and physical factors such as geography, religion, clothing, means of subsistence, and complexion.[39] Orthodox writers took it for granted that humans were ultimately of one family descended from Adam and Eve, an idea later known as monogenesis.[40] Dissenters proposed the existence of humans before Adam (Pre-Adamites), multiple acts of creation, or several groups of survivors following the grounding of Noah's ark to account for observed differences between peoples. Later known as polygenists, the most radical argued that such differences were enough to distinguish between multiple species and that human development was impervious to environment, education, or evolution. Notions of human antiquity strengthened arguments for monogenesis and polygenesis, because both factions needed vast tranches of time to explain the variation from a single stock or, more controversially, the emergence of multiple human species.

Crucially, accepting human antiquity gave rise to new methodologies for understanding human difference. One of the biggest problems facing scholars of prehistory was the lack of available evidence. How could anyone know how our ancestors had lived from a few stone tools and fossilized remains? Even if material traces could be found, how could that tell us anything about the social and cultural lives of our ancestors? Sir Francis Palgrave's lament exemplified the profound sense of an unknowable prehistory:

> You can no more judge of their age than the eye can estimate the height of the clouds: the shapeless masses impart but one lesson, the impossibility of recovering by induction any knowledge of the speechless past. Waste not your oil. Give it up, that speechless past; whether fact or chronology, doctrine or mythology; whether in Europe, Asia, Africa or America; at Thebes or Palenque, on Lycian shore or Salisbury Plain: lost is lost; gone is gone forever.[41]

Palgrave began his passage with the ancient Celts but ended by proclaiming global attempts to reconstruct ancient human history futile. Lubbock acknowledged the difficulty of reconstructing ancient history but advised scholars wishing to confute Palgrave to look around, since so many "savage or semi-savage tribes live in the same manner even at the present day."[42] Peter Mandler's contribution to this volume, "Looking Around the World," explores the political, particularly imperial, possibilities created by equating ancient people with modern Aboriginal peoples.[43] Here, it is worth noting Lubbock's claims that "memorials of antiquity" were usually made as "interesting vignettes, not as historical pictures."[44] Thus, "the Van Diemener [Aboriginal Tasmanian] and South American" were "to the antiquary what the opossum and the sloth are to the geologist."[45] The synchronic alignment of ancient humans with living peoples was made explicit by the book's full title: *Prehistoric Time: As Illustrated by Ancient Remains and the Manners and Customs of Modern Savages*. Lubbock's comparative method made modern colonized peoples into the foundations of prehistoric archaeology by claiming that they were living exemplars of ancient human society.

Drawing comparisons between ancient human remains and modern "savages" was hardly unique to Lubbock. In the first few weeks after the discovery of the Neanderthal skull of 1857, its novelty was partly established by how unlike it was to any living peoples, even those assumed to be most undeveloped. Yet, within five years Huxley's *Man's Place in Nature* (1863) used the image of an Aboriginal skull superimposed with the remains of the "ape-like" Neanderthal (fig. 3). For Huxley, the difference was such that a "small additional amount of flattening and lengthening with the corresponding increase of the supraciliary bridge, would convert the Australian braincase into a form identical with that of the aberrant fossil."[46] The racist conclusion that Aboriginal peoples were a Stone Age people trapped in a state of arrested development continues to resurface to this day. Crucially, the comparative method

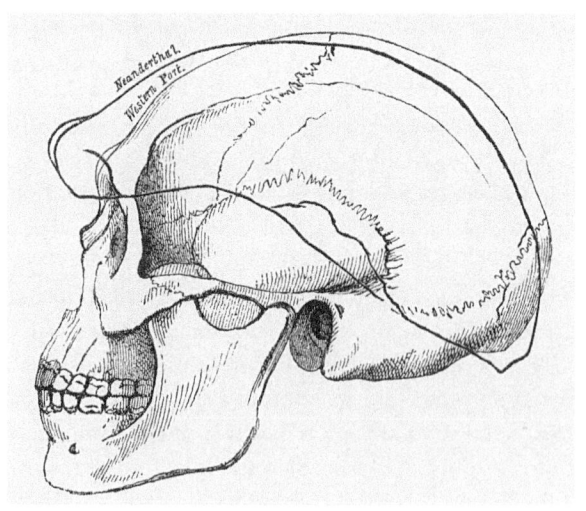

FIGURE 1.3. An aboriginal skull superimposed with a Neanderthal skull fragment, from T. H. Huxley's *Man's Place in Nature* (1863). Author's collection.

was no panacea, since it was possible to find dissimilarities as well as similarities between ancient and modern humans.[47] Arguments that prehistoric human development bore significant similarities to ostensibly "primitive" humans in the present raised a number of possibilities for the human future. For some, if humans were not extended kin but developing along the bifurcating paths, perhaps this entailed rethinking European imperial rule, as explored in Peter Mandler's essay in this volume, "Looking Around the World." The most disturbing interpretations of the past revolved around making predictions regarding modern Aboriginal peoples from ancient human extirpations.

Vanishing Peoples

Throughout the nineteenth century, scientists, politicians, military officials, travelers, missionaries, painters, and writers all contributed to creating and perpetuating the belief that some peoples were doomed to die. Many proposed that any concurrence of "prehistoric" and "modern" peoples in the same territory would prove fatal to many peoples worldwide. In this scenario, indigenous peoples were expected to recede into the past and fade from human memory as the colonizers invaded. Vast demographic depletions were noted as many societies found themselves ravaged by the new diseases, loss of land, and warfare that came with imperial expansion. The violence sparked considerable discussion about the kinds of political activity that were appropriate for "civilized" nations. Significantly, such dispossession had been noted by early modern writers, such as las Casas; however, a new urgency emerged in the nineteenth century.[48]

Humans were seen as both the agents and subjects of extinction. Prehistoric annihilations were conceived of as antecedents to the loss of modern peoples. Extinction was explained through a variety of mechanisms. Permanent physical disappearance was the most dramatic. Prehistoric extirpations were often argued to be the result of migration and conquest.[49] In Denmark, Lubbock noted that others had proposed there was "so sharp and well-marked a distinction between the tumuli of the Stone Age and those of the Bronze period, that the use of bronze might be considered as having been introduced by a *new race of men*, [who had] rapidly exterminated the previous inhabitants . . . and were all together in a much higher state of civilisation."[50] Later the archaeologist William Boyd Dawkins argued in *Early Man in Britain* (1880) that when Paleolithic races encountered Neolithic invaders, "there were the same feelings between them as existed in Hearne's times between the Eskimos and the Red Indian, terror and defenceless hatred being, on the one side, met by ruthless extermination on the other." Dawkins drew direct comparisons with ancient and modern exterminations to explain why cavemen were "gradually driven from Europe, without leaving any mark on the succeeding peoples either in blood or in manners and customs."[51] Ancient losses were quickly invoked in explaining the emergence of new humans and the nature of interethnic encounters, as well as in framing indigenous dispossession in settler colonies.

Alternatively, diachronic theories of human social development promoted specific forms of cultural loss as desirable routes to ostensible social progress. In the prevalent four-stages taxonomy, people were classified on the basis of their modes of subsistence. As people passed from depending on hunting to pasturage, agriculture, and finally commerce, societies were said to become increasingly complex and civilized; envisioned progress necessarily entailed loss.[52] The same fate was often said to be faced by the modern dispossessed. Most famously, in 1869 William Lanney, problematically argued to be the last Aboriginal Tasmanian man, passed away. Just seven years later, Trukanini, ostensibly the last Aboriginal Tasmanian woman, followed in his wake.[53]

Responses to human endangerment ranged from conservationist protection to exterminationist violence. Across this spectrum, visions of the human past competed. Many conservationist arguments depended upon maintaining and protecting synchronic forms of human difference. These projects were predicated on a persistent Edenic state in which humans existed in harmony with the environment, whether as precontact communities or "noble savages." Conservationism was inextricably tied to broader moral and religious philanthropic movements, such as the Aborigines' Protection Society (est. 1836).[54] The society believed that an accurate assessment of the condition of colonized peoples would "show the absolute necessity of adopting immediate measures for their protection and preservation."[55] The society's paternalistic, missionary imperialism largely failed to live up to the hopes of founding members; the campaigning continued until the society merged with the Anti-Slavery Society to form the Anti-Slavery and Aborigines Protection Society in 1909. Notoriously, anthropologists responded by calling for global ethnographic salvage, since the "death of every old man brings with it the loss of knowledge never to be replaced."[56]

Alternatively, by the later nineteenth century, human endangerment could be promoted as the inevitable consequence of evolution through interracial competition. Darwin's *On the Origin of Species* (1859) was infamously opaque regarding humanity's origins. Immediately, combative debates arose regarding human evolution. The long-awaited publication of Darwin's views on human evolution in *The Descent of Man* (1871) further naturalized and racialized extinction as a feature of intercultural contact: "Extinction follows chiefly from the competition of tribe with tribe, and race with race." Darwin used human groups as units of natural selection. For him, the perceived differences in social and cultural development between groups determined who might survive; the only hope for peoples he perceived to be weaker lay in the aid of a "deadly climate."[57] In his writings, extermination was frequently conflated with evolution through natural selection and was even proposed as the necessary corollary of civilized expansion. The hunter, military officer, and conservationist Frederick Courteney Selous served in both the 1893 and 1895 Ndebele wars. His published accounts prophesy that the Ndebele (Matabele) would soon lose their homeland, just as the peoples of the Cape Colony had already lost theirs to white settlers: "Matabeleland is doomed by what seems a law of nature to be ruled by the white man, and the black man must go, or conform to the white man's laws, or die in resisting them. It seems a hard and cruel fate for the black man, but it is a destiny which the broadest philanthropy cannot avert." For Selous, British colonists were akin to "irresponsible atom[s] employed in carrying out a pre-ordained law" that had operated since "organic life was first devolved upon the earth—the index of all law which Darwin has aptly termed the 'Survival of the Fittest.'"[58] It is significant that a colonial officer viewed settler colonial conflicts as empirical proof of a new natural "law." Likewise, many Victorians imagined the expansion of their settler colonies as evidence of their own superiority.

Modern historians acknowledge the deadliness of nineteenth-century settler colonialism, but the emphasis has shifted to new questions—most importantly, is settler colonialism predominantly, or even inherently, genocidal?[59] Within genocide studies, some of this research has been aimed at identification and prevention. Establishing which forms of past violence constitute genocide has been valuable in bringing the extreme violence of many settler societies into focus.[60] Meanwhile, there have been attempts to identify the circumstances under which genocide has occurred (often at contested frontiers) and to order the stages of escalation (bouts of conflict, planned dispossession, and genocidal "final solutions") as a means of guiding contemporary measures for genocide prevention.[61] As a means of understanding settler colonialism in nuanced historical terms, this has created difficulties. For example, there is a genuine risk of seeing frontier violence in teleological terms, characterizing settler colonial expansion as a series of predictable stages rather than contingent processes, losing a sense of regional specificity, and implying that categorizing past losses as genocide will reveal how historical actors conceived of their own actions. The concept of genocide has yielded valuable avenues for understanding settler colonialism, but it is always worth asking how we can use the term with the greatest historical nuance *and* political effectiveness. We know that demographic loss was naturalized as endemic extinction and, worse still, used to justify extermination; nonetheless, we need far more comparative work on settler colonies, violence, and the making of political policy over the *longue durée* and beyond modern European empires to draw broader conclusions.[62] In this vein, histories of sovereignty and land rights, the hierarchies of value associated with different means of subsistence, the mechanisms and strategies for rationalizing dispossession, humanitarianism, the role of science in informing political policy, and the lasting impact of such debates all offer potentially fruitful avenues.

: : :

Competing visions of human pasts sparked intriguing debates in nineteenth-century Britain. Excavations of "petrified" human bones and ancient tools, in conjunction with the new epistemological status of extinction, raised urgent questions regarding our ancestors' antiquity, descent, and evolution. These questions became the central theoretical concerns for emerging disciplines such as archaeology and anthropology. Disciplinary fissures have contributed to distinct literatures on human descent, evolution, and antiquity. Exploring nineteenth-century notions of human pasts transcends such fragmentation and recaptures the shock caused by the first suggestions of antiquity and evolution. Accepting human antiquity entailed reconceptualizing our pasts in diachronic and synchronic forms. Human history stretched considerably further into "deep time" than commonly accepted; it was no longer conflated with the traces of literate peoples, and the earliest Europeans were seen anew as lowly and brutish. New methodologies were invented to recover the lives of our ancestors. The wealth of material allowed for a profusion of human pasts to burst forth from excavations of the earth. Disturbingly, the shadow of ancient extinctions loomed large over many modern peoples, and past losses were used to rationalize the violence and dispossession in settler societies such as North America, South Africa, and Australia. More broadly, indigenous peoples, like museum objects, were increasingly likely to be reclassified as endangered relics of the past. In developmental and civilizational terms, early Europeans were equated with modern indigenous peoples. Hunter-gatherers had long been used to illustrate the earliest stages of subsistence; yet, as human histories extended into "deep time," modern, usually colonized, peoples were recast as "living fossils" of embryonic development and exemplars of the formerly "savage" nature of Europeans. Thus, older stadial and civilizational classifications of social and cultural development persisted into the later nineteenth century and became relevant to new debates regarding human endangerment. Recovering the extraordinary cacophony of theories regarding our ancestors and their legacies is crucial for histories of race, nation, and empire. As human pasts were rewritten, the geography of the modern world offered new possibilities. Traveling became time-traveling.

Notes

1. These concerns are brought together in Adrian Desmond and James Moore, *Darwin's Sacred Cause: Race, Slavery and the Quest for Human Origins* (London: Allen Lane, 2009).
2. On the nation, see Chris Manias, *Race, Science, and the Nation: Reconstructing the Ancient Past in Britain, France and Germany* (London: Routledge, 2013); on geohistory, see Martin J. S. Rudwick, *Worlds Before Adam: The Reconstruction of Geohistory in the Age of Reform* (Chicago: University of Chicago Press, 2008), and Ralph O'Connor, *The Earth on Show: Fossils and the Poetics of Popular Science, 1802–1856* (Chicago: University of Chicago Press, 2008); on colonial policies, see Ian McNiven and Lynnette Russell, *Appropriated Pasts: Indigenous Peoples and the Colonial Culture of Archaeology* (Lanham, MD: AltaMira Press 2005).
3. Matthew R. Goodrum, "The History of Human Origins Research and Its Place in the History of Science: Research Problems and Historiography," *History of Science* 47 (2009): 337–57. Manias, *Race, Science, and the Nation*, provides a welcome example of what might be achieved, especially when including Europeans theorizing about their own pasts.
4. Handbill, 1840, Geological Society of London, LDGSL/547, image 10–03. The text was routinely reproduced in newspaper advertisements such as "EXHIBITION of a HUMAN FOSSIL," *Morning Post*, June 9, 1840, front page.
5. "Human Fossil, Alleged to be Ante-Diluvian," the *Colonist* (Sydney), November 7, 1840, 3.
6. *Morning Post*, June 9, 1840, front page.
7. Mark V. Barrow, *Nature's Ghosts: Confronting Extinction from the Age of Jefferson to the Age of Ecology* (Chicago: University of Chicago Press, 2009).
8. Martin J. S. Rudwick, *Bursting the Limits of Time: The Reconstruction of Geohistory in the Age of Revolution* (Chicago: University of Chicago Press, 2005), 244.
9. Helen Cowie, *Conquering Nature in Spain and Its Empire, 1750–1850*

(Manchester: Manchester University Press, 2011); and Barrow, *Nature's Ghosts*, 39–42.

10. Barrow, *Nature's Ghosts*, and Rudwick, *Bursting the Limits of Time*, especially figs. 5.1, 7.4 and 8.3.

11. For the most succinct summary, see Rudwick, *Worlds Before Adam*, 407–22. For more detailed accounts, including the later nineteenth century, see A. Bowdoin von Riper, *Men Among the Mammoths: Victorian Science and the Discovery of Human Prehistory* (Chicago: University of Chicago Press, 1993).

12. Rudwick, *Worlds Before Adam*, 407.

13. Colin Kidd, *The Forging of Races: Race and Scripture in the Protestant Atlantic World, 1600–2000* (Cambridge: Cambridge University Press, 2006).

14. Marianne Sommer, "An Amusing Account of a Cave in Wales: William Buckland (1784–1856) and the Red Lady of Paviland," *British Journal for the History of Science* 37, no. 1 (2004): 53–74. Buckland was not a literalist interpreter of the Bible, but he did use theories that were compatible with short biblical chronologies.

15. Rudwick, *Worlds Before Adam*, 228–32.

16. Rudwick, *Worlds Before Adam*, 407–22.

17. Schmerling had published a paper on the finds in 1831, but this appears not to have been picked up outside Belgium (Rudwick, *Worlds Before Adam*, 412).

18. Van Riper, *Men Among the Mammoths*, 62–73.

19. "A HUMAN FOSSIL," *Morning Post*, June 10, 1840, 8.

20. "Human Fossil, Alleged to Be Antediluvian," *Patriot*, June 11, 1840, 8.

21. William Buckland, note written on a promotional handbill for the exhibition, cited in Francis Buckland, *Curiosities of Natural History*, 4th ed. (New York: Rudd & Carleton, 1860), 104.

22. Drawing of the "Human Fossil," Geological Society of London, LDGSL/547, image 10–04.

23. The society and its collections are both now housed in Chambers Street in the National Museums of Scotland. Unless otherwise stated, this history of the museum is based on *The Scottish Antiquarian Tradition: Essays to Mark the Bicentenary of the Society of Antiquaries of Scotland 1780–1980*, ed. A. S. Bell (Edinburgh: John Donald, 1981). In particular, see R. B. K. Stevenson's essays, "The Museum, Its Beginnings and Its Development, Part One: To 1858," 31–85; and "The Museum, Its Beginnings and Its Development, Part Two: The National Museum to 1954," 142–211, in the same volume.

24. Marinell Ash, "'A Fine, Genial, Hearty Band': David Laing, Daniel Wilson and Scottish Archaeology," in Bell, ed., *The Scottish Antiquarian Tradition*, 86–113.

25. Daniel Wilson, *Synopsis of the Museum of the Society of Antiquaries of Scotland* (Edinburgh: Society of Antiquaries of Scotland, 1851), x.

26. Wilson, *Synopsis*, 1.

27. Peter Rowley-Conwy, *From Genesis to Prehistory: The Archaeological Three Age System and Its Contested Reception in Denmark, Britain, and Ireland* (Oxford: Oxford Univeresity Press, 2007), 39.

28. Suhm, for instance, divided Scandinavian history into in a "dark age" from Odin I to Odin II, a "fabulous age" from Odin III to c. 800 CE, and all subsequent time as a "historical age." See Rowley-Conwy, *From Genesis to Prehistory*, 28–29

29. Rowley-Conwy, *From Genesis to Prehistory*, 48–81.

30. On Europeans, see Manias, *Race, Science, and the Nation*.

31. D. Schaaffhausen, "On the Crania of the Most Ancient Races of Man. With Remarks, and Original Figures, Taken from a Cast of the Neanderthal Cranium. By George Busk, FRS," *Natural History Review* 1 (1861): 155–75, 155; Michael Hammond, "The Expulsion of the Neanderthals from Human Ancestry: Marcellin Boule and the Social Context of Scientific Research," *Social Studies of Science* 12, no. 1 (1982): 1–36; and Paige Madison, "The Most Brutal of Human Skulls: Measuring and Knowing the First Neanderthal," *British Journal for the History of Science*, 49, no. 3 (2016): 411–32.

32. See Manias, *Race, Science and Nation*.

33. Van Riper, *Men Among the Mammoths*. On Lubbock's reception, see Clive Gamble and Theodora Moutsiou, "The Time Revolution of 1859 and the Stratification of the Primeval Mind," *Notes and Records of the Royal Society* 65, no. 1 (2011): 43–63.

34. The Ethnological Society of London (est. 1843) and the Anthropological Society of London (est. 1863) have primarily been associated with the wranglings between ostensibly monogenist ethnologists and polygenist anthropologists. Recent research has shown this division to be a caricature. See Efram Sera-Shriar, *The Making of British Anthropology, 1813–1871* (London: Pickering and Chatto, 2013), especially the treatment of Hunt and Huxley.

35. See Van Riper, *Men Among the Mammoths*, and Manias, *Race, Science, and the Nation*, for an indication of the extent of this engagement and the possibilities for further work.

36. Nancy Leys Stepan, *The Idea of Race in Science: Great Britain, 1800–1960* (London: Macmillan, 1982).

37. Sadiah Qureshi, *Peoples on Parade: Exhibitions, Empire and Anthropology in Nineteenth-Century Britain* (Chicago: University of Chicago Press, 2011), 217.

38. R[obert] Dunn, "Archaeology and Ethnology: Remarks on Some of the Bearings of Archaeology upon Certain Ethnological Problems and Researches," *Transactions of the Ethnological Society of London* 5 (1867): 305–17, 307.

39. Roxann Wheeler, *The Complexion of Race: Categories of Difference in Eighteenth-Century Culture* (Philadelphia: University of Pennsylvania Press, 2000).

40. Kidd, *Forging of Races*. The ostensible shift to theories of innate differences has been attributed to many factors, including debates over enslavement, the diminishing importance of climatic theory, and the emergence of a new scientific racism. See Stepan, *Idea of Race in Science*; Seymour Drescher, "The Ending of the Slave Trade and the Evolution of European Racism," *Social Science History* 14 (1990): 415–50; and Hannah F. Augstein, *Race:The Origins of an Idea, 1760–1850* (Bristol: Thoemmes, 1996). On why the broader accep-

tance of polygenism has been overstated, see Peter Mandler, "The Problem with Cultural History," *Cultural and Social History* 1, no. 1 (2004): 94–117.

41. Francis Palgrave, *The History of Normandy and of England*, 4 vols (London, John W. Parker & Son, 1851–1864), 1:469–70. The latter part of this quotation from "Give it up . . ." onward appeared in John Lubbock, *Pre-Historic Times, As Illustrated by Ancient Remains, and the Manners and Customs of Modern Savages* (London: Williams and Northgate, 1865), 1.

42. Lubbock, *Prehistoric Times*, 122.

43. McNiven and Russell, in *Appropriated Pasts* (2005), argued that such "antiquation," although well known by historians, had not "previously been identified as an expressly colonial tenet that legitimated European colonial expansion, settlement and dispossession" (8). In contrast, they call for as much attention to be paid to the entanglement of archaeology and empire as has been devoted to anthropology. See, for example, Henrika Kuklick, *The Savage Within: The Social History of British Anthropology, 1885–1945* (Cambridge: Cambridge University Press, 1991); and Nicholas Thomas, *Colonialism's Culture: Anthropology, Travel, and Government* (Cambridge: Polity Press, 1994).

44. Lubbock, *Prehistoric Times*, 1.

45. Lubbock, *Prehistoric Times*, 336.

46. Thomas H. Huxley, *Man's Place in Nature* (London: Williams and Norgate, 1863), 155–56.

47. Chris Manias, "The Problematic Construction of 'Palaeolithic Man': The Old Stone Age and the Difficulties of the Comparative Method, 1859–1914," *Studies in History and Philosophy of Science Part C: Studies in History and Philosophy of Biological and Biomedical Sciences* 51 (2015): 32–43.

48. John Hausdoerffer, *Catlin's Lament: Indians, Manifest Destiny, and the Ethics of Nature* (Lawrence: University Press of Kansas, 2009); and Sadiah Qureshi, "Dying Americans: Race, Extinction and Conservation in the New World," in *From Plunder to Preservation: Britain and the Heritage of Empire, 1800–1950*, ed. Astrid Swenson and Peter Mandler (Oxford: Oxford University Press, 2013), 269–88.

49. On the complicated debates about the nature of ancient Europeans, see Manias, *Race, Science, and the Nation*.

50. Lubbock, *Pre-Historic Times*, 99; emphasis added.

51. William Boyd Dawkins, *Early Man in Britain and His Place in the Tertiary Period* (London: Macmillan, 1880), 243. Such accounts of conquest and extermination are striking, but we must remember that displacement, intermingling or coexistence were all recognized responses to prehistoric interethnic encounters. See the treatment of the Iberians in chapter 4 of Dawkins, *Early Man in Britain*, 309–41; and Chris Manias, "'Our Iberian Forefathers': The Deep Past and Racial Stratification of British Civilization, 1850–1914," *Journal of British Studies* 51 (2012): 910–35.

52. The most developed analysis of the four-stages model remains Ronald Meek, *Social Science and the Ignoble Savage* (Cambridge: Cambridge University Press, 1976).

53. Aboriginal Tasmanians have consistently been forced to argue that they have not been exterminated, as settlers claimed. See Rebe Taylor, "Genocide,

Extinction and Aboriginal Self-Determination in Tasmanian Historiography," *History Compass* 11 (2013): 405–18; and Tom Lawson, *The Last Man: A British Genocide in Tasmania* (London: I. B. Tauris, 2014). The most extensive work on endangered races remains Patrick Brantlinger, *Dark Vanishings: Discourse on the Extinction of Primitive Races, 1800–1930* (Ithaca, NY: Cornell University Press, 2003), which is the basis of my current research.

54. James Heartfield, *The Aborigines' Protection Society: Humanitarian Imperialism in New Zealand, Fiji, Canada, South Africa, and the Congo, 1837–1909* (London: Hurst, 2011).

55. UK Government, *Report of the Parliamentary Select Committee on Aboriginal Tribes (British Settlements): Reprinted, with Comments, by the 'Aborigines Protection Society'* (London: William Ball, 1837), xi.

56. William Rivers, "Report on Anthropological Research Outside America," in *Reports on the Present Condition and Future Needs of the Science of Anthropology*, ed. W. H. R. Rivers, A. E. Jenks, and S. G. Morley, Carnegie Publication 200 (Washington, DC: Carnegie Institution, 1913), 5–28, 7. In the later nineteenth century, calls for fieldwork were tied to polemical debates on how best to secure the future of anthropology. By caricaturing and rejecting the work of predecessors who had not worked *in situ* for long periods, anthropologists claimed a new kind of expertise and, by the mid-twentieth century, had largely succeeded in both institutionalizing the methodology and establishing it as a demarcation criterion for good practice. See George W. Stocking, *After Tylor: British Social Anthropology, 1888–1951* (London: Athlone, 1995).

57. Charles Darwin, *The Descent of Man, and Selection in Relation to Sex*, 2 vols. (London: John Murray, 1871), 1:238.

58. Frederick Courteney Selous, *Sunshine and Storm in Southern Rhodesia: Being a Narrative of Events in Matabeleland both before and During the Recent Native Insurrection up to the Date of the Disbandment of the Bulawayo Field Force* (London: Rowland Ward, 1896), 67. Importantly, Selous misattributed "Survival of the Fittest"; it actually was coined by Herbert Spencer.

59. Stephen Howe, "Colonising and Exterminating? Memories of Imperial Violence in Britain and France," *Histoire@Politique* 11, no. 2 (2010): 1–18; and Patrick Wolfe, "Settler Colonialism and the Elimination of the Native," *Journal of Genocide Research* 8, no. 4 (2006): 387–409.

60. Ben Kiernan, *Blood and Soil: A World History of Genocide and Extermination from Sparta to Darfur* (New Haven, CT: Yale University Press, 2007).

61. Benjamin Madley, "Patterns of Frontier Genocide, 1803–1910: The Aboriginal Tasmanians, the Yuki of California, and the Herero of Namibia," *Journal of Genocide Research* 6, no. 2 (2004): 167–92. Whether such approaches prove effective for prevention remains to be seen. It is frustratingly easy to be disheartened, given the ethnic cleansing and genocide we know is taking place in the present day from the Rohingyas in Burma to the Yazidis in Iraq.

62. Ned Blackhawk, *Violence over the Land: Indians and Empires in the Early American West* (Cambridge, MA: Harvard University Press, 2008) is a stunning exemplar for such work.

2

Looking Around the World

Peter Mandler

Over the course of the nineteenth century, the conceptions of both time and space lengthened, and these two conceptions were connected. With their discovery of the antiquity of human origins, the Victorians came to see their own history as a long gallery encompassing every possible stage of human development, from savagery to barbarism to the high civilization they felt they enjoyed in the present. At the same time, they were looking around the world and discovering the same dizzying array of human types, from those peoples widely held to be at the lower stages of development—the Aboriginal peoples of Australia, the so-called Hottentots or Bushmen of Southern Africa, the scattered hunters and gatherers of Polynesia—through the nomadic native peoples of North America, African and Central Asian pastoralists, the settled villagers of South Asia and China, to "high" cultures similar to their own. Putting these two sequences together, they found their own story in time being played out in space as well.

The Victorians' construction of these sequences was neither accidental nor innocent. The antiquity of human origins was their "discovery"—based on finds of human fossils and of the material culture of "prehistoric" cultures (the word *prehistoric* was coined for this purpose in 1851)—but their "discovery" was not only about new things; it was equally about new ways of understanding these new things. The ground was prepared by the geology of Charles Lyell and the evolutionary biology of Charles Darwin. It was fertilized by doubts about the literal truth of Revelation and by a new faith in human power—the power of technology, of print, education, and science, of self-determination and self-improvement—and by a rampant confidence in the superiority of their own moral, social, and economic arrangements. The encounter with "backward" peoples in the spread of Victorian political and economic dominance around the world (of empire, in a word, though not only the formal political kind) provided a vast amount of new evidence about the stages of civilization; equally, however, the insertion of this evidence into a single "social-evolutionary" sequence, with the Victorians themselves as the consummation, provided an ingenious set of novel warrants for their dominance. At the same time, seeing the "backward" peoples of the world as one's own ancestors placed oneself in a disturbing and unexpected relation to them. They could no longer be so easily kept at a distance; they were, as one London journalist put it in 1871, "[u]nder our feet," for there, "while we walk the streets of this great metropolis, lies the history of human civilisation."[1] The construction of the social-evolutionary sequence at mid-century therefore both rationalized and problematized empire by posing once again the question raised by the slave-trade abolitionists a few generations earlier about the African—"Was he not a man and a brother?"

Stages of Civilization

It is not surprising that the Victorians arrayed the vast stores of new information about the peoples of the world along a social-evolutionary sequence, because their Enlightenment predecessors had already done that for them, without the help of "prehistory." "Stadial" theories of a progression from savagery to civilization can be traced further back into the early modern period, but the Enlightenment's belief in the universality of reason and progress motivated a much closer and more careful documentation of the stages and explanation of how and why different cultures passed through them. Ethnographic information increasingly available from far afield was for this purpose more valuable and more vital than unreliable historical material about Europeans' own distant past. The Seven Years' War—tellingly also dubbed the French and Indian War—drew attention in the 1750s to the great diversity of Native

American cultures. James Cook's voyages in the Pacific between 1766 and 1779 revealed another, completely unexpected world of diversity amongst the Polynesians. The world was becoming a laboratory. As Edmund Burke wrote to William Robertson, one of the Scottish Enlightenment figures who put this information to fullest use, "the great Map of Mankind is unrolled at once, and there is no state or gradation of barbarism, and no mode of refinement which we have not at the same moment under our view; the very different civility of Europe and of China; the barbarism of Persia and of Abyssinia; the erratick manners of Tartary and of Arabia; the savage state of North America and of New Zealand."[2] Yet the Enlightenment's interest in the stages of development was not very historical: their value lay in the unfolding of the present to which they bore witness. How one got from one stage to the next was entirely "conjectural," as Dugald Stewart admitted in 1793. Neither was it truly ethnographic: the material gathered from the Americas and the Pacific simply filled in the "gradations" that had been established on first principles. This was, as Uday Mehta has observed, "anthropological minimalism." A less minimalist approach would depend on more sustained contact with these strange peoples and with the problems of their management that trade, Christianization, and empire raised, as the eighteenth century turned to the nineteenth.[3]

The new agendas that Europeans brought to their contact with non-European peoples all involved to some extent, as Jeffrey Cox has described with regard to the missionary enterprise, the "defamation of the other" to bend them to European designs. In many cases they entailed a sudden shrinkage of the historical imagination, as the "states and gradations of barbarism" became a problem for Europeans to deal with rather than a curious glimpse of conjectured previous phases of European life. Whether this loss of historical imagination rendered non-Europeans more distant—dehumanized them—is debatable. In their relatively early phases many of these imperial projects aimed rather to *re*humanize non-Europeans in more recognizable European forms. Thus missionaries were impatient to drag "savages" into civilization, or at least into the light of God, as quickly as possible. Utilitarian reformers, too, were desperately impatient with "savagery" and "barbarism" and thought that the customs that held so many of the peoples of the world back from progress needed to be broken up—by force if necessary. The Enlightenment optimism about progress as a natural, historical process gave way to a fierce desire to accelerate the passage of time, exhibiting either an anxiety about whether progress truly was natural, or a humanitarian zeal for reason and freedom, or both.

Both are evident in James Mill's *History of British India*, among the most defamatory of all contemporary European accounts of a non-European people, dripping equally with contempt for the stationary state of the Indians and with enthusiasm for their progressive reconstruction. The two are connected—Mill would not have been nearly so contemptuous had his zeal for reform not been so great. In fact, as John Burrow has pointed out, Mill retained the stadial language of his Enlightenment predecessors; it was simply that he now focused on the failure to progress as a problem to be solved. "By conversing with the Hindus of the present day, we, in some measure, converse with the Chaldeans and Babylonians of the time of Cyrus," he wrote, and "with the Persians and Egyptians of the time of Alexander." The problem to be solved was that the Indians had, unlike the Greeks, not progressed beyond the silly superstitions of the Chaldeans and Babylonians. This cruder dichotomization of barbarism and civilization presented the British with their civilizing mission.[4]

It is now conventional to argue that the remorseless effect of this imperial encounter over time was to supersede the "conjectural" historical imagination of the Enlightenment and to replace it with a more brutal project of control, subjection, and extermination. Empire, it is said, is intrinsically utopian—it seeks to replace one way of life (which it doesn't understand) with another (its own)—and like all such utopian projects it is bound to be disappointed, over and over again.[5] Its only recourse is to repression in the short term and violent suppression in the longer term. Thanks to this line of analysis, we now have some very vivid and detailed accounts of this process in different settings. Richard Price has described the "implosion" of missionary utopianism about the prospects of civilizing the Xhosa in the Eastern Cape of South Africa in the 1840s and the resulting policy of breaking up Xhosa culture by force in a series of increasingly bloody wars from the 1850s.[6] Catherine Hall tells a similar story of growing disappointment and frustration with the aftermath of slavery abolition in Jamaica. In these circumstances, the missionaries' understanding of African slaves as "infants," "no longer locked in another time, archaic African time or the pre-modern time of slavery," and thus, after liberation, free to "enter the present," slips quickly into a perception of Africans as *permanently* infantile, incapable of maturation, as the freed slaves stubbornly resist their subjection to European routines of landless labor and civic subordination. The inevitable result was resistance, in the form of the Morant Bay uprising of 1865, and repression, with Governor Eyre's policy of summary arrest and execution of the ringleaders.[7] Most notoriously, the Sepoy Mutiny of 1857 in India, the ultimate disappointment of the civilizing mission, is said to have crystallized a "crisis of liberalism" and established the necessary preconditions for the more overtly exploitative imperialism of the post-1870 period.[8] These successive failures of the civilizing mission—forming an "arc of disillusionment," as Mantena nicely puts it—are said to have been registered as much "at home" as in the empire, decisively consigning subject peoples in the British public's imagination to the distant past from which they could only very slowly, if ever, be retrieved, or, worse, knocking them out of European time altogether by biologizing them as alien species.[9]

Was this "arc of disillusionment" as uniform as this analysis implies, and did it always bend in the same direction? Nicholas Thomas has wisely observed that if empire is intrinsically "presumptuous," we ought not to mirror it in presuming too quickly that we know what it was about.[10] The Victorians' belief in history, especially as its arc culminated in their own civilization, was a powerful one and not so easily destroyed by setbacks in difficult and various situations. As is also said, if "all empire is local," we must attend to the particularities of those various situations to understand the local responses to the shifting fortunes of the civilizing mission.[11] Thus, even in those cases of spectacular failure already cited, the "men on the spot" cast their responses in very different forms. In the Eastern Cape, the violence wrought upon the Xhosa was justified as *enabling* the civilizing mission. As William Holden, author of *The Past and Future of the Kaffir Races*, put it in 1866, "the sword must first—not *exterminate* them, but—break them up as tribes and destroy their political existence," and when "set free from the shackles by which they are bound, civilisation and Christianity will no doubt make rapid progress among them, for they are a noble race, no wise deficient in mental capacity."[12] In contrast, in the aftermath of the Sepoy Mutiny, the widespread response (in conditions where breaking up and destroying "traditional" society was clearly impractical) went in nearly the opposite direction—the perceived need was to understand traditional society and to redeploy it in the service of order and stability; thus the rise of the "ethnographic state," "driven by the belief that India could be ruled using anthropological knowledge to understand and control its subjects," as Nicholas Dirks has said.[13] In one setting "the creation of a colonial knowledge system involved a transition from knowledge to ignorance"—the replacement of the missionaries' "open knowledge system" with "fixed stereotypes" and palpable simplifications—but in the other setting James Mill's crude attempt to erase caste gives way to an ambitious attempt to harness it to colonial imperatives.[14]

In other locales the civilizing mission was not seen as so spectacular a failure, or a failure at all. In West Africa, the successful eradication of European slave trading—a persistent concern of the British, who, after their own abolition of slavery in 1833 maintained a naval squadron off the West African coast to suppress other people's trade in humans[15]—sustained optimism about the civilizing mission. As Philip Curtin has observed, "The dominant British attitude toward Africa became more conversionist than ever," until the 1870s. Better information about Africans in their indigenous environments rendered more subtle their placement in "conjectural" histories of development, such that Ashanti pastoralism was now placed above rather than below the stage of settled agriculture, and a better understanding of tribal societies in Africa caused their peoples to be put above the "savages" of North America or the South Pacific.[16] Fresh debates broke out as to whether commerce, or Christianity, or anti-slavery coercion was the best instrument of the civilizing mission.[17] David Livingstone, for example, experienced the disillusionment with the Xhosa that led him to support the wars against them in the 1850s, but in no way abandoned "the original [evangelical] dream of a rapid Christianization of the world," and simply moved on to fresh pastures in the interior to pursue that dream.[18]

In the South Pacific, that debate seemed to have been settled by the widespread conversion of indigenous peoples to Christianity, clearly setting them on the path out of "savagery." Where resistance was met, a racial alibi was always available—thus the "demonization" of the darker Melanesian peoples, especially after the killing of the missionary John Williams in the New Hebrides in 1839, and the merciless genocide of the Aboriginal peoples of Tasmania. But where, as in Tahiti or Fiji, indigenous peoples—under pressure or for endogenous reasons, or both—saw advantages in adopting new gods, missionary triumphalism could prevail, and the progressive histories of the civilizing mission remained dominant to the end of the century.[19] It is certainly not the case, as Brantlinger contends, that "[p]ublic support for missionary societies declined around mid-century."[20] While organized support for anti-slave-trade activity inevitably declined—with the decline of the slave trade—the focus of public support simply shifted to new targets. Anti-slavery opinion shifted to East Africa, where public support for the efforts of David Livingstone actually peaked after his death in 1873. Missionary activity shifted from the Caribbean and the South Pacific, where it had been deemed successful, to new frontiers in Central Africa and East Asia. Fueled by new recruits from a "wave of evangelical religious enthusiasm" after 1857, and increasingly from America, the missionary presence in the non-Western world continued to grow until the 1920s.[21]

As this fresh wave of missionary enthusiasm attests, empire is not *only* local. Conditions in the metropole were different and were not necessarily subject to the "arc of disillusionment." As Richard Price grants, "a vision of British imperialism that fused moral purpose with the civilizing functions and values of commerce" was "terribly important to British culture in Britain," though "it said very little about the conditions of missionary existence in the empire itself"; thus the coincidence of disillusionment in the Eastern Cape just "at the moment when the missionary effort in Britain was being installed as the moral center of Britain's imperial identity."[22] And at the same mid-century moment, as John Lubbock notes, "suddenly a new light has arisen in the midst of us" that forced a different kind of reconsideration of the temporal sequence of the "civilizing mission." This was the discovery, "under our feet," of a new history of human civilization, "not on the sandy plains of the Nile or the Euphrates, but in the pleasant valleys of England and France, along the banks of the Seine and the Somme, the Thames and the Waveney."[23]

Prehistory and Civilization

The conjectural sequence of human civilization assembled by the Enlightenment may have been tested by the stubborn failure of colonized peoples in Africa and Asia to travel along the sequence in the way and at the speed desired by missionaries and other civilizers. But as we have seen, it was not abandoned, and during the first decades of the nineteenth century, archaeology was unearthing new sources of evidence that might be used to revivify and reinforce it. As Sadiah Qureshi shows in her essay in this volume, a new, longer history of humanity was suggested as early as 1833 by the discovery and identification of early human skulls and bones at Engis by P. C. Schmerling. By the time of the much more extensive discovery in the Neanderthal in 1856–57 and at Brixham Cave in 1858, there was already, as Qureshi describes, something of "a consensus on human antiquity," at least among scientific experts. This consensus was given more than paleontological value by the contemporary findings of Boucher de Perthes (in France in the 1840s) and the likes of Thomas H. Bowker and Langham Dale (in South Africa in the 1850s and 1860s) that early humans had a culture—characterized by stone tools, early forms of pottery, early forms of metallurgy—similar to that traditionally associated with nonwhite "savages," and that this culture was common to early peoples in far-flung corners of the world. But it was only as the much longer timescales of geological and biological time provided an increasingly accepted framework within which to understand these discoveries, after about 1860, that they were put together to form the sequence that confirmed and extended the conjectural histories of the Enlightenment, establishing that the histories of Western civilization did indeed have a "prehistory" that paralleled the present state of many non-Western peoples. As John Lubbock pointed out, this not only cast in doubt biblical theories of "degeneration" (under pressure of sin) from the ideal state of the Garden of Eden, it also threatened traditional (biblically influenced) understandings of European civilization's "origins" in the early civilizations of the Middle East, and drew attention back instead to the Enlightenment's conjectures about how civilization emerged from a much baser state of "savagery."[24] Thus arose the "new light" that Lubbock identified in 1865.

The 1860s were a decade of frenetic activity in the new field of "prehistory," which was widely advertised in public prints. As early as 1858 the leading learned societies of London had set up a "Cave Committee" to coordinate a concerted campaign of excavations in likely sites around Britain. The first international paleoethnological conference was held in Neuchâtel in 1866, and the first international prehistorical conference took place in Paris in 1867. The British Museum began to mix its early European artifacts ("antiquities") with non-European artifacts of a similar complexion ("ethnography") in 1866.[25] As Qureshi shows, especially in this initial phase of prehistorical debate, many possibilities were considered and aired about early human development, including conjectures about multiple human species and their possible survival into historic time. But most influentially, a series of best-selling works explaining early human origins and development through prehistory were published that systematized and popularized the idea that all human societies had started in "savagery" and passed through a series of civilizational sequences in the process that has been dubbed "social evolution." The most important of these works in the English-speaking world were Lubbock's *Pre-Historic Times* (1865), Edward Tylor's *Researches into the Early History of Mankind and the Development of Civilization* (1865), Lubbock's *The Origin of Civilisation and the Primitive Condition of Man* (1870), Tylor's *Primitive Culture* (1871), and Lewis Morgan's *Ancient Society* (1877). Promiscuously mixing evidence from prehistoric archaeology and contemporary ethnography, this body of work reestablished firmly in educated society the idea of a sequence of development, almost invariable, that had been practically programmed into the human constitution and that thus characterized equally some peoples in prehistory and other peoples in what Europeans considered to be historic time.[26]

By mixing the evidence of European archaeology and non-European ethnography, the new prehistory undoubtedly reinforced the traditional Christian belief in "monogenesis": that is, that all human societies (of all "races") were descended from a common stock. Asians and Africans were not alien species, but members of a single human species. That message was carried loud and clear by all the main social-evolutionary thinkers, who were determined to overthrow the biblical chronology but not necessarily to undermine Christian faith; Morgan, for one, thought the innate tendency of all humans to progress through the stages was a sign of "the plan of the Supreme Intelligence."[27] However, as is generally recognized, the long extension of human chronology entailed in the elaboration of prehistory offered much room for disagreement about the divergence of human stocks after their common origins. Just as different experiences of empire produced different understandings of the civilizing mission, so the study of prehistory produced diverse interpretations of the current state of peoples around the world. There was much debate over why Europeans had attained the highest states of civilization first, and why non-Europeans had, to various degrees, been held back. Unlike scientists today, the Victorians seldom agreed about the operation of biological evolution in historic time. It was possible to assert, as did Herbert Spencer, that the persistence of the savage state in historic time was evidence that savage peoples had, for whatever reason, failed to evolve physically in ways conducive to civilization, and thus that the civilizing mission could not apply to them—it could not substitute for the work of physical evolution achieved by civilized peoples. This assertion was given further credibility by persistent belief, even among "Darwinists," in the inheritance of acquired characteristics, or "use-inheritance": the Lamarckian idea that changes in human culture could be inherited, or, in Spencer's words, that "mental development [was] a process of adaptation to social conditions, which are continually remoulding the mind and are again remoulded by it."[28] Thus Spencer held that "inferior" races were locked mentally in a past from which they could only emerge at a glacial pace; others took this "Social Darwinist" conclusion to mean that "superior" races were likely to overrun, and perhaps extirpate, these "inferior" races before they had a chance to evolve to keep up.[29]

This was naturally a common view amongst the physiologists,[30] who focused on physiological distinctions, but it was not the view held by the most popular figures of prehistory, the "culturalists" as we might call them, led by Lubbock, Tylor, and Morgan. The culturalists assumed that humans were basically all alike—"We have the same brain, perpetuated by reproduction, which worked in the skulls of barbarians and savages in by-gone ages," in Morgan's words—and that, as a result, they tended to develop similar ideas and inventions when faced with similar environments.[31] This explained the many independent inventions of similar tools, rituals, and institutions that could be found around the world. The innate abilities of all humans to learn, and to pass on their accumulated learning through culture, meant that all peoples tended to progress also along roughly similar lines. Thus the culturalists were confident that they could construct a sequence of development, common to all peoples (taking into account the effects of different environments and exogenous contacts with peoples in other stages of development),[32] similar to the Enlightenment's stadial history but no longer conjectural. This confidence they based not so much on the archaeological evidence, which had extended the human timescale with which they worked but was fragmentary and difficult to interpret, as on the tremendous accumulation of ethnographic material that had built up since the late eighteenth century, which could be inserted into that timescale. "Our acquaintance with remote tribes is now so vastly extended," as one reviewer congratulated Tylor, "that we are completely delivered from a large number of misapprehensions which, till recently, rested on seemingly good authority."[33]

Since all human inventions were seen equally to be subject to the laws of development, the culturalists' coverage was broad and ambitious, ranging from the development of primitive kin systems into barbarous clans and modern states to the evolution of animistic views of natural phenomena into the great world religions. But the most popular social-evolutionary sequence—less controversial, perhaps, and also easier to document from both archaeological and ethnographic evidence—was technological. Elaborating on the Three Age model devised by the Danish antiquarian C. J. Thomsen, Lubbock popularized the idea that all human societies progressed from the Stone Age (subdivided by Lubbock into the Palaeolithic and the Neolithic) to the Bronze Age to the Iron Age.[34] By extending the timescale backwards and giving more precision to the Enlightenment's vague ideas about "savagery" and "barbarism," the three-age model highlighted the "distinctively human" qualities of the stages before civilization, showing the "savage" also as, in the words of another social evolutionist, J. F. McLennan, "a tool-user, an artist, a thinker, an ingenious craftsman."[35]

Did this backward extension of human history make "living savages" feel more or less like "men and brothers"? It has been strongly argued that the lengthening of the timescale of development distanced "savages" from civilization by drawing attention to their profound backwardness and querying in sinister ways their failure to develop. "Stone Age" was, in this view, an epithet even more defamatory than "savage" or "primitive." In this way the new schema set the capstone on the "arc of disillusionment" and justified increasingly brutal treatment of "living fossils" that were so far from civilization.[36] But that was not how the culturalists themselves presented their arguments. As their principal targets were "degenerationist" arguments based on biblical authority, they laid emphasis on the progressive nature of humans. All the peoples of the world alive in their own time, they argued, showed signs of progressive development from the Palaeolithic, and, given how very far distant the Palaeolithic was proving to be, even the most "primitive" were closer to moderns than to their origins.[37] Evidence of degeneration, Tylor wrote, usually arose when a more advanced civilization had disrupted the development of a less advanced people—he cited the effects of the slave trade on Africans and described the much-slandered Bushmen as "the persecuted remnants of tribes who have seen happier days."[38]

The culturalists' emphasis on humans as "tool-user[s], artist[s], thinker[s], ingenious crafts[men]," solving problems presented by their environment, was certainly meant to dignify them all, regardless of their stage of development. McLennan defended Australian aborigines against the libels of their oppressors—why should not the inventors of the boomerang be viewed just as Galileo, Newton, and the authors of the Reform Act were viewed?—and compared their social institutions to those developed by the "ancient nations" upon which European civilization was built.[39] Tylor's similar comparisons between the ancient Swiss, the Aztecs, the Ojibwa, and the Zulu were surely intended rather to rehumanize than to dehumanize the latter. "It is no more reasonable," he wrote, "to suppose the laws of mind differently constituted in Australia and in England, in the time of the cave-dwellers and in the time of the builders of sheet-iron houses, than to suppose that the laws of chemical combination were of one sort in the time of the coal-measures, and are of another now. The thing that has been will be; and we are to study savages and old nations to learn the laws that under new circumstances are working for good or ill in our own development."[40] The technological emphasis of the culturalists had the added advantage of insulating them somewhat from contemporary debates about the moral status of "savages." While defending the rational, functional responses of tool-makers to their environments, on the whole they were more cautious about judging the morals of "primitive" societies.[41]

These arguments could therefore have many different implications for the civilizing mission. For those following the "arc of disillusionment," especially the "man on the spot," they could be used to condemn "savages" targeted by empire as living fossils. Regardless of whether or not the West's superiority was inborn, it could be argued that colonized peoples' failure to progress so far doomed them to marginalization, even extinction, as Qureshi shows. But the same arguments could equally be used to justify civilization, either by suasion or by force—a dramatic acceleration of social development, as implied in the utopian projects of earlier evangelicals and utilitarians, now given a firmer, "scientific" justification in proofs of the common mental equipment of all peoples. And they could also be used to justify protection of peoples in earlier stages of development from the disruptive impact of culture contact, as Tylor seemed to be implying in his defense of the Bushmen. All of these uses were, of course, supportive of empire in its various manifestations, but they imply very different kinds of imperial policy—some reillusioning rather than disillusioning, others starting up entirely new lines of thought about the capabilities and futures of "primitive" peoples.[42] And, as the culturalists well knew (and often said), their study was still in its infancy; firm conclusions for policy were hard to make. That accounts in part for their own ambivalence about empire. They did not think "primitive" peoples were stalled in the past, and they certainly looked hopefully to the diffusion of the blessings of civilization to all, sometimes (as in Lubbock) with expectations as utopian as their evangelical forebears; but they were less certain than those utopians about how, or how quickly, those blessings might be diffused. The highest stages of technological development, they knew, gave a godlike power to Europeans that tended to disrupt the stately progression of social evolution, for better or ill.

A telling illustration of the plasticity of social-evolutionary thought can be found in the ideas of Henry Maine. Maine was not strictly speaking a social evolutionist; he was a jurist whose field of study was limited to comparisons between the development of legal ideas in Europe and India. As Lubbock and McLennan complained, he was insufficiently interested in ethnography to get a truly comprehensive grasp of primitive societies, but as Maine would have countered, his interests as a colonial administrator were more practical than theoretical.[43] Nor did he believe so teleologically as did the social evolutionists in a necessary sequence of progressive stages; as he wrote to Darwin in 1883, "There is nothing in the recorded history of society to justify the belief that . . . the same transformations of social constitution succeeded one another everywhere, uniformly if not simultaneously."[44] Nevertheless, his ideas about *a* stage of social development—in his famous phrase, "from status to contract"—were generally interpreted in the context of social-evolutionary thought, and, when applied, given a bewildering variety of interpretations.

Mantena, who coined the phrase, sees his thinking as integral to the "arc of disillusionment"—by evoking so vividly the primitive conditions of customary society, characterized by kinship ties, the village-community, and a degree of stability and cohesion that suggested "immobility," he further distanced "primitive custom" from modern civilization. In Mantena's view, Maine's dichotomization of barbarism and civilization made his schema less stadial even than the Enlightenment's, and cast Indian society so far into the past as to make it indefinitely uncivilizable—one kind of "alibi of empire."[45] On the other hand, she grants that, by affiliating customary society in India with customary society in Ancient Greece and Rome, Maine is recognizing Indians not so much as "fossils" but as "the 'living past' of Europe," asserting "deep affinity" as well as "radical difference." This affinity could be used by utilitarians to warrant not stasis but forced civilization—a different alibi. It could also be used, as Mantena acknowledges, to facilitate reform by developing hybrid modes of governance that recognized both Indian particularity and the demands of modernity, thus bridging Indian custom and modern civilization; this was probably Maine's own preferred interpretation, as he saw himself as the lawgiver on the pattern of the Roman lawyers who bridged "ancient usages . . . with the legal ideas of our own day."[46] This role did give reformers a chance to assist the "quickening" of progress without disrupting the stability of native society.[47] Beyond this, as Mantena is more reluctant to acknowledge, a better understanding of Indian custom could raise the possibility of a specifically Indian path to modernity. Maine's depiction of the village community was eagerly taken up by European land reformers, inspired also by John Stuart Mill's late speculations on alternative forms of land ownership, who were critiquing the unilinear models of progress implied in classical political economy. As Mill himself wrote in his review of Maine's *Village-Communities*, Maine's researches into Indian "custom" might well be found

> to prove, not that institutions and ideas belonging to past times have been unduly prolonged into an age to which they are unsuitable, but that old institutions and ideas have been set aside in favour of others of comparatively modern origin. . . . The question is opened whether the older or the later ideas are best suited to rule the future; and if the change from the one to the other was brought about by circumstances which the world has since outgrown—still more if it appears to have been in great part the result of usurpation—it may well be that the principle, at least, of the older institutions is fitter to be chosen than that of the more modern, as the basis of a better and more advanced constitution of society.[48]

In other words, it was by no means clear to Mill and his followers that the "modernity" of present-day European civilization was the only or the best form for India or for Europe either. Indeed, the village community was also deployed by early Indian nationalists arguing for "India as a dominant, evolving civilisation," crafting its own version of representative government and modern liberty.[49] That is not what Maine had intended at all.

The Salience of Difference

There is room for debate as to how far Mill at this stage had moved away from a unilinear understanding of development more characteristic of the social evolutionists, though his willingness to contemplate progressive development upon non-European lines was not by any means unique. But this turn in his thinking does illustrate the fertility and unpredictability of the effects of the renewed interest in ethnography that the social evolutionists had triggered by affiliating "primitive" peoples in the present to earlier stages of European development. Social evolutionism incited closer study of primitive peoples who had previously been marginalized or, in the "anthropological minimalism" of the Enlightenment, the object more of fantasy and speculation than of serious consideration. In some locales, it supported colonialism and the "ethnographic state"; in others, their interests diverged. In Britain it birthed the modern discipline of anthropology; Tylor held the first designated university position in this field from 1884. Tylor himself, while holding to the orthodox view of unilinear development, showed enough openness of mind to acknowledge that further study might yield hitherto unimagined complexities and diversities.[50] He was right.

If there is an intrinsic tendency for empire, with its ulterior motives of control, to defame the Other, there is a countervailing tendency for anthropology to identify with the Other. Tylor's students were decreasingly willing to rely on the secondhand reports of missionaries and colonial administrators, and by the end of the century, they were poised to initiate a new tradition in ethnography—participant observation—which at least offered the possibility of the observer being drawn into the thought-world of the observed, rather than vice-versa. By looking around the world in search of people like themselves, only in an earlier stage of development, the Victorians could end up learning to appreciate the salience of difference.

Notes

1. *Examiner*, May 27, 1871, 536–37, from a review of Edward B. Tylor's *Primitive Culture*, 2 vols. (London: John Murray, 1871).

2. Ronald L. Meek, *Social Science and the Ignoble Savage* (Cambridge: Cambridge University Press, 1976), remains the standard work; the quotation from Burke is at 173.

3. Uday S. Mehta, "Liberal Strategies of Exclusion," in *Tensions of Empire: Colonial Cultures in a Bourgeois World*, ed. Frederick Cooper and Ann Laura Stoler (Berkeley: University of California Press, 1997), 80.

4. J. W. Burrow, *Evolution and Society: A Study in Victorian Social Theory* (Cambridge: Cambridge University Press, 1966), 43–53; for other arguments that pair Mill's defamatory and reforming instincts, see Javed Majeed, *Ungoverned Imaginings: James Mill's "The History of British India" and Orientalism* (Oxford: Clarendon Press, 1992), 127–28, 132–3, 135–40; Nicholas B. Dirks, *Castes of Mind: Colonialism and the Making of Modern India* (Princeton, NJ: Princeton University Press, 2001), 32–37; and Theodore Koditschek, *Liberalism, Imperialism, and the Historical Imagination: Nineteenth-Century Visions of a Greater Britain* (Cambridge: Cambridge University Press, 2011), 76–87. For the comparisons of the Indians with the ancient Middle Eastern peoples, see James Mill, *The History of British India*, 5th ed., 6 vols. (London; James Madden, 1858), 1:113, 116n, 118, 237. But it has become common to stress the defamation without rooting it in the project of reform, seeing ideas of progress as simply "alibis of empire": see, for example, Uday Singh Mehta, *Liberalism and Empire: A Study in Nineteenth-Century British Liberal Thought* (Chicago: University of Chicago Press, 1999), 67–75, 89–94; Jennifer Pitts, *A Turn to Empire: The Rise of Imperial Liberalism in Britain and France* (Princeton, NJ: Princeton University Press, 2005), 127–33; Karuna Mantena, *Alibis of Empire: Henry Maine and the Ends of Liberal Imperialism* (Princeton, NJ: Princeton University Press, 2010), esp. 184.

5. Richard Price, *Making Empire: Colonial Encounters and the Creation of Imperial Rule in Nineteenth-Century Africa* (Cambridge: Cambridge University Press, 2008), 6–7.

6. Price, *Making Empire*, 128–29.

7. Catherine Hall, *Civilising Subjects: Metropole and Colony in the English Imagination, 1830–1867* (Cambridge: Polity, 2002), 120–38, 186, 205–7; see also James Epstein, "Freedom Rules/Colonial Fractures: Bringing 'Free' Labor to Trinidad in the Age of Revolution," in *The Peculiarities of Liberal Modernity in Imperial Britain*, ed. Simon Gunn and James Vernon (Berkeley: University of California Press, 2011), 53.

8. See, for example, Thomas R. Metcalf, *Ideologies of the Raj* (Cambridge: Cambridge University Press, 1994), ix–x, 42–43, 48; Hall, *Civilising Subjects*, 54; Mantena, *Alibis of Empire*, 1. In this analysis, the failure of the civilizing mission is seen as just as much of an "alibi of empire" as the civilizing mission itself.

9. Mantena, *Alibis of Empire*, 2, 18, 35–36, 55, 87, 187; Price, *Making Empire*, 156; Hall, *Civilising Subjects*, 339, 398, 436; Patrick Brantlinger, *Dark Vanishings: Discourse on the Extinction of Primitive Races, 1800–1930* (Ithaca, NY: Cornell University Press, 2003), 2, 6, 69, 127–30.

10. Nicholas Thomas, *Islanders: The Pacific in the Age of Empire* (New Haven, CT; London: Yale University Press, 2010), 297.

11. Kathleen Wilson, *The Island Race: Englishness, Empire and Gender in the Eighteenth Century* (London: Routledge, 2003), 213, n. 74. This perspective is nicely captured by histories of the civilizing mission "from below," such as those by many of the contributors to David Lambert and Alan Lester, eds., *Colonial Lives Across the British Empire: Imperial Careering in the Long Nineteenth Century* (Cambridge: Cambridge University Press, 2006).

12. Cited by Price, *Making Empire*, 178, and cf. 179, where Price himself concludes that "the old universal humanitarianism of evangelical culture was *still* valid," but only if the Xhosa were broken so as not to be Xhosa any longer.

13. Dirks, *Castes of Mind*, 43–44.

14. Price, *Making Empire*, 156; Dirks, *Castes of Mind*, 51.

15. This undertaking is finally given the historical treatment it deserves in Richard Huzzey, *Freedom Burning: Anti-Slavery and Empire in Victorian Britain* (Ithaca, NY: Cornell University Press, 2012).

16. Philip D. Curtin, *The Image of Africa: British Ideas and Action, 1780–1850* (London: Macmillan, 1965), 389–91, 415–20.

17. Curtin, *Image of Africa*, 420, 428–31; Huzzey, *Freedom Burning*, 123–28, 135–37.

18. Price, *Making Empire*, 142–34; and see John L. Comaroff, "Images of Empire, Contests of Conscience: Models of Colonial Domination in South Africa," in Cooper and Stoler, eds., *Tensions of Empire*, 163–97, on the varieties of colonialism that proliferated in the region at this period.

19. Thomas, *Islanders*, 114–21, 203–5, 253–54, 283; Brantlinger, *Dark Vanishings*, 127–38, 142–45.

20. Brantlinger, *Dark Vanishings*, 87.

21. Cox, *British Missionary Enterprise*, 145–52, 172–74, 182–87, 267.

22. Price, *Making Empire*, 142–43.

23. John Lubbock, *Pre-Historic Times, as Illustrated by Ancient Remains*,

and the Manners and Customs of Modern Savages (London: Williams and Norgate, 1865), 268.

24. Lubbock, *Pre-Historic Times*, 268, 275; Tylor, *Primitive Culture*, 1:52–3. On the South African discoveries, see Saul Dubow, *A Commonwealth of Knowledge: Science, Sensibility, and White South Africa* (Oxford: Oxford University Press, 2006), 105–8. As David Gange has pointed out, traditional understandings of civilizational origins in the ancient Middle East kept their hold in religious circles at least until the discovery of prehistoric evidence there too in the 1890s (*Dialogues with the Dead: Egyptology in British Culture and Religion, 1822–1922* [Oxford: Oxford University Press, 2013], 24–30, 153).

25. Donald K. Grayson, *The Establishment of Human Antiquity* (New York: Academic Press, 1983); A. Bowdoin Van Riper, *Men Among the Mammoths: Victorian Science and the Discovery of Human Prehistory* (Chicago: University of Chicago Press, 1993); Chris Manias, *Race, Progress and the Nation: Sciences of the Ancient Past in Britain, France and Germany, 1800–1914* (London: Routledge, 2013), chapter 5.

26. The best account of the emergence of the idea of "social evolution" remains Burrow, *Evolution and Society*.

27. Lewis H. Morgan, *Ancient Society*, ed. Leslie A. White (1877; Cambridge, MA: Harvard University Press, 1964), 468. Most of the social evolutionists, Lubbock excepted, came from radical, nonconformist backgrounds: Tylor was from a Quaker family; Morgan and J. F. McLennan were Presbyterians.

28. Herbert Spencer, "The Comparative Psychology of Man" (1876), in *Essays: Scientific, Political & Speculative*, 3 vols. (London: Williams and Norgate, 1891), 1:351–70; see, for context, George Stocking, *Victorian Anthropology* (New York: Free Press, 1987), 140–43, 224–25.

29. See also Tony Bennett, "Habit, Instinct, Survivals: Repetition, History, Biopower," in Gunn and Vernon, eds., *Peculiarities of Liberal Modernity*, 102–18.

30. See, for example, William B. Carpenter, *Principles of Mental Physiology* (London: William Benjamin, 1874), 228–29; Henry Maudsley, *The Physiology of Mind* (London: Macmillan, 1876), 161, 214–18.

31. Morgan, *Ancient Society*, 59; see also Tylor, *Primitive Culture*, 144; [J. F. McLennan], "The Early History of Man," *North British Review* 50 (1869): 516–49, 522.

32. Edward B. Tylor, *Researches into the Early History of Mankind and the Development of Civilization* (1865), ed. Paul Bohannan (Chicago: University of Chicago Press, 1964), 3–4, 137–44.

33. Henry Calderwood, from a review of Tylor's *Primitive Culture*, *Contemporary Review*, December, 1871, 210–11.

34. Van Riper, *Men Among the Mammoths*, 184–95; Manias, *Race, Progress and the Nation*, chapter 5.

35. [McLennan], "Early History of Man," 523.

36. Mehta, *Liberalism and Empire*, 107–8; Brantlinger, *Dark Vanishings*, 164–65, 171–81; Koditschek, *Liberalism, Imperialism*, 211–20, but cf. 216 on Lubbock, and 231–33 on Maine.

37. Tylor, *Researches*, 239–40; Lubbock, *Pre-Historic Times*, 474; Morgan, *Ancient Society*, 16–18; and see John McNabb, *Dissent with Modification: Human Origins, Palaeolithic Archaeology and Evolutionary Anthropology in Britain, 1859–1901* (Oxford: Archaeopress, 2012), 54–59. Conversely, it was also now clear how long even Europeans had spent in "savagery," a point nicely made by Chris Manias, "Contemporaries of the Cave Bear and the Woolly Rhinoceros: Historicizing Prehistoric Humans and Extinct Beasts, 1859–1914," in *Historicizing Humans: Deep Time, Evolution, and Race in Nineteenth-Century British Sciences*, ed. Efram Sera-Shriar (Pittsburgh: University of Pittsburgh Press, 2018).

38. Tylor, *Primitive Culture*, 43.

39. [McLennan], "Early History of Man," 531–33.

40. Tylor, *Primitive Culture*, 5–6, 144.

41. Lubbock tended to be more censorious: see *Pre-Historic Times*, 462–72, and especially John Lubbock, *The Origin of Civilisation and the Primitive Condition of Man*, ed. Peter Riviere (Chicago: University of Chicago Press, 1978), 257–74, orig. pub in 1870; Tylor was less so: see *Primitive Culture*, 24–29.

42. This point is well made by Adam Kuper, *The Reinvention of Primitive Society: Transformations of a Myth* (London: Routledge, 2005), 10–11. For further argument along these lines, see Andrew Sartori, "The British Empire and Its Liberal Mission," *Journal of Modern History* 78 (2006): 623–42.

43. Lubbock, *Origin of Civilisation*, 1–2; Burrow, *Evolution and Society*, 232; Stocking, *Victorian Anthropology*, 150–56.

44. Cited in Kuper, *Reinvention of Primitive Society*, 13.

45. Mantena, *Alibis of Empire*, 2–6, 35–36, 86.

46. Mantena, *Alibis of Empire*, 131–32, 145, 158, 171–72, 177. In developing my own understanding of Maine's position, I owe a great deal to discussions with my M.Phil. student Zuleika Cheatle-Conte.

47. This is denied by Mantena, *Alibis of Empire*, 169; cf. Koditschek, *Liberalism, Imperialism*, 231–33.

48. J. S. Mill, "Mr. Maine on Village Communities," *Fortnightly Review*, 1 May 1871, 544, 549–54.

49. C. A. Bayly, *Recovering Liberties: Indian Thought in the Age of Liberalism and Empire* (Cambridge: Cambridge University Press, 2012), 105, 131, 142, 160, 164–67, 170–73, 181–84, 201, 211; for other Indian uses of the social-evolutionary "progress narrative," see Koditschek, *Liberalism, Imperialism*, 235, 271–75, 280, 287, 290, 296, 302–3. Andrew Sartori makes an even stronger claim that for many contemporary land reformers, both British and Bengali, "custom" could be integrated into liberal-universalist understandings of political economy that showed "traditional" Indian society to be already "modern" by any definition: see *Liberalism in Empire: An Alternative History* (Berkeley: University of California Press, 2014).

50. Tylor, *Primitive Culture*, 5, 24–29.

3

The World Beneath Our Feet

Adelene Buckland

One line of insight would have done more than all those lines of description.
—Virginia Woolf[1]

Magical beings abound in geological writing of the late eighteenth and early nineteenth centuries. Humphry Davy's *Consolations in Travel* (1830), for instance, sees its protagonist voyage into outer space with a "Genius" somewhat "above" the human being "in power and intellect." The Genius helps that protagonist overcome "the mortal veil" of his "senses" to see a glimmer of the universe that ordinarily lay beyond human understanding.[2] Later, a mysterious (though human) "Unknown" appears, echoing many of the insights of the "Genius," to describe "the early changes and physical history of the globe" (Davy, 132).[3] Lord Byron's closet drama *Cain* (1822) imagined a similar interplanetary journey to witness a vast prehuman earth history—this time in the company of Lucifer.[4] Gideon Mantell's 1838 *Wonders of Geology* deployed a "higher intelligence from another sphere" to thread fragments of geological evidence into a sweeping story of the earth's history that, it is implied, could only be imagined by the merely human observer. And John Mill's *The Fossil Spirit; or, a Boy's Dream of Geology*

(1858) uses an Eastern fakir's reincarnations to make the geological past visible to a single human consciousness. Each of these magical beings enables speculation on the power and nature of an almost unimaginably long span of time.

As it was constituted by the Geological Society of London, formed in 1807, geology was at the height of fashion in Britain in the early decades of the nineteenth century, underpinned by a strictly empirical rhetoric. This empiricism raised questions about the spatiotemporal limitations of humans as geological observers: if the earth was not merely a few thousand but perhaps millions of years old, then most of its history predated human life. Nobody had seen the processes of the early earth in action. At the same time, much evidence for geological history was locked in an inaccessible underground, stretching thousands of meters beneath the earth's surface. Magical beings were useful figures by which to imaginatively string together the fragments of evidence that remained of this elusive past, giving shape to the extrahuman perspectives that might be required to rightly comprehend a history so vast and so mysterious. Through these supernatural figures, writers and geologists allowed themselves to speculate on the origin, causes, and connections of things that remained out of empirical reach. Because such beings were so obviously fictional, their appearances in geological texts enabled this speculation without sacrificing their authors' commitment to inductive rigor.

But there is a more ignominious magical geological being, the gnome, who features equally commonly in these texts to play with the disjunction between human scales of perception and the scales and depths of geological time. Possibly drawn from the Greek *genomos*, meaning "of the earth," two-footed gnomes possessed *no* special perceptive powers beyond the power to move through rock as easily as a human might move through air. But this mythical power made gnomes useful "virtual witnesses" of worlds buried so deep underground, or so far in the past, as to be inaccessible to human eyes.[5] Fantastical as they may seem, gnomes structured imaginary encounters with invisible or lost geological objects in empirical terms. And they offered writers and geologists a means of reflecting on the potentially schismatic relationship between human beings and the geological earth. While some magical beings represented an effort to imagine the earth as seen from *above*, to achieve a total view of the earth not possible from its surface, gnomes suggested an attempt to understand the past from *within*, to view earth history from the inside out. Encapsulating the spatial and temporal shortcomings of being human and living on (or in) a deep and ancient earth, gnomes offered eighteenth- and nineteenth-century writers a concrete image for the methodological difficulties of generating knowledge about the earth that was both empirical and visionary, stretching far across time and space.

A Being of the Nether World

In the most famous geological work of the century, *Principles of Geology* (1830–33), Charles Lyell set out to rid his science of the supernatural thinking he thought had dogged its advancement. Drawing on David Hume's *Treatise of Human Nature*, Lyell distinguished between two kinds of imagination as they related to scientific work. The first was a kind of reason—the belief that objects continue to exist when we can't see them. Nonetheless, such a belief is "natural and necessary in the human mind," and without it, many attempts at scientific reasoning would be impossible.[6] One example of this was *vera causa* reasoning: that "all inferences from experience suppose . . . that the future will resemble the past," and, by implication, that the past can be understood by its *resemblance* to the present—and not, as many geologists then believed, by its difference from it.[7] The famous subtitle to *Principles of Geology* exemplified such reasoning, drawing on James Hutton's methods, as described by John Playfair in the 1802 *Theory of the Earth*, to produce "an attempt to explain the former changes of the earth's surface by reference to causes now in operation." For Lyell, this method would offer the most powerful tool for making the past visible.

This was important because the method of most geologists writing in the 1820s was to examine rocks and fossils preserved across millennia in order to build maps of the earth's strata. This approach was felt to be rigorously empirical, simply analyzing the extant evidence. For Lyell, however, this did not take enough account of the ways in which that evidence had been lost or distorted, or was otherwise incomplete. Nonetheless, as Lyell wrote in *Principles*, this approach suggested that the earth had begun as a fiery, uninhabitable, molten mass, slowly cooling to support increasingly complex forms of life and culminating in a period of geological quiescence that had seen "the creation of man" (Davy, 137). Earth, indeed, could be imagined to have been created specifically for human beings, who had appeared only after millions of years of geological *progress* toward that goal. In this version of events, the odd volcano or flood could be considered a sort of spasm, reminding humans of more violent earlier epochs. But Lyell argued that this story of geological progress was produced by the second kind of

Humean imagination, an irrational imagination that had nothing to do with things seen, producing absurd fictions such as "winged horses, fiery dragons and monstrous giants."[8] Lyell implicitly compared this view of earth history with the delusional imagination of men "in an early stage of advancement."[9] Such men viewed "occurrences afterwards found to belong to the regular course of events . . . as prodigies," and "ascribed" natural processes to the intervention of "demons, ghosts, witches, and other immaterial and supernatural agents" (1:76). Supernatural beings had no place in Lyell's geology.

The critical thing here, however, is that Lyell also emphasized with renewed vigor just how much of earth history was invisible to human beings. Not only did the underground *represent* a past predating human memory, and not only did its physical location render it inaccessible, but there was also a third, more serious problem. For Lyell, the record of the rocks and fossils, even should it ever be completely reconstructed, was radically incomplete. The chances of any species being fossilized and then surviving millennia was vanishingly rare. Humans simply had too small a sample to know earth history from the rocks and fossils alone. Just because the older rocks seemed to contain only marine and freshwater fossils, for instance, did not mean that only aquatic species had lived in those times. It was simply an index of the fact that deposition and fossilization were much more likely to happen under water, preserving greater numbers of those species. There were also emerging a handful of exceptions to the progressionist view, so that Lyell could argue that the whole narrative was not to be trusted. In addition, the apparent lack of fossils in the oldest rocks did not mean they predated life on earth. Untold aeons and species may have existed before the formation of granite, the supposed first rock, but the evidence for them had been lost in the millennia that had since elapsed. For Lyell, the earth was of indefinite age and had been subjected to constant climatic variation around a mean rather than gradual cooling. And geological progress was an apparition built on a fantasy of omniscience, a wrongheaded belief that the rock record could be reliably reconstructed.

So, faced with the incompleteness of the geological record, what were geologists to do? Lyell had a solution. Geologists would need to invest the seemingly trivial action of present geological processes—the meanderings of rivers through the land, the laborious construction of coral reefs, for instance—with a significance it would take imagination to comprehend. Instead of looking at relics of earth history and describing them until a story of earth history emerged, description itself would need to be cast in a new frame. And, to make his methodological point felt, Lyell called on the services of a gnome, inviting his readers to descend with him deep into the underground, to a world beyond human gaze, in the company of "A being, entirely confined to the nether world," a "'dusky melancholy sprite,' like Umbriel" (1:82): the gnome who travels to the Cave of Spleen in Alexander Pope's 1714 poem *The Rape of the Lock*. Lyell's gnome is "never permitted to 'sully the fair face of light,' and emerge into the regions of water and of air" (1:82). And he is a bad geologist. He can't see erosion, or weathering—geological processes that occur on the earth's surface. He can't see the deposition of new strata beneath the waters. Entombed in the underground, he constructs a fantastical earth history, correlating his infernal home at the earth's core with the present (just as humans assume the surface strata on which they live to be the most recently deposited), and the more distant strata with the past. Starting with the molten, fiery masses deep in the "bowels of the earth," the gnome, moving toward the surface, would first see masses of granite and basaltic rocks, shot through with metallic veins. A little later, molten and metamorphosed rocks containing a few fossils might indicate that life had begun to exist when they were deposited. Later still, the rocks would become more neatly layered, more fossiliferous. Based on this journey, the gnome might with seeming reasonableness conclude that the early globe had been populated by myriad forms of life, but had become progressively hotter and less hospitable. One day, he might assume, "the whole globe shall be in a state of fluidity and incandescence" (1:83). The gnome, using the same methods as "human philosophers," nonetheless frames "theories the exact converse" of theirs—a narrative of decline rather than progress (1:82). Confined to the earth's depths, unable to think beyond the direct evidence of his senses and his own limited spatiotemporal position, the gnome produces nonsense. The point was underscored through literary allusion: Pope's Umbriel travels to the underworld inside his heroine's spleen, where he sees "bodies chang'd to various forms": "Men prove with child, as pow'rful fancy works,/And maids turn'd bottles, call aloud for corks."[10] The underworld reveals diseased, feminine imagination, producing bizarre logics and chimerical monsters. The implication was that human philosophers, trapped on the surface of the earth, were doing the same.

Occult Spirits

Lyell's allusion to Pope and his use of the gnome to help envision processes and spaces beyond human perception, has been discussed by other critics. What has not been explored is a quite conscious, poetic-geological tradition bridging Pope and Lyell, in which gnomes repeatedly occur as vehicles for the imaginative exploration of the underground. Lyell perhaps attempts to terminate this tradition, for his gnome's insights into earth history are predicated on the inadequacies of his vision. But the tradition begins with Pope, who explained his source for Umbriel in his "dedicatory epistle" to *The Rape of the Lock*, claiming to have borrowed his gnomes and sylphs from *Le Comte de Gabalis*, a book written in 1670 by Abbé N. de Montfaucon de Villars and translated into English in 1680, which satirized what Pope calls "the Rosicrucian doctrine of spirits."[11] This "doctrine" was associated with a secret society of Rosicrucian alchemists and healers following the two early-seventeenth-century manifestos for the Fraternity of the Rose Cross.[12] The doctrine was derived from the Renaissance philosopher and alchemist Paracelsus, who argued that Aristotle's four "sublunary" elements comprising the world beneath the heavens—fire, air, water and earth—were tenanted by elemental beings: salamanders, sylphs, nymphs, and gnomes. Challenging the long-held belief that all sublunary spirits were fallen angels, Paracelsus advocated the firsthand study of nature, arguing that elementals were embodiments of the invisible processes by which inanimate matter was made to move, think, breathe, and live. Though invisible to human eyes, the elementals were actually part of Paracelsus's argument for detailed, empirical attention to nature rather than adherence to received doctrine. From their inception, gnomes and other elemental beings, then, represented the imaginative underpinnings even of empirical science, the special and secret processes required to see the earth rightly. In the subsequent Rosicrucian tradition, powers of special sight were accorded to those admitted to the sect, who invented special rituals to help them "see" the otherwise-invisible elementals. Rosicrucian characters, indeed, fill the pages of Gothic novels, including Horace Walpole's *Castle of Otranto* (1764), Thomas Love Peacock's Gothic parody *Nightmare Abbey* (1818), Percy Bysshe Shelley's *St. Irvyne; or, the Rosicrucian* (1811), William Godwin's *St. Leon: A Tale of the Sixteenth Century* (1799), Mary Abigail Brooks's *Zophiel; or, the Bride of Seven* (1833), and Edward Bulwer-Lytton's *Zanoni* (1842). When Pope uses the term "Rosicrucian," he refers only to the Paracelsian "elementals" and not the subsequent tradition of the Rose Cross, but in the literary-scientific texts that referenced him, gnomes continue to be imagined as technologies for revealing to the mind's eye a past, dwelling in an underground, that human beings could not see.[13]

In 1785, then, the MP John Sargent published a verse-drama entitled

The Mine, crediting Pope's "inimitable poem" for having "made the system of the Rosicrucians . . . familiar to every one." Partly, this "system" is archaic and superstitious, helping set his scene in the mysterious underground: "So congenial is it to the human mind," Sargent wrote, "to associate the idea of supernatural agents with darkness and the wonders of the subterraneous world, that such superstitious notions are generally found to prevail among the people who inhabit the mine countries."[14] On the other hand, the inaccessibility of the underground gave a fantastical quality even to seemingly dispassionate scientific accounts of it. In one of his many footnotes to the poem—a technique typical of the "philosophical verse" genre to which the poem belongs—Sargent cited Richard Watson's *Chemical Essays* (1782) to compare human attempts "to explore the internal structure of the earth, by digging small holes in its surface," with a gnat trying "to investigate the internal formation of the body" of an elephant with only "its slender proboscis" as a tool (Sargent, 75). As a result, the "undiscovered space" of the internal earth was "filled" not only with the superstitions of miners but with "the conjectures of speculative men," including Robert Boyle, Thomas Burnet, and Buffon. Sargent's goblins are not only the supernatural beings of local mining lore, then, but agents of geological change and illuminators of the underground for the human scientific mind.

At first, Sargent's gnomes perform geological processes in Miltonic language: they pour deluges, raise volcanoes to create islands, and work until "riven mountains from their base are hurl'd / And elemental wars convulse the world" (Sargent, 9)—an echo of book 4 of *Paradise Lost*, in which the fallen angels tear up the mountains: "So Hills amid the Air encountered Hills / Hurl'd to and fro with jaculation dire, / That under ground, they fought in dismal shade." The line is also echoed in Pope's *Epistle to Dr. Arbuthnot* (1735), in which "Pit, Box, and gall'ry in convulsions hurl'd / Thou stand'st unshook amidst a bursting world."[15] But the gnomes also have a more specific role to play in the story, in which a young woman, Juliana, follows her husband into the mines in order to care for him after his wrongful imprisonment. Her husband (ludicrously) fails to recognize her because of the darkness and because she is wearing a veil. Sargent adds a further Miltonic subplot, this time drawn from *Comus*, in which a woman uses free will and rational thought to withstand a series of magical attempts to seduce her: Sargent quotes Milton's "No Goblin or swart Fairy of the Mine / Hath hurtful power o'er true virginity" (Sargent, 60). But in Sargent's poem Juliana is not assailed by the underground beings—the gnomes—but by a would-be seducer and criminal named Conrad. The gnomes are sent to rescue her.

More specifically, the gnomes are sent by the underground queen to "Raise before" Juliana's "ravish'd sight / The pageant of eternal night; / And let her fancy-kindled soul attain / The unknown wonders of our boundless reign" (Sargent, 28),—mineral wonders as they had been classified by the Swedish naturalist Karl Linnaeus. Sargent admits that Linnaeus's attempts to classify minerals according to their external, visual characteristics were clumsy, and endorses chemical analysis as "more accurate." But he requires the *visual* resplendence of Linnaeus's underground to perform Juliana's "ravishing."[16] His gnomes therefore divide the mineral kingdom into its three Linnaean classes: *Petrae* (earths and stones), presided over by "barren Petra, giant queen" surveying her "dun realms" from a throne of "Imperial Granite" (28–29); *minerae* (salts, sulfurs, and metals) including "Phosphor," glowing green to "counterfeit the beam of day," and the "electric flame" of amber (29–30); and *fossilia* (fossils, earthy soils, and concretions), revealing "fairy scenes" and "heavenly visions" "wreath'd in rocky fold," antlers "branching in wrinkled stone" (30–31). Each category is paradoxical: the bloom of "barren" rocks, the "darkness visible" of *minerae*, the life-in-death of *fossilia*. Sargent reinstates the sensuous in the mineral, drawing out the vitality and brilliance of its seemingly invisible, seemingly inert properties through the poetic agency of the gnomes.

Juliana's gaze on the mineral kingdom is entirely passive. She does not seek out the secrets revealed to her, but has them presented to her "ravish'd sight" while dreaming. Sargent draws on the multiple meanings of the verb *to ravish* here. Rape is more than implied: indeed, Juliana is fortified to withstand Conrad's threat by an *alternative* seduction of her senses by the gnomes. Other kinds of theft and threat prevail: *to ravish* has a now-rare sense "to plunder, rob, steal from" (*OED*), and the parliamentary Sargent hopes that, after reading his poem, his readers might want to "unveil the yawning mountain's store," imitating the gnomes' geological "unveiling" by plundering the earth and its treasures for national economic gain. Since Conrad also seeks to "unveil" Juliana, the poem displaces potential guilt for its imperialist and acquisitive agenda, the plunder of the earth, by transferring it to this sexual subplot: Juliana remains veiled and is saved from rape precisely because she is "ravished" by the sight of the "unveiled" secrets of the underworld. Sexual plunder is displaced by imperial acquisition.

This is possible because of a third sense of *ravishing*, "to remove (a person) from earth, esp. to heaven," and "to transport (a person, the mind, etc.) with the strength of some emotion; . . . to entrance, captivate, or enrapture" (*OED*). Juliana is threatened with physical "dragging away" but is instead rewarded with mental transportation, removed "to heaven," enraptured. The spatial connotations matter, for Sargent's gnomes are the opposite of Lyell's: while the Umbriel of *Principles* is trapped in the underground, unable to see beyond his actual field of vision, Sargent's gnomes enact for Juliana the imaginative transportation across time and space that overcomes her physical entrapment, weaving "Visions of celestial dye" from the very rocks that bind her. The gnomes make manifest otherwise invisible aesthetic and mineral properties in the underground. As Juliana is inspired by the gnomes to survive the darkness, so her readers are inspired by its paradoxical brilliance to penetrate and mine it. The gnomes give concrete form to the imaginative (moral and scientific) transcendence required to *see* the underground. Such "ravishing" is procured almost against the seer's will, moreover, so that the earth is presented as an aesthetic (and economic) gift to the observer, who is situated in a space between body and mind, between what can be seen, and what must be imagined.

Veils and Vases

The Mine is now largely forgotten, but in its day several poets expended considerable effort in puffing it. Charlotte Smith, author of the well-known geological poem "Beachy Head," William Hayley, Anna Seward, and Erasmus Darwin were a powerful literary coterie representing a "generation of favourite poets," according to William Hazlitt in 1818. Each praised *The Mine* as a triumphant national achievement, and Sargent as a poet worthy of comparison to Milton. Indeed, the frontmatter to the second edition of Darwin's famous poem *The Botanic Garden* contained a sonnet comparing Sargent, who makes "with magic light the mineral kingdoms shine," with Darwin, bidding "all the vivid plants with passion glow."[17] The first part of Darwin's poem, *Economy of Vegetation*, is made up of four cantos representing each of the Aristotelian elements, each presided over by Paracelsian beings, including the gnome-powered second canto, "Earth." And the second part of the poem, *The Loves of the Plants*, would render Linnaeus's taxonomy of the vegetable kingdom in verse, just as Sargent had done for the mineral kingdom.[18]

The key difference is that Darwin's poem is cosmological. His underground is not a self-contained world beneath the surface of the earth, but is part of what Darwin calls "the transmigrating Ens," "the perpetual circulation of matter," for which he drew on Hutton's "oeconomy of nature" in which continents were wasted to the ocean floor, melted by the earth's internal fires, and erupted to the surface in endless soil-producing cycles.[19] Darwin's geology exists almost solely to produce soils in which the plants in his "botanic garden" would grow, helping the early world to "dissolv[e]" and "distil" (Darwin, 2.93) the shells of its most ancient forms of life and to "dispart" (2.118) salt from earths and sulphurs, regenerating the earth; and, at the end of the canto, resuming this "vernal toil" (2.542), to power the pleasurable processes by which all life, especially human, was sustained.[20]

Hutton's geology had trumpeted the status of man as an "intellectual being."[21] "The globe of this earth is evidently made for man," he had written, for "He alone, of all the beings which have life upon this body, . . . is capable of knowing the nature of the world, . . . and he alone can make the knowledge of this system a source of pleasure and the means of happiness."[22] "Man," indeed, "is not satisfied, like the brute, in seeing things which are; he seeks to know how things have been, and what they are to be," extending his line of sight into the imaginary territories of past and future. But, confronted by the earth's immense age, the human mind fell short of fulfilling its desires. The perpetual natural cycle in which continents were raised and ravaged represented "an idea that does not easily enter into the mind of man in its totality," the earth's monuments appearing "like a thing that is imaginary."[23] Mountains and other monuments like the pyramids appear "imaginary" because they are too abstracted, too vast, for their relation to the concrete, tangible, events that created them to be comprehended. Hutton reminds his readers that these are events to which they are "daily witness," but which seem so inconsequential that they are too easily forgotten.[24] This was of critical importance to Lyell, too, for whom common geological processes could make and remake worlds over aeons, and for whom the difficulty was encouraging his readers to imagine this process.

Like Sargent's gnomes, Darwin's would work in Paracelsian fashion to embody and represent the invisible properties of nature, making it "shine" for human eyes.[25] But they are also shaped by this more directly Huttonian sense that the problem for the geologist is less that he can't *see* the earth's past and more that he hasn't found a method for imagining it. "The Rosicrucian doctrine of Gnomes, Sylphs, Nymphs, and Salamanders," Darwin wrote, "was thought to afford a proper machinery for a Botanic poem; as it is probable, that they were originally the names of hieroglyphic figures representing the elements" (Darwin, vii). "Hieroglyphic figures," the elementals represent an early human attempt to know the world through personification: "earth" or "air" endowed with properties made concrete and knowable through the action of the beings that represent them. For Darwin, this is what poetry does for the earth. In the "Interludes" throughout the poem, Darwin writes repeatedly on this theme. "The Poet writes principally to the eye," he argues, while "the Prose-writer uses more abstracted terms" (Darwin, Interlude 1, 48). Simile, he writes, should be used to "bring the scenery before" the "eye" (Interlude 2, 93); poetry makes "sentiments and passions visible . . . by describing . . . the effects or changes, which those sentiments and passions produce upon the body" (Interlude 2, 135). Poetry helps infer causes from effects in the manner described by David Hume, performing the crucial connecting work between what we can and can't see. While John Locke had argued for human exceptionalism, because humans alone possessed the power of abstract thought, Hume maintained that all ideas, *including* abstractions, were copies or impressions of our sensory experience. In this tradition, Darwin carves out a practical role for poetry, which reconnects abstract thought with the sensory experience that made it possible, bringing thought back into palpable connection with the body.

The point at which the gnomes begin to assume agency in his poem, then, is exactly the point at which human beings enter the story of the earth. Humans watch "self-born fires the mass fermenting glow" (Darwin, 2.275) in earth's "deep-sepulchred," "rock-ribbed ponderous vault" (2.274), clays "ductile" (2.277) in the heat and spreading over the earth, and then they gaze as the world's first potters mould and heat the same clay in like fashion to make teapots and plates. Later, gnomes turn potters, reflecting human action to produce a vision of the earth as moulded clay:

> GNOMES! as you now dissect with hammers fine
> The granite-rock, the nodul'd flint calcine,
> Grind with strong arm the circling chertz betwixt,
> Your pure Ka-o-lins and Pe-tun-tses mixt;
> O'er each red saggar's burning cave preside,
> The keen eyed fire-nymphs blazing by your side;
> And pleased on Wedgwood ray your partial smile,
> A new Etruria decks Britannia's isle (2.297–303)

The text here describes, and is illustrated with, an image of Darwin's friend Josiah Wedgwood's Portland vase, a dramatic reconstruction of an Etruscan masterpiece. Darwin's gnomes dissect granite and grind chertz, transforming stone into clay as the earth once melted rocks, and as the humans once moulded them to sculpture (itself in imitation of the gnomes' grindings of chertz). Humans and the earth act in cyclical concert, each imitating the other. And, for Darwin, Wedgwood's achievement is *poetic*. He turns the *idea* of the vase, which had been found shattered into pieces, into a real object to be seen and touched, just as the gnomes distil the fragments of the earth into malleable clay. The achievement of the Portland vase, for Darwin, is that it makes the abstract idea of the vase, and of Etruscan pottery, satisfyingly concrete. And, lest this new vase become in itself an abstracted artistic ideal, Darwin's poem reminds us of the work it took to achieve it, of the mixing and grinding of raw matter it took to make it whole.[26]

Darwin takes this a step further, moreover, by drawing human beings into the soil, making them part of the circulating matter his canto describes. The climactic episode of the canto follows this description of the vase, centering on the inexplicable vanishing of Cambyses II's army in the Persian desert as he attempted to invade the Siwa Oasis in the sixth century BCE. Darwin saw this invasion as an obscene act of imperialist aggression powered by an army of slaves, the brutality of which had been repeated by the Mexican and Peruvian empires (history was cyclical) but had also recently been challenged by the American and French Revolutions (history was also progressive). Most importantly, Darwin fills in historic mystery—nobody knows what happened to the army—with the direct action of the gnomes, who "outstretch" their "unnumber'd arms . . . o'er the guilty wretch" Cambyses (Darwin, 2.433–34), withholding food and water from the earth to create famine, and then unleashing a sandstorm, "inhum[ing]" the army's "struggling limbs" (2.490) under "one great earthy Ocean" (2.494). The gnomes literally move matter into and out of human hands (or withhold it), moving clay from the depths of the earth into the potter's wheel and back again. And they move human beings into and out of the world of matter, murdering them and drawing them into the earth, consigning slavery (and starving bodies) to an underworld where they will be made to feed "each bursting bud" anew. The gnomes do not unveil the secrets of history in Darwin's poem: the gnomes *are* its mysterious workings

Darwin's favorite poet, Mark Akenside, described the imagination as "a middle place between the organs of bodily sense and the faculties of moral perception," and Darwin's elementals also give tangible shape to the human imagination by occupying precisely this middle ground, uniting thought and experience in a single figure.[27] They dramatize interconnections between tiny pieces of sensory data and what seem to be (but never are) original thoughts, situating human beings both at the summit of geological history, uniquely situated in time and space to comprehend its immensity, and *within* it, their thoughts reworking sensory data in imitation of and subjection to the geological recycling of matter. There are no "unveilings" in Darwin's poem, unlike Sargent's, in which human beings are occupants of the surface, only trespassing on the deeps. In Darwin, the human being is never fully situated above ground, hoping to penetrate its secrets. Instead, human beings are constituent elements of the processes they seek to describe: processes constantly, unrelentingly on the move.

Kings of the Underworld

Darwin's poetry was later perceived as repellent by the Romantic poets (though several admitted to youthful intoxication with his dazzling heroic couplets), and, as the Terror unfolded in France, a parody of Darwin's libertarian politics wrecked his position as the leading poet of his generation.[28] Both grand cosmological thinking, with its tendency to overreach the evidence, and the materialist and evolutionary tendencies of Darwin's work were unfashionable in early nineteenth-century Britain. Nonetheless, "Rosicrucian" poems of Darwinian reflection lived on, though with changed emphasis. Eleanor Anne Porden's *The Veils; or, The Triumph of Constancy* (1815), is one such poem, inspired in part by Porden's attendance at Humphry Davy's Friday Evening Discourses at the Royal Institution. Porden refers to the "Rosicrusian [sic] doctrine, which peoples each of the four elements with a peculiar class of spirits" and on which "the machinery" of the poem "is founded." This is, she notes, "a system introduced into poetry by Pope, and since used by Darwin, in the Botanic Garden."[29] Furthermore, Porden likens her (male) protagonist's sadness both to Belinda's mourning "for her ravish'd hair" (Porden, bk. 3, line 17) after Umbriel crashes his "bag of passions, sobs, and sighs" (bk. 3, line 19) upon her, and to Juliana's sadness as she descends to "Idria's mine to tend her banish'd love" (bk. 3, lines 122–29).

Nonetheless, Porden claims her elementals "differ" from theirs, in that they represent "the different energies of nature, exerted in producing the various changes that take place in the physical world" (Porden, viii). But these "energies" are more malevolent than they are in Sargent or Darwin. Three ladies each have veils stolen by elemental kings they are condemned to marry unless, within a year, they and their human lovers can recapture the veils. The central relationship is between Maria, whose veil was stolen by the gnome king, Albruno, and her would-be lover, Henry. Albruno is surrounded by allegorized rocky troops, such as the "six primitive rocks: granite, porphyry, marble, serpentine, schist and sienite" (bk. 2, line 64), here dressed "in purple armour" (bk. 2, line 303) or vests with a "silvery lustre" (bk. 2, line 309) according to their appearances in nature. The allegorizing of the rocks is a new technique, not borrowed from Darwin or Sargent, and it invests them with new power. Not only is the gnome king opposed to human desires (unlike Sargent's and Darwin's human-aiding gnomes), but the underground itself is out to defend him.

Porden's subterranean world is relentlessly gloomy, enveloped by "the thick gloom of everlasting night" (Porden, bk. 2, lines 45–56), entirely lacking the refulgence of Sargent's mine. In part, this is because her gnomes are antagonists, "Fiends that delight in hurricane and storms" (bk. 2, line 52), who throw off but "a transient brightness," flashing so quickly it seems "but the lightning's gleam" as they pass (bk. 2, lines 47, 48, 50). Only when Henry accidentally scratches a piece of lead ore, causing an explosion, do "The strata now proclaim their watry birth" (bk. 3, line 275): "The strata now erected, now deprest, / Now disappearing, now again confest, / The events of unrecorded time declare, / Sad monuments of elemental war" (bk. 3, lines 283–86). The rocks only "proclaim," "confes[s]" and "declare" their histories briefly, by the light of human accident. The process of "unveiling" is more accidental and aggressive, and less sensuous and assured, than it was in Sargent's mine.

When Henry reaches the veil, it is in a cavern, set in a crystal vase, and he can't even see it directly. Instead, he sees it reflected in "a wondrous mirror . . . / That both illumines and reflects the cave, / Like that strange gem instinct with life and light / that self-suspended shines and chases night" (bk. 3, lines 138–41). This mirror dispels not the physical darkness of the underground but the moral darkness of its beholders, revealing "the bosom's inmost thoughts" (bk. 3, lines 144–47), showing the disguised Henry "restor'd to native" (human) "form" (bk. 3, lines 141–43). The poem turns scientific discovery into romance, a quest to unveil nature's secrets. What is unveiled, finally, in "rays of solar light" whose "lucid forms [are] too fine for mortal sight" (bk. 3, lines 148–49), is the hidden heroic soul of the scientific observer—an idea drawn from Davy's descriptions of the heroic man of science. Crucially, the quest is not to unveil the natural world, but to recover the veil itself. The veil's owner possesses (but preserves, rather than penetrates, at least at first) the heroine's virginity; he also owns the veil behind which nature's secrets are guarded—secrets "too fine for" most "mortal sight." *The Veils* reflects less a desire to explain the world than the desire to own the *right* to explore it, to decide who should be let into its mysteries and who should be kept out. In the much more cautious post-Revolutionary climate in Britain, it offers a cautious delineation of the correct moral spirit required to see the world accurately, and a new sense that the underground contains elusive treasures that might shock or control human life rather than yield its treasures at human bidding. You can be right inside, or underneath, the objects of your desire, but, without this crucial ingredient, still you might not *see* them. And they will no longer reveal themselves to humans at their pleasure.

While it is not known if Lyell read Porden, he would certainly have been familiar with another (anti-Darwinian) imitation of Darwin written in the late 1810s: *King Coal's Levee*, an attempt to lay out the stratigraphic and geographic order of the rocks in England and Wales in verse, written by Northumbrian poet John Scafe. Like Porden, Scafe allegorizes the rocks, and his poem is full of deliberately groan-inducing puns ("Marquis SLATE . . . was a testy wight; / Would *split* with his best friends upon occasions slight"). It is an aristocratic *jeu d'esprit* drawing on William Roscoe's story of a party of insects and animals, *The Butterfly's Ball, and the Grasshopper's Feast* (1802). As such, its plot is whimsically slight: the only thing that happens in its 1145 lines is the crashing of King Coal's party by a group of "Plebeian PEBBLES," escorted off the premises at the king's behest by a group of thuggish gnomes. The gnomes have no other geological, narrative, or symbolic purpose, comically reduced to little more than doormen. Indeed, in direct contrast with *The Botanic Garden*, the poem does not attempt to explain geological *processes* at all: it is little more than a comic list of the strata, its fun deriving from jokes about their arrangement and distribution rather than by the underlying history of the world that had produced them.

Nonetheless, the book attracted scandalous notice after it was published in a second edition in 1819, with forty pages of appended geological notes written by two Oxford geologists, William Daniel Conybeare and William Buckland, to be used as a teaching aid for undergraduates (including, at that moment, the young Lyell). Simultaneously, however, the poem was almost republished as a seditious pamphlet by a group of radicals in Yorkshire and Lancashire, on the strength of which it was offered as evidence to the Privy Council at Oxford for Buckland and Conybeare's revolutionary dispositions. For, once the pebbles are thrown out of the levee, they reappear united under the wrathful Giant Gravel, who predicts the imminent demise of King Coal: "Prophetic vision sees a *Davy* aid / Earth's daring sons thy kingdom to invade," Gravel proclaims. Davy's safety lamp, (supposedly) lighting up the underground with greater efficiency and safety, would enable the working of ever-deeper coal seams. In this poem, this means the destruction of King Coal, who would eventually be fully mined, and the survival of the unmineable pebbly rabble. Bringing light to the underground, bringing knowledge to the darkness, could mean overthrowing kings, quite literally undermining the existing social and natural order.

The poem and its reception prompted an ironic response from the two Oxford geologists, whose anonymously published "Critical Dissertation on *King Coal's Levee*," immediately and ironically proclaimed the poem "as forming an epoch in the history of Geology."[30] Not only that, the poem was to be read as proclaiming a "new school" of poetry, the school of "'the unideal'" (Buckland and Conybeare, 48), which was to be antithetical to Darwin's glittering style: "Misled by false taste," previous poets of science (i.e., Darwin) "have availed themselves of the fascination of dignified and harmonious verse, and of appropriate and witty allusions to scientific principles or discoveries in nature," they wrote, dividing the reader's attention "between the sense and the poetry" with their copious footnotes (48–49). While Sargent's lengthy notes followed his poem, so that you had to turn to the back of the book in order to find them, Darwin and Porden more typically had pages so densely annotated with notes that they were more footnote than poem. In many ways this form precisely enacted the tension between the transcendental view of Creation potentially afforded by science and the detailed empirical work by which it might be produced.[31] By contrast, Buckland and Conybeare laughingly remarked that readers would be better off reading their (separately written and printed) notes than bothering with the poem at all, since Scafe's poetry had helpfully "emancipate[d] poetry from its cumbrous connexion with sense" (50), "nobly and completely rescuing it from the manacles of meaning" (49). In this new interpretation, nothing about the poem is "dignified," "harmonious," or "appropriate." All its "meaning" is reserved for the geological appendix, which enumerates the strata and their locations in England and Wales. And so "the narrative of the poem" is said to make no sense either: it is neither a political allegory like Darwin's, nor (also like Darwin's) a geological story of cosmological scope: "To what former or future revolutions of the globe, or to what facts in Geology it refers, we are utterly at a loss to conjecture" (59–60). But therein lies its value. Buckland and Conybeare laughingly announce the poem as a bunch of discombobulated descriptions, telling no story either of its own or of earth history, and in doing so they dispel any idea of its revolutionary potential (or of the revolutionary potential of geological knowledge in any sense). Like the best geologists, Scafe describes the rocks and the rocks alone, his poem taking its form from their mangled and chaotic arrangements.

This goes hand in hand with the fact that the gnomes' illuminating powers in Sargent and Darwin are replaced with the Davy lamp, which lights up the darkness in a literal rather than metaphorical sense. The lamp implies that there is no special route to seeing: geology is descriptive, not speculative. No longer in need of imaginative reconstruction, this kind of geology can be at least imagined as safely empirical, productive of lists of strata rather than grand cosmological visions. Though Buckland's storytelling prowess was legendary, and he certainly did evince theoretical speculations on the nature of the earth, in a potentially dangerous climate he and Conybeare seized on *King Coal's Levee* as a chance to advertise their science as a purely descriptive enterprise. Armed with the Davy lamp, the man of science could fantasize about penetrating the secrets of the darkness, liberated from the necessity of imaginative or narrative extensions of the evidence he now saw plainly in the deeps. This fantasy of total penetrability enables an articulation of the stratigraphic project of reassembling history from remains with which Lyell was to take such issue in *Principles*. The facetious tone of both the poem and the critical dissertation remind us, however, that even for Buckland and Conybeare, this remained a quite conscious fantasy.

Sojourners on the Surface of the Planet

It has long been recognized that Lyell was a child of the Scottish Enlightenment, as indebted to Hume and to Hutton as to Darwin before him. He shares with these writers a sense of the limited ability of the human imagination to accurately perceive the world around it, and of the imagination, used rightly, as a nonetheless vital tool of reason. What has been less strongly articulated is just how deeply Lyell was indebted to the empiricist rhetoric of men like Buckland, his teacher at Oxford in the years of the Scafe controversy, and to the Geological Society of London, with its public calls for geology to be a science of fieldwork and stratigraphic mapping. His rejection of the progressionist synthesis of that research was largely due to his fear that progressionist narratives could lead to evolutionary ones, so that he shared with Buckland and Conybeare a resistance to large-scale explanatory mechanisms that would make sense of the earth's past in a single sweep. And his antipathy to evolution was in part a horrified reaction to the suggestion that the human was not a special being whose intellectual powers rendered him closer to god than the apes.[32] Lyell does not want his human beings quite as much of the earth as Darwin, either Erasmus or Charles. And Lyell's gnome has more in common with the malevolent, even useless gnomes of Porden and Scafe than with the enlightening gnomes of Sargent and Darwin, intensifying the sense of an underground whose secrets could only with difficulty be disinterred, whose stories could only cautiously be reconstructed. Lyell simply extended these difficulties to the world above ground as well, stressing the imaginative dimensions of the *work* of empirical science even as he avoided the transcendent perspectives assumed by forebears like Davy.

These tensions are in play as he writes on the restorative power of the *vera causa* geological imagination. Accepting the "uniformity" of the earth's processes and intensities across history he writes, "the deficiency of our information" on "the most obscure parts of the present creation" is "removed":

> For, as by studying the external configuration of the existing land and its inhabitants, we may restore in imagination the appearance of the ancient continents which have passed away, so we may obtain from the deposits of ancient seas and lakes an insight into the nature of the subaqueous processes now in operation, and of many forms of organic life, which, though now existing, are veiled from our sight. Rocks, also produced by subterranean fire in former ages at great depths in the bowels of the earth, present us, when upraised by gradual movements, and exposed to the light of heaven, with an image of those changes which the deep-seated volcano may now occasion in the nether regions (1:165–66).

Lyell restores to human sight both "the appearance of the ancient continents" and the "many forms of organic life" currently living but hidden beneath the sea, drawing on the biblical sense of "veil," "with reference to the next world" (*OED*). Rocks made doubly obscure by their location underground and their prehuman formation (the world behind the veil) are comparable to those rocks already lifted "and exposed to the light of heaven." In effect, Lyell eradicates the spatial binary of depth and surface altogether: the surface, it turns out, is all there is, though it is both deeper and higher than previously imagined. As such, the present earth is the *perpetual* scene of unveiling, bathed in light, in which every ancient rock or fossil is "an image" of a rock or fossil forming in the present but lost in "the nether regions"; every present process is an echo of a once-lost past or a prophecy of a process yet to come. This language of the heavens, combined with the language of subterraneous fire, endows the human being with celestial—yet still empirical—powers of vision.

The key word here is *insight*, "the fact of penetrating with the eyes of the understanding into the inner character of hidden nature of things; a glimpse or view beneath the surface; the faculty or power of thus seeing" (*OED*). The word offers a spatial view of a "power" of "seeing" both visionary and penetrative, unveiling "the hidden nature of things," its intuitions garnered through movement "beneath the surface." In this respect, Lyell's gnome generates a comedy of empiricism: his gnome is able to see the subterranean deeps, but fated only ever to describe them. What he lacks, quite specifically, is *insight*, a sight situated both *within* the body (and therefore real)—and also inside or beneath the earth in a literal sense—but which also requires mental and imaginative traveling beyond it.

This is clearly drawn from Hume and Hutton. But it equally resists the Darwinian implication that the human being can be folded into the history he hopes to describe. It seeks to find a way back to a "higher" view of the world without ever standing quite so far above it as Byron's Lucifer, Mantell's "higher consciousness," or Davy's "Genius"—by maintaining a view that remains, in essence, *human*. Neither a geology of the hellish past, nor a geology of heavenly omniscience, Lyell's geology claimed to be a true geology of the earth, holding the limitations of earthly perspective in view. But its central trick was to fold those subterraneous deeps and heavenly heights into one view, all the time reminding us of their essential unknowability.

Blindness and Insight

In *Blindness and Insight*, Paul De Man famously argued that critical insight is very often produced by blind spots and myopias of which critics themselves are unaware. It has sometimes been assumed that our Enlightenment and nineteenth-century forebears were uncritical of the vagaries of descriptive language, naive in their faith that a simple empirical view of the world would yield its secrets, and blindly anthropocentric in their view of the story of humanity and its relationship to "nature." And yet Lyell and the tradition of geological writers I have explored here produced a powerful and influential view of our earth and the ways in which we inhabit, occupy, and comprehend it that was built *precisely* on this recognition that vision (whether deep, intimate acquaintance with the body of the earth or an almost Godlike or omniscient view of the world from a great height) required an acknowledgment that the position and power of human observers were necessarily limited and indeed controlled by the vagaries of the earth itself. As Lyell put it, "We are mere sojourners on the surface of the planet, chained to a mere point in space, enduring but for a moment of time," but "the human mind is not only enabled to number worlds beyond the unassisted ken of mortal eye, but to trace the events of indefinite ages before the creation of our race, and is not even withheld from penetrating into the dark secrets of the ocean, or the interior of the solid globe." Elemental beings embodied this tension—of occupying both surface and depth at the same time, being both blind and insightful, knowing and unknowing.

"Insight" was just one solution to the imaginative problems posed by the effort to render invisible and lost historic worlds comprehensible via empirical methods in this period. But the concentration of multiple forms of unknowability in the underground—its inaccessibility, its darkness, its radical incompleteness (at least in Lyell's account of it) as a record of the past, and its association with other lost, supernatural worlds—reminds us that nineteenth-century explorations of the past often meant debating the nature and limits of the human imagination. The challenges this presented to the Victorians cannot be understood in rigidly disciplinary terms. I have defined "geological writing" very loosely here for this reason, ranging across texts both "literary" and "scientific" as they cross and borrow from one another and take one another seriously. And *insight* is a term deliberately designed to evoke literary, aesthetic, religious, and scientific sensibilities all at once. It captures the ways in which, in the nineteenth century, the fantastic loomed large and lingered long in even the most seemingly straightforwardly descriptive and empirical encounters with a past whose intangibility was so keenly and so sharply felt. And it reminds us that, for the Victorians who inherited Lyell's geological vision, the past, and the ways in which it could be understood, very rarely offered certain ground.

Notes

1. Virginia Woolf, *The Captain's Deathbed* (1950), 108.
2. Humphry Davy, *Consolations in Travel; or, The Last Days of a Philosopher* (London: John Murray, 1830), 18, 42. Further references are cited parenthetically with page numbers in the text: e.g. (Davy, 132).
3. See James A. Secord, *Visions of Science* (Oxford: Oxford University Press, 2015), chap. 1, for a detailed discussion of Davy's imaginary beings. Jan Golinski devotes a chapter to the "Genius" in *The Experimental Self* (Chicago:

University of Chicago Press, 2016), chap. 2, where he also discusses the "Unknown" as "a projection of the genius he [Davy] aspired to be" (72).

4. See Ralph O'Connor, "Mammoths and Maggots: Byron and the Geology of Cuvier," *Romanticism* 5, no. 1 (1999): 26–42, on Byron's paleontological imagery.

5. On virtual witnessing and the "literary technologies" of science, see Shapin and Schaffer, *Leviathan and the Air-Pump: Hobbes, Boyle and the Experimental Life* (Princeton, NJ: Princeton University Press, 1985; repr. 2011), esp. 60–69.

6. David Hume, *A Treatise of Human Nature: Being an Attempt to Introduce The Experimental Method Of Reasoning Into Moral Subjects* (London: John Noon, 1739–40), T1.4.7.4, SBN 266.

7. David Hume, *Essays, Moral and Political* (London: A. Millar, 1748), E4.21, SBN 37.

8. Hume, *Treatise of Human Nature*, T1.1.3.4, SBN 10.

9. Charles Lyell, *Principles of Geology*, 2 vols. (London: John Murray, 1830–33): 1:76. Subsequent references are given parenthetically in the text: e.g., (Lyell, 1:78).

10. Alexander Pope, *The Rape of the Lock: An Heroi-comical Poem*, 2nd ed. (London: Bernard Lintoot, 1714), canto 4, 33.

11. Pope, *Rape of the Lock*, 15.

12. See Frances Yates, *The Rosicrucian Enlightenment* (London: Routledge, 1972; repr. 2001) for a classic account of the history of Rosicrucianism, with a controversial argument about its foundational significance in the Scientific Revolution.

13. Here, I am indebted to Jan R. Veenstra, "Paracelsian Spirits in Pope's *Rape of the Lock*," in *Airy Nothings: Imagining the Otherworld of Faerie from the Middle Ages to the Age of Reason*, ed. Karin E. Olsen and Jan R. Veenstra (Brill: Leiden, Netherlands 2013), 213–39.

14. John Sargent, *The Mine: A Dramatic Poem* (London: Cadell, 1785), 58. Subsequent references are given parenthetically in the text by page number; e.g., (Sargent, 75).

15. Alexander Pope, *An Epistle from Mr. Pope to Dr. Arbuthnot* (London: J. Wright, 1735), l, 87–88.

16. On Linnaeus and chemical classification, see particularly Rachel Laudan, *From Mineralogy to Geology: The Foundations of a Science, 1650–1830* (Chicago: University of Chicago Press, 1985), 21–35, 70–86; also Lisbet Koerner, "Carl Linnaeus in His Time and Place," in *Cultures of Natural History*, ed. Nick Jardine, James A Secord, and E. C. Spary (Cambridge: Cambridge University Press, 1996), 145–621, at 45; and Daniel Merriam, "Carolus Linnaeus: The Swedish Naturalist and Venerable Traveler," *Earth Sciences History* 23, no. 1 (2004): 88–106. On chemistry in this period more generally, see Jan Golinski, *Science as Public Culture: Chemistry and Enlightenment in Britain, 1760–1820* (Cambridge: Cambridge University Press, 1992).

17. "Polwhele," in Erasmus Darwin, *The Economy of Vegetation* (London: J. Johnson, 1791), frontmatter.

18. On *The Botanic Garden* and its sources and contexts see Patricia Fara,

Erasmus Darwin: Sex, Science & Serendipity (Oxford: Oxford University Press, 2012), and Paul A. Elliott, *Enlightenment, Modernity and Science: Geographies of Scientific Culture and Improvement in Georgian England* (London; New York: I. B. Tauris, 2008), 48–76; for the best full-length study of Darwin's poetry, see Martin Priestman, *The Poetry of Erasmus Darwin: Enlightened Spaces, Romantic Times* (Farnham, Surrey: Ashgate, 2013). I am especially indebted to Priestman's sense of the spatial organization of Darwin's poems and his detailed discussion of Rosicrucianism and its links to Darwin through both the Royal Society and Freemasonry in chaps. 8 and 10. On Darwin and Linnaeus, see especially Janet Browne, "Botany for Gentlemen: Erasmus Darwin and "The Loves of the Plants," *Isis* 80 (1989): 592–621; and Londa Schiebinger, "The Private Life of Plants: Sexual Politics in Carol Linnaeus and Erasmus Darwin," in *Science and Sensibility: Gender and Scientific Enquiry, 1780–1945*, ed. Marina Benjamin (Oxford: Blackwell, 1991).

19. Darwin, *The Economy of Vegetation*, 2.584. Subsequent references are given parenthetically in the text: e.g., (Darwin, 2.93).

20. On pleasure in Darwin, see Noel Jackson, "Rhyme and Reason: Erasmus Darwin's Romanticism," *Modern Language Quarterly* 70, no. 2 (2009): 171–94.

21. James Hutton, *Theory of the Earth; or, An Investigation of the Law Is Observable in the Composition, Dissolution and Restoration of the Land upon the Globe* (1788); repr. in G. W. White and V. A. Eyles, *James Hutton* (Darien: Hafner, 1970), 286.

22. See also James Hutton, *Theory of the Earth, with Proofs and Illustrations*, 2 vols. (Edinburgh: William Creech, 1795), 116–17. On Darwin's geology, see Noah Heringman, *Romantic Rocks: Aesthetic Geology* (Ithaca, NY: Cornell University Press, 2004), 191–227.

23. Hutton, *Theory of the Earth*, 295.

24. Hutton, *Theory of the Earth*, 295.

25. See Heringman, *Romantic Rocks*, on the gnomes and their ability to make visible invisible scientific processes "and to attach them to concrete agencies as yet inaccessible to 'rigorous' science" (223).

26. Maureen McNeil quite rightly also notes that this is a "stylised picture of labour," attributing the work of grinding rocks to the magical agency of the gnomes and focusing attention on Wedgwood as the assembler of these magical forces(*Under the Banner of Science: Erasmus Darwin and His Age* [Manchester: Manchester University Press, 1987]), 20. My attention is rather on the relationship between artistic production and the formation of the world suggested by this passage.

27. [Mark] Akenside, *The Pleasures of the Imagination: A Poem in Three Books* (London: R. Dodsley, 1744), 3.

28. Erasmus Darwin, "Loves of the Triangles," *Anti-Jacobin; or, Weekly Examiner*, April 16, April 23, and May 7, 1798.

29. Eleanor Anne Porden, *The Veils; Or the Triumph of Constancy* (London: John Murray, 1815), vii–viii. Subsequent references are given parenthetically in the text: e.g., (Porden, bk. 3, 1ine 17).

30. [Buckland and Conybeare], "Critical Dissertation on *King Coal's*

Levee," 47, appended to John Scafe, *A Geological Primer in Verse* (London: Longman, 1820), 49–68. Subsequent references are given parenthetically in the text: e.g. (Buckland and Conybeare, 48).

31. Thanks to Ralph Pite for suggesting this line of argument to me.

32. See James A. Secord, "Introduction," in Charles Lyell, *Principles of Geology*, ed. James A. Secord (London: Penguin Books, 1997).

Part Two: Origins

4

Ad Fontes

Simon Goldhill

Benjamin Disraeli had plenty of reasons to be leery of the privileges of genealogy, but it is still something of a surprise to see the verve with which he expressed mockery of contemporary reconstructions of national history, because of their fantasies, ideological projections, and silences: "If the history of England be ever written by one who has the knowledge and the courage, and both qualities are equally requisite for the undertaking, the world would be more astonished than when reading the Roman annals by Niebuhr. Generally speaking, all the great events have been distorted, most of the important causes concealed, some of the principal characters never appear, and all who figure are so misunderstood and misrepresented, that the result is a complete mystification."[1] *Sybil*, the novel in which this historiographical sniping occurs, is obsessed precisely with the restoration of family fortunes, the genealogical imperative, the authorization of class by history, the reestablishment, indeed, of Sybil to her true, inherited identity. None the less, Disraeli, founder of the Young England group and about to become a major parliamentary figure, cynically likens the history of England to the early history of Rome, not as the paradigm of the growth of empire, as one might expect, but as a parallel opportunity for the sort of debunking that made Niebuhr a red rag to bullish conservative historians. Niebuhr had shown how Livy's history of early Rome could not provide a secure basis for a true account, but contained stories that ought to be classed as no more than legends or myths: they were unreliable sources. So, too, claims Disraeli, if a historian had both the intellectual and social nerve to risk such an enterprise, the stories of early England, which formed such a key element in our national narrative, could be shown to be no more than a "complete mystification"—the result of misunderstanding and misrepresentation.

One aim of this essay is to show how this self-positioning of Disraeli provides a telling insight into a particular, dominant aspect of nineteenth-century traveling in time and its specific relation to and disavowal of a foundational scholarly rhetoric of the Renaissance. Disraeli's self-positioning here is undoubtedly complex.[2] The narrator's voice of even a "state of the nation" novel is not necessarily coextensive with the writer's convictions—rather, it is a performative act of self-positioning. He is writing as a public figure, a Jewish convert to Anglicanism, whose popular novels were seen as part of his flashy and louche personality, and whose Jewishness was never fully disavowed. Disraeli knowingly worked his public persona. This combination of a narrator's self-formation, a challenging critique of the political myths of genealogy, and the willful commitment to the rhetoric of restoration through the return to a glorious previous era, sets Disraeli in a particularly intricate and conflicted negotiation of a personal and national past.

To return to the sources—*ad fontes*—was the battle-cry of humanist scholarship of the Renaissance. The phrase is biblical—"As the deer longs for the sources (*ad fontes*) of the waters, so my soul longs for you, O Lord" (Ps. 42:1–2)—but Erasmus made it his own, and it became a watchword of the new learning. In his *De Ratione Studii ac Legendi Interpretandique Auctores* (1511), Erasmus writes that "First and foremost, one must hasten to the sources (*ad fontes*) themselves, the Greeks and the ancients." For, indeed, "all scholarship is blind without Greek learning." He declares ringingly: "I affirm that with slight qualification the whole of attainable knowledge lies enclosed within the literary monuments of ancient Greece. This great inheritance I will compare to a limpid spring of undefiled water; it behoves all who are thirsty to drink and be restored."[3] The appropriation of the language of the Gospels and the Hebrew prophets is even more shocking than the overheated claim for the embracing omniscience of the texts of the Greek past (scarcely tempered by the caution of "with slight qualification"). The heavily freighted metaphors of the pure waters of religious truth and divine inspiration—Christ's call to his own message—have become here the injunction to learn Greek, to return to the "fountain" of Western culture. By contrast, for Erasmus's enemies, learning Greek could thus be derided as the "font of all evil,"[4] and even as a "heresy"[5]—an accusation equally shocking in its extremism, granted that the New Testament had been written in Greek. Within the battles of the Reformation and the Counter-Reformation, an understanding of Greek was seen as a sign of the dangerous, critical, new knowledge that threatened the established values and ways.[6] Yet both sides in what was a physically as well as theologically violent debate, agreed that at the source lay truth. To go back *ad fontes* was to establish not just an origin, but an established, authorized foundation of knowledge.

To return *ad fontes* in the nineteenth century, however, meant a different sort of journey. Erasmus was certain that the return to the source provided a foundational truth. For many scholars of the nineteenth century, the shock of making such a journey was that it produced a destabilizing sense of an ever-receding past, an always-contentious anxiety about what value might be ascribed to the earliest evidence, and an increasing uncertainty about the very recognition of an origin. For them, to go back *ad fontes* produced not the limpid waters of truth so much as the slime and mud of the evolutionary morass, or the abyss of geological time, or a profusion of competing textual authorities. Even the journey back to the Garden of Eden—the spiritual aim of the ascetic practices of the early Church—in the nineteenth century became a passionate disagreement between the proponents of monogenesis and the challenging theorists of polygenesis. If Adam was not the foundational source of the human race, what price the word of God?[7] This essay claims that the nineteenth-century exploration of the journey *ad fontes* fundamentally inverted the commitments of Renaissance scholarship in a way that goes to the heart of the modern concept of temporality and its narratives.

There are some familiar contexts for this claim, which need a very brief sketch before I turn to the biblical and classical material that form the focus of my argument. It has become a commonplace that the nineteenth century experienced a sea-change in the understanding of historical time. Geology was the discipline that, above all, revolutionized the Victorian comprehension of deep time. Geology produced a model of the long history of the material world that challenged in particular the possibility of traditional Christian chronology. In addition, its invention of stratigraphy offered a way of conceptualizing, representing, and articulating the past that, in combination with later evolutionary science, made cyclical or nondynamic natural history harder to maintain.[8] This new awareness of the "abyss of time," which made James Ussher's celebrated chronological precision seem so dated, developed hand-in-hand with a different awareness of the present moment. A new, insistent language of the "signs of the times"—"of today," "of the period," "of the moment"—is only one typical gesture of the historical self-consciousness that marks out the self-aware proclamations of modernity in the nineteenth century. As the past became longer, the present became faster. To talk of oneself as located within a specific moment in history is a paradigmatic positionality of the nineteenth-century subject.[9]

Thomas Hardy gives a particularly vivid impression of this historical subjectivity as he looks back from 1893 to the Great Exhibition of 1851: "For South Wessex, the year [1851] formed in many ways an extraordinary chronological frontier or transit-line, at which there occurred what one might call a precipice in Time. As in a geological "fault," we had presented to us a sudden bringing of ancient and modern into absolute contact, such as probably in no other single year since the Conquest was ever witnessed in this part of the country."[10] For the nostalgic, fictional, rural world of South Wessex, the Great Exhibition is as disruptive an event as the Norman Conquest. It appears as a dividing line in history and in the self-consciousness of change—a true "precipice in Time," as Hardy memorably terms it, a precipice that is vertiginous to look over. This break is expressed as a "geological 'fault.'" The inverted commas around the word *fault* mark it as a trendy technological term—performing the rupture of old and new with a word typographically marked out as newfangled. It is inevitably geology that provides the language for this new sense of historical time. The shock of the exhibition appears as "a sudden bringing of ancient and modern into absolute contact." This is not just a recognition of the excitement of seeing the artifacts of ancient history in a strikingly modern iron and glass building, but also an image of an immediate and pressing clash between past and present as conceptual frames: a cognitive dissonance across time. To look back to 1851 is to recognize how looking back to ancient times is as destabilizing as looking over a precipice.

The combination of geology and giddiness in the face of time is iconic. Faults, folds, fractures, the crushing metamorphosis of fossils produce a destabilizing sense of a damaged record in nature itself.[11] James Playfair, who popularized Hutton's revolutionary geological theories and provided the epigraph to Charles Lyell's seminal *Principles of Geology* (1830–33), remembered visiting rock formations with Hutton and the effect it had on him: "Revolutions still more remote appeared in the distance of this extraordinary perspective. The mind seemed to grow giddy by looking so far into the abyss of time; and while we listened with earnestness and admiration to the philosopher who was now unfolding to us the order and series of these wonderful events, we became sensible how much farther reason may sometimes go than imagination can venture to follow."[12] Playfair, dizzy before the abyss of time, records how his imagination, that prime Romantic faculty, could not in fact travel as far back into time as the reason of the philosopher (as he terms the physician, chemist, and geologist Hutton). Lyell himself is even clearer about the threat of such a perspective. "'In the economy of the world,' said the Scottish geologist, 'I can find no traces of a beginning, no prospect of an end'"—and this challenge to the grounding Christian narrative of the six days of creation and the end of time prompts from Lyell a beautiful, awestruck reflection on how "the imagination was first fatigued and overpowered by endeavouring to conceive the immensity of time" and the concomitant fact that "no resting place was assigned in the remotest distance."[13] Geology's threat is to remove the possibility of a return *ad fontes*, which, as Lyell notes with careful reticence, made the geologist appear "with unhallowed hand, desirous to erase characters already regarded by many as sacred." It is not by chance that Lyell in the *Principles* quotes Niebuhr as an inspiration: just as Niebuhr questioned the privileged authorities of the past, the sacred texts of classics, in a way that, in the eyes of his contemporaries, would lead to a challenge to scripture, so Lyell's analysis of material evidence of the past was deeply threatening.[14] As Ruskin beautifully put it, "If only the Geologists would let me alone, I would do very well, but those dreadful Hammers! I hear the clink of them at the end of every cadence of the Bible verses."[15] Much as Darwin's theory of evolution challenged the perfection of man made in God's image with a theory of continuous and continuing change, so Lyell's geology challenged the solidity of the earth with a narrative of boundless transformation. The Bible loves its genealogies from Adam; Darwin exclaims by contrast "What an infinite number of generations, which the mind cannot grasp, must have succeeded each other in the long roll of years!"[16] It became increasingly harder by the end of the nineteenth century to reconcile geology, biology, and theology—not least because the still desired return *ad fontes* was a journey to be recognized now as necessarily falling short.

Different time-scales became, consequently, an integral factor in the development and regulation of academic disciplines in the nineteenth century. Chronology as a science has a long history, with a particular flourishing in the Renaissance, when scholars of the stature of Joseph Scaliger made serious contributions to the debate.[17] By the end of the nineteenth century, however, it was no longer possible for any serious scholar even to try to integrate the chronology of the generations of the Bible with geological time, nor to use the earliest stories of Greece or of the Bible as an uncontested framework for history. By the end of the century, the idea of the universal flood seemed, even to a broader public, contrary to scientific evidence: there was now a long prehistory for humans that had left no solid philological record. Evolutionary biology, theology, history, and geology worked with different ideas of the generations of man, and increasing professionalization and institutionalization of the academic disciplines through the century helped formalize these distinctions as constitutive firewalls. The disciplines as disciplines worked with multiple, unreconciled temporal frames.

Yet theology and classics, along with mathematics, remained privileged disciplines of the university and the education system in general in England, and had a strong presence also across Western Europe.[18] In mid-century Britain, more than 60 percent of the curriculum in elite schools was dedicated to learning Greek and Latin, with the dual rationale of training the young both in the languages of Christianity and in the culturally privileged excellence of the classics.[19] The nineteenth century also witnessed the most strident and embracing arguments about the social purchase of religion and its institutions since the Reformation, a contest not just between religious groups but about the very role of religion in society. From the evangelical fervor in the first decades of the century through the political fights over emancipation and Erastianism to the loss-of-faith novels at the end of the century, Christianity—its role in the nation and in each person's life—was a dominant force in the public sphere in Britain.[20] As with the classics, Christianity certainly offered the potential for revolution—the cityscape of Britain was transformed by religious building programs; evangelical Christians labored to alleviate the poverty of the slums; missionary movements spread across the world.[21] Yet, as with the classics, the institutions of the Anglican Church were fully integrated into the power structures of the country and could stand for them. For many, rejecting the church of their parents was a revolutionary option.

Christianity and the classics marched in tandem, then, in their authority and insecurity. Christianity had developed in antiquity in the Roman Empire in conscious opposition to the social values around it, and yet it had taken over both the empire and the philosophies within it, as it became the empire's dominant ideological and institutional force. This long, precarious, and often violent negotiation between Greco-Roman culture and Christianity took shape in the nineteenth century as the promotion of the necessary foundations of *Bildung*, the education into and performance of the values of civilization. What pasts counted and how different models of the past could be reconciled were shared public concerns of the nineteenth century.

These three tensions, then—between giddying deep time and the insistent (self-)awareness of the present moment in history; between competing temporal scales, which inform the foundations of competing disciplines; and between the privileged classical and the privileged Christian pasts—together provide a framework for what follows. In the remainder of the essay, I investigate through different genres of intellectual inquiry the move back toward an original source—the journey *ad fontes*—and the different forms it takes in the nineteenth century. In each case, I argue, especially with regard to classical antiquity and the Bible, that we find a telling revision of the scholarship of the past—in particular, of the humanistic scholarly turn—which opens a crucial vista onto the strategies and import of nineteenth-century intellectual traveling in time. I begin this investigation, however, with a case of paraded certainty.

Constantine Tischendorf, who brought to Europe the Codex Sinaiticus, the oldest known manuscript of the New Testament, was explicit that his discovery was designed, from the very moment when he planned his adventures in the Middle East, to be a significant contribution to polemics about the status of the Gospels. He refers explicitly to Erasmus's celebrated and provocative work on the New Testament in dismissive terms, although Erasmus was a hero of Tischendorf's: "The first editions of the Greek text, which appeared in the sixteenth century, were based upon manuscripts which happened to be the first to come to hand."[22] Tischendorf is correct that Erasmus's edition was not based on a complete or even a critical survey of extant manuscripts. Here, immediately, Tischendorf marks a strategic step away from earlier scholarship. Lachmann, a hero of philology in nineteenth-century Berlin, had formalized a method for evaluating the history of manuscript transmission—stemmatics—which demanded a full collection and collation of manuscripts, and an evaluation of the value of each witness with regard to such a model.[23] Not all ancient witnesses are equally valid as sources. The hope was not the impossible dream of returning to the author's text but to reconstruct an archetype. As we will see, both the term "reconstruction" and the gap between an archetype (which may not be singular) and the point of origin are significant.

Tischendorf aims from the start to be up-to-date and scientific. He is clear that critical history had left many Christians "in painful uncertainty as to what the Apostles had actually written."[24] He desired to work not from the *textus receptus*, as other scholars since the sixteenth century had, but to discover "the most ancient manuscripts." He had searched the European libraries with notable successes, but his trip to the East was explicitly to hunt down "some precious manuscripts slumbering for ages in dust and darkness."[25] His self-representation as an adventurous hunter of lost treasures is fully fleshed out in his self-aggrandizing story of discovery. What he promises the reader in an "age in which attacks on Christianity are so common" is nothing less than "a full and clear light as to what is the real text of God's word written, and to assist us in defending the truth by establishing its authentic form."[26]

The Codex Sinaiticus is a marvelous discovery, for sure, fully deserving its mythic narrative. It is our oldest exemplar of the Gospels. But it cannot quite provide the authentic and real word of God that Tischendorf so grandly announces. There are two interconnected reasons why the return *ad fontes* stalls. First, the manuscript itself contains some twenty-three thousand corrections. Their extent is unparalleled in other ancient manuscripts. These changes are early, and are made mainly by six hands; they not only include routine overwriting of faded letters or corrections of spelling, but also the insertion of omitted lines and words, changes in wording, and even the deletion of material. In some cases phrases are deleted by one corrector and reinserted by another. At the very least, Codex Sinaiticus demonstrates "how early Christian groups and individuals read and altered the text."[27] That is, it shows how already in the fourth century, the text was corrupt, alterable, altered, and subject to dissent. It is at least a step or two away from the authentic, real text. Our oldest manuscript cannot provide the pure waters of truth but already needs evaluation, collation, and correction.

How, then, to get back to the original? Tischendorf—and this is the second reason why his return to the sources stalls for us, if not for him—argues a very surprising and very weak case, which is painfully exposed by his translator's naive praise for its certainty. Any text accepted as authentic in the fourth century could not have been written only in the third century, he argues, because the later Christians would have been too close to the production of such a forgery to accept it as true; by the same token, any text written in the second century could not have been accepted by anyone in the third. In a hop and a skip, we are back "almost, if not quite, to the days of the Apostles."[28] The complex and conflicting evidence about the production and circulation of the Apocryphal Gospels, the Synoptic Gospels, and the Gospel of John, in separate and competing religious groups, is reduced to a declaration of a single and direct line of descent. The journey *ad fontes* is marked by its fantasy of simplicity. Yet even so, it cannot finally repress or conceal the "almost, if not quite" (non)arrival at the source.

Such rhetoric is endemic in the textual scholarship of the period, and we must remember that from the heartland of *Altertumswissenschaft* in German-speaking lands, textual criticism was established across Europe as the pinnacle of learning in the most privileged of academic fields, the classics. The desire of textual criticism is evident and often expressed: to reach the very text as written by the author himself. The corruption of manuscripts over time requires their emendation, and the criterion for such emendation can only be the authentic and original form of the author's written words. Yet it is also clear that there can be no autograph manuscript extant from antiquity. And the vast majority of our manuscripts are copies from many hundreds of years after the author's death. Consequently, the authentic, authorized very words must always involve a projection, an imagination, and an evaluation of what might have been. Hence Lachmann's aim is not more than the reconstruction of the archetype, a postulated source, some way from the autograph, from which the extant manuscripts can be argued to derive, stemmatically. There is always a gap between the desire to reach *ad fontes* and its possible fulfillment. Scholars bridge this gap with varying levels of the rhetoric of certainty and varying markers of the closeness of the "almost, if not quite." But the gap always remains.

Textual criticism as a science of the nineteenth century, like Disraeli's *Sybil*, is thus indebted to its genealogical imperative, its family trees of stemmatics. Like Christian teleological genealogies, its procedures and its ideology require a single point of origin, its Adam, in the author's autograph. But textual critics found themselves destined always to work with the messy, complex business of evaluating multiple cousins, uncles, brothers, and surrogates many years down a fragmented line. The earliest evidence is already open to multiple interpretation and the uncertainty of epigonal transformation.

It is thus worth comparing how interpolation in the holy texts is handled. In the Reformation, emendation of the texts of the Gospels, especially the case of the so-called Johannine Comma, was a battleground on which people could be burnt to death for their opinions.[29] Herman van Flekwyk died at the stake in 1569 with Erasmus's deletion of the Comma on his lips: "I have heard, that Erasmus, in his *Annotations* upon that passage, shows that this text is not in the Greek original."[30] The Anabaptist van Flekwyk, faced with the mortal threat of the Inquisition, cites Erasmus as his shield of truth. He is prepared to die on the grounds of Erasmus's textual criticism, just as the inquisitor is prepared to use the text's authenticity as a lethal argument against heresy. In this brutal contest of faith, the return *ad fontes* is a matter of mortal import and absolute truth. By contrast, the Codex Sinaiticus ends the Gospel of Mark at chapter 16, verse 8 with the closing formula *euangelion kata Markon: The Gospel according to Mark*. As scholars itemized, the roughly contemporary Codex Vaticanus, the Codex Bobbianus, one hundred or more earliest Armenian manuscripts, and the earliest Georgian manuscripts also lack the final lines of the Gospel as printed in the *textus receptus* and the King James Bible. For the most of the major textual critics of the nineteenth century—followed by almost all scholars today—this absence, coupled with the significant evidence that major early fathers did not know this ending, was conclusive proof that the traditional concluding eleven verses of the Gospel of Mark (16:9–20) were a later addition—an early addition, but an addition nonetheless. That is, the traditional text of the Latin bible, hallowed by long usage and familiarity, declared authentic—representing the best traditions of the church—by the Council of Trent, was an interpolation. Yet this "false ending" of Mark continued to be used in liturgy. It was printed in smaller letters in a parallel column in the Revised Version of the Bible by Westcott and Hort, but printed still. The issue of interpolation was vigorously debated—and sometimes still is—by scholars (and hated by evangelicals); but earnest divines, textual critics, and laymen who accepted that it was a proven interpolation did not necessarily feel that their faith was undermined or compromised. Renaissance scholars were committed to the existence of *fontes* that could be attained and thus committed to the idea of reaching an authentic text to die for. The return to the earliest manuscripts of the Bible in the nineteenth century produced not just a recognition of the variety of textual traditions and the need for a revised version, but also a sense of unattainable distance and necessary uncertainty—for which no one was prepared to die.

In the nineteenth century, then, the Holy Grail of an authentic original was repeatedly appealed to in the rhetoric of textual emendation, but necessarily remained an absent ideal, a projection of imaginative reconstruction, always behind the curtain of actual, real variations of manuscripts. Such a tension within the rhetoric of authenticity was also made strikingly explicit in the most physical of reconstructions, the movement to restore the churches of England, which transformed the appearance of the city and countryside. In the nineteenth century, fully seven thousand of the nation's ten thousand churches were restored, generally in line with the principles of the Gothic revival.[31] Many restorations were nothing less than full-scale redesigns, and, once again, the challenge to familiar modes of religious experience produced heated debate, eventually resulting in the heritage movement. The case of the Church of the Holy Sepulchre in Cambridge (the Round Church) is paradigmatic.[32] When the building was restored in 1841, the bell-tower was removed and replaced with a pitched roof. The Gothic windows were taken out and replaced with narrow Norman arches. The north aisle was completely rebuilt and extended to make it now the same length as the chancel. A south aisle was added to create symmetry. The gallery inside and the staircase to reach it were removed. The result of the restoration was not just the systematic removal of the signs of the fifteenth century and other later periods, but also the fulfillment of a design that no period had ever seen—a Victorian idealized imagination of a supposed Norman blueprint. With a sanguine sense of consensus, which would turn out to be wholly misplaced, the young Edward Augustus Freeman comfortably asserted that "No one probably objects to the destruction of the superincumbent storey, and the restoration of the nave to its original state."[33] Like textual emendation, restoration takes us back to the desired, fantasized "original state."

Yet the process by which this return *ad fontes* is achieved is expressed with disarming straightforwardness by Freeman: "The building is to be simply brought back to what we know or reasonably suppose from analogy to have been its original state."[34] Like so many claims of simplicity, this statement is layered with complex ideological appeals and assumptions. Restorationists aim at "the original state"—by which is not necessarily meant the first construction of the building, it turns out, but the idealized form of its structure as it existed in the mind of the medieval designers, even and especially when they did not have the technology or the wherewithal to construct that ideal. Restoration is "to seize the opportunity to render the building as nearly as possible what it must have been in the mind's eye of its original designer."[35] Hence the weasel slide in Freeman's credo between "what we know" and what we "reasonably suppose from analogy." This allows space for the claim, in the case of the Church of the Holy Sepulchre, for example, that symmetry *would have been* the aim of the designer (by analogy), although the church had not previously ever been built like that (we know). So, Eugène Emmanuel Viollet-le-Duc, the high priest of restoration, defined his practice thus: "To restore an edifice means neither to maintain it, nor to repair it, nor to rebuild it; it means to re-establish it in a completed state, which may in fact never have actually existed at any given time."[36] This is a remarkably clear statement of the ideology of restoration: it sets out to create what may never have existed at any given time—a constructive act of imaginative idealism. Indeed, *The Ecclesiologist* reveled in "the logical possibility of working a building into a state of *abstract perfection*" through a process of creative destruction.[37] Freeman's expression "brought back" is a disingenuously passive veil for this reworking, which calmly conceals the destructive effort required for abstract perfection.

Where so many textual critics found it necessary to disavow or veil the impossibility of ever reaching the original that authorized their practice, for the ecclesiastical architectural restorationists, the original was *already* an imaginative construct, which may never had had physical existence. Freeman himself was deeply influenced by Niebuhr and by Thomas Arnold, Niebuhr's translator. So, "Style [in architecture] is seen to have a certain equivalency with language. Drawing on Niebuhrian philologic methodology, architectural style, like language, is interpreted by Freeman as an elemental form of cultural expression, one that is understood to embody the unique experience and character of a people."[38] Thus while architecture may not seem to have the cosmic scope of geology, it yet leads to issues of national identity and the profoundest impact of culture on society, just as philology may seem to involve no more than niceties of textual transmission, but can invoke the deepest theological importance.

Tischendorf describes the crisis of faith produced by textual scholarship as a "painful uncertainty as to what the Apostles had actually written." For many, the work of critical history initiated by Wolf in Homeric scholarship, and continued by Niebuhr in Roman history and David Strauss on biblical texts produced not just uncertainty, but a collapse of faith. As Bishop Wilberforce thundered in already too late anticipation, "The alarming question is . . . whether the human mind, which with Niebuhr has tasted blood in the slaughter of Livy, can be prevailed upon to abstain from falling next upon the Bible."[39] So Renan, whose *Life of Jesus* instrumentally stoked the painful uncertainties, wrote iconically, "My faith has been destroyed by historical criticism, not by scholasticism nor by philosophy."[40] Mrs. Humphry Ward's novel *Robert Elsmere* sold more than two hundred thousand copies in America and was a *cause célèbre* for a decade after its appearance in 1888. She too had her faith undermined by ancient history: "What convinced *me* finally and irrevocably was two years of close and constant occupation with materials of history in those centuries that lie near the birth of Christianity."[41] It is, she concluded, with Renan, "*the education of the historic sense* which is disintegrating faith."[42] In *Robert Ellsmere*, the hero, a young man in holy orders, goes through a similar process of historical study resulting in a painful loss of faith—and the novel hit a nerve with the public. *Ben Hur*, the best-selling novel of the period, was constructed as a rejoinder to *Robert Ellsmere*. The novel, as a genre, became the place where the origins of Christianity could be imaginatively restored.[43]

My final example takes us back to a man whom many have regarded as the father of modern critical thinking about classical antiquity, Friedrich August Wolf. Wolf's *Prolegomena to Homer* was published in Latin in 1795, but its influence was felt throughout the nineteenth century, and most directly by Lachmann, whose theorizing of textual criticism became the standard model for the field.[44] Wolf studied the inconsistencies within the Homeric texts, at the level of linguistic variation, manuscript variation and, above all, through the secondary tradition of scholia (scholars' comments collected on the margins of the manuscripts) and the indirect traditions of quotation. He added to this a historical analysis from which he concluded that writing was not available as a technology at the time when the poems of Homer were composed. This led him to argue that the poems were composed orally, over several generations and sites, committed to memory—with all the flaws of memorizing and repetition—edited first in what has become known as the Pisistratid recension, in the early fifth century BCE in Athens—and then subject to scholarly intervention in Hellenistic Alexandria from the third century BCE onward, which provided testimony to long-standing variations in traditions of transmission. It is hard to recover the shock of this argument, so familiar have many of these arguments become. But shocking the case was, not least because it challenged the image of the genius Homer at the fountainhead of Western civilization, and threatened the unitary and privileged status of the *Iliad* and the *Odyssey* as the foundation-stones of the grandest genre of them all, epic poetry. Wolf had been much influenced by the biblical criticism of Eichorn, and what was at stake in this threat to the status of the poetry of antiquity was felt to be similarly transformational.

Yet what is crucial for my argument here is how Wolf's *Prolegomena* also reveals the same pattern we have seen elsewhere—of a return to the sources producing a recognition that such a journey cannot be fulfilled except in the imagination of a kind of idealist reconstruction: "I gave up hope, then, that the original form of the Homeric Poems could ever be laid out save in our minds, and even there only in rough outline."[45] For Wolf, we can by dint of scholarly work "restore the poems to a form that is neither unworthy of them nor inconsistent with the canons of learned antiquity"—that is, "what contented Plutarch, Longinus, or

Proclus," the Homeric text as read in later antiquity, hundreds of years after the first instantiations of Homeric epic. But "if we demand the bard in simon-pure condition . . . we will have to take refuge either in empty prayers or in unrestrained license in divination."[46] This is true also of the scholia, the ancient commentators, where opinions of different scholars from different periods are mixed together and often unattributed: "From the time when the Aristarchan reading became the transmitted text, . . . new emendations and annotations were composed and attached to it in particular, with the omission, in general, of the first authors of the readings, except perhaps where they disagreed among themselves."[47] Like the annotations and vowel pointings of the Hebrew Bible, the Masorah, integral to the production of the received text of the Pentateuch, so the "Homeric scholia are bodies of material that were finished off by different processes of sifting at different periods."[48] So even for what appears to be the most strongly established recension, the Aristarchan, overlays of further readings and corruptions give a confused and difficult version of Homer. That Wolf ends by explicitly likening the text of Homer to the text of the Hebrew Bible underlines—as many feared—that Wolf's work was felt to be aimed at the sacred and the order that stems from the sacred. As Elizabeth Barrett Browning wrote in *Aurora Leigh*, wittily capturing that fear:

> Wolf's an atheist,
> And if the *Iliad* fell out, as he says,
> By mere fortuitous concourse of old songs,
> We'd guess as much, too, for the Universe.

Wolf's *Prolegomena* set the agenda for Homeric criticism for the next century or more. Not only did he establish the scholarly tools that critics would develop and utilize, but also—and, for my purposes here, most tellingly—he made it necessary for generations of scholars to argue not just about interpolations and consistency of narrative, which in Homeric scholarship had a long tradition, for sure, but equally and necessarily about the scene of composition itself.

It is a familiar critical claim that the return to the source or origin is an authorizing gesture that has considerable power in the rhetoric of history, politics, genealogy, and ideology. It has long been recognized, too, that Victorian science, epitomized by the evolutionary biology of Darwin or the geology of Lyell, constructed a worrying destabilization of such a rhetoric of authority, particularly with regard to the originary stories of the Bible, and the role of classical antiquity and the biblical tradition as the foundations of Western cultural value, not least by the postulation of "deep time," a perspective of dizzying scope. Yet this essay has argued that we need to recognize also a different, specific rhetorical strategy of nineteenth-century writing, which takes classical and biblical philology as its privileged model, and which takes shape against a Humanist idealism. This strategy may parade the traditional rhetoric of the authority of the origin, but repeatedly stalls in its journey toward it. It recognizes the impossibility of reaching the goal of absolute authenticity, the source itself, and aims rather at a fictionalized, idealized reconstruction, as a response to new confusions and possibilities of an intellectual traveling back in time. Where Erasmus sought the very words of the Gospel, and martyrs died with his emendations on their lips, even Tischendorf does not proceed beyond the "almost if not quite." Viollet-le-Duc restores what "may never have existed" in the name of an idealized model of the past. To journey toward the source was, for these Victorian writers, to face the uncertainty of multiple authorities and different textual traditions. Caught between the brassy assertions of certainty and the vertiginous effects of deep time, textual critics, architectural theorists, theologians, writers of fiction, and ancient historians found a practical, historical positionality from which they could still imagine themselves to see the *fontes* in the distance, even if they could no longer drink from their pure waters.[49]

Notes

1. Benjamin Disraeli, *Sybil; or, The Two Nations* (London: Henry Colburn, 1845), 14.

2. Charles Richmond and Paul Smith, eds., *The Self-Fashioning of Disraeli, 1818–1851* (Cambridge: Cambridge University Press, 1998); J. P. Parry, "Disraeli and England," *Historical Journal* 43 (2000): 699–728.

3. Desiderius Erasmus, *De Ratione Studii ac Legendi Interpretandique Auctores*, trans. Brian McGregor, in *Collected Works of Erasmus: Literary and Educational Writings, 1–2*, ed. Craig R. Thompson, vol. 24 (Toronto: University of *Toronto* Press, 1978), 24:669.

4. This is the view of Baechem, cited by Erika Rummel, *Erasmus and his Catholic Critics I: 1515–1522* (Nieuwkoop: De Graaf, 1989), 139.

5. Desiderius Erasmus, *Antibarbari*, trans. Margaret Mann Phillips, in Thompson, ed., *Collected Works of Erasmus*, 23:32. See also Desiderius Erasmus, *Erasmi Epistolae*, ed. Percy S. Allen, 12 vols. (Oxford: Clarendon Press, 1906–58) 4:400–11 [*ep* 1167].

6. Simon Goldhill, *Who Needs Greek? Contests in the Cultural History of Hellenism* (Cambridge: Cambridge University Press, 2002), 14–59.

7. See Colin Kidd, *The Forging of Races: Race and Scripture in the Prot-*

estant World, 1600–2000 (Cambridge: Cambridge University Press, 2006); Colin Kidd, *British Identities Before Nationalism: Ethnicity and Nationhood in the Atlantic World, 1600–1800* (Cambridge: Cambridge University Press, 1999); George Stocking, *Race, Culture and Evolution: Essays in the History of Anthropology* (Chicago: University of Chicago Press, 1982), and *Victorian Anthropology* (New York: Free Press, 1987), 247–52; John P. Jackson and Nadine M. Weidman, eds., *Race, Racism and Science: Social Impact and Interaction* (New Brunswick, NJ: Rutgers University Press, 2006), 35–59. See also Sadiah Qureshi, "Looking to Our Ancestors," chap. 1 in this volume..

8. Charles Coulston Gillispie, *Genesis and Geology: A Study in the Relations of Scientific Thought, Natural Theology, and Social Opinion in Great Britain, 1790–1850* (Cambridge, MA: Harvard University Press, 1951); James A. Secord, *Controversy in Victorian Geology: The Cambrian-Silurian Dispute* (Princeton, NJ: Princeton University Press, 1986); Martin J. S. Rudwick, *Scenes from Deep Time: Early Pictorial Representations of the Prehistoric World* (Chicago: University of Chicago Press, 1992), *Bursting the Limits of Time: The Reconstruction of Geohistory in the Age of Revolution* (Chicago: University of Chicago Press, 2005), and *Worlds Before Adam: The Reconstruction of Geohistory in the Age of Reform* (Chicago: University of Chicago Press, 2008); Adelene Buckland, *Novel Science: Fiction and the Invention of Nineteenth-Century Geology* (Chicago: University of Chicago Press, 2013).

9. Reinhart Koselleck, *Futures Past: On the Semantics of Historical Time*, trans. Keith Tribe (Cambridge: Cambridge University Press, 1985).

10. Thomas Hardy, *The Fiddler of the Reels and other Short Stories, 1880–1900*, eds., Keith Wilson, Kristin Brady, and Patricia Ingham (London, 2003), 191.

11. Buckland makes this argument in *Novel Science*.

12. James Playfair, "Biographical Account of the Late Dr James Hutton, F. R. S. Edin.," *Transactions of the Royal Society of Edinburgh* 5 (1805): 73.

13. Charles Lyell, *Principles of Geology*, 2 vols. (London: John Murray, 1837), 1:63.

14. Buckland, *Novel Science*, 119–21.

15. John Ruskin, *The Letters of John Ruskin: 1827–1869*, vol. 36 of *The Works of John Ruskin*, ed. Edward Cook and Alexander Wedderburn (London: Allen, 1909), 115.

16. Charles Darwin, *On the Origin of Species* (London: John Murray, 1859), 287.

17. Anthony Grafton, "Dating History: The Renaissance and the Reformation of Chronology," *Daedalus* 132, no. 2 (2003): 74–85, and *Joseph Scaliger: A Study in the History of Classical Scholarship II: Historical Chronology* (Oxford: Clarendon Press, 1993).

18. Christopher Stray, *Classics Transformed: Schools, Universities, and Society in England, 1830–1960* (Oxford: Clarendon Press, 1998); Sheldon Rothblatt, *The Revolution of the Dons: Cambridge and Society in Victorian England* (Cambridge: Cambridge University Press, 1968), and *The Modern University and Its Discontents: The Fate of Newman's Legacies in Britain and*

America (Cambridge: Cambridge University Press, 1997); Sheldon Rothblatt and Bjorn Wittrock, eds., *The European and American University since 1800: Historical and Sociological Essays* (Cambridge: Cambridge University Press, 1993); Amanda Anderson and Joseph Valente, *Disciplinarity at the Fin de Siècle* (Princeton, NJ: Oxford University Press, 2002); Andrew Abbott, *Chaos of Disciplines* (Chicago: University of Chicago Press, 2003); William Clark, *Academic Charisma and the Origins of the Research University* (Chicago: University of Chicago Press, 2006).

19. Simon Goldhill, *Victorian Culture and Classical Antiquity: Art, Opera, Fiction and the Proclamation of Modernity* (Princeton, NJ: Princeton University Press, 2011), 1–19.

20. See, for example, W. O. Chadwick, *The Victorian Church*, 2 vols. (London: SCM Press, 1966–70); Timothy Larsen, *Contested Christianity: The Political and Social Contexts of Victorian Theology* (Waco, TX: Baylor University Press, 2004); Stewart J. Brown, *Providence and Empire: Religion, Politics and Society in the United Kingdom, 1815–1914* (Harlow, UK: Pearson Education, 2008); Boyd Hilton, *The Age of Atonement: the Influence of Evangelicalism on Social and Economic Thought, 1785–1865* (Oxford: Oxford University Press, 1986); Michael Wheeler, *The Old Enemies: Catholic and Protestant in Nineteenth-Century British Culture* (Cambridge: Cambridge University Press, 2006); David W. Bebbington, *Evangelicalism in Modern Britain: A History from the 1730s to the 1980s* (London: Unwin Hyman, 1989); Norman Vance, *The Sinews of the Spirit: The Ideal of Christian Manliness in Victorian Literature and Religious Thought* (Cambridge: Cambridge University Press, 1985).

21. See Peter Mandler's chapter in this volume; see also Brian Stanley, *The Bible and the Flag: Protestant Missions and British Imperialism in the Nineteenth and Twentieth Centuries* (Leicester: Apollos, 1990); Andrew Porter, *Religion versus Empire? British Protestant Missionaries and Overseas Expansion, 1700–1914* (Manchester: Manchester University Press, 2004); and Brown, *Providence and Empire*.

22. Constantine Tischendorf, *When Were Our Gospels Written?* (London: Religious Tract Society, 1867), 15–16.

23. Sebastiano Timpanaro, *The Genesis of Lachmann's Method*, ed. and trans. Glenn Most (Chicago: University of Chicago Press, 2005); surprisingly undervalued in James Turner, *Philology: The Forgotten Origin of Modern Humanities* (Princeton, NJ: Princeton University Press, 2014).

24. Tischendorf, *When Were Our Gospels Written?*, 17.

25. Tischendorf, *When Were Our Gospels Written?*, 20.

26. Tischendorf, *When Were Our Gospels Written?*, 36.

27. David C. Parker, *Codex Sinaiticus: The Story of the World's Oldest Bible* (London: British Library, 2010), 3. The number of corrections is from this work..

28. Tischendorf, *When Were Our Gospels Written?*, v.

29. See Grantley Robert McDonald, *Raising the Ghost of Arius: Erasmus, the Johannine Comma and Religious Difference in Early Modern Europe* (PhD diss., University of Leiden, 2011), 1–142.

30. Robert Wallace, *Antitrinitarian Biography*, 3 vols. (London, E. T. Whitfield, 1850) 2:272–80, at 277. On the Annotationes, see Erika Rummel, *Erasmus' Annotations on the New Testament: From Philologist to Theologian* (Toronto: University of Toronto Press, 1981).

31. Chris Brooks and Andrew Saint, eds., *The Victorian Church: Architecture and Society* (Manchester: Manchester University Press, 1995), especially Chris Brooks, "Introduction," 1–29, and "Building the Rural Church: Money, Power and the Country Parish," 51–81; Astrid Swenson, *The Rise of Heritage: Preserving the Past in France, Germany and England, 1789–1914* (Cambridge: Cambridge University Press, 2013); Simon Goldhill, *The Buried Life of Things: How Objects Make History in Nineteenth-Century Britain* (Cambridge: Cambridge University Press, 2014), chap. 5.

32. James F. White, *The Cambridge Movement: The Ecclesiologists and the Gothic Revival* (Cambridge: Cambridge University Press, 1962).

33. Edward A. Freeman, *Principles of Church Restoration* (London: J. Masters, 1846), 14. See Swenson, *The Rise of Heritage*; Goldhill, *The Buried Life of Things*.

34. Freeman, *Principles of Church Restoration*, 5.

35. A Looker On, "Theory of Restoration," *The Builder* 28 (1870): 649–50, 650.

36. Eugène-Emmanuelle Viollet-le-Duc, *Dictionnaire raisonné de l'architecture française du XIe au XVIe siècle*, 10 vols. (Paris: B. Bance, 1854–68) 8:14.

37. "Church Restoration. [Review of] *Principles of Church Restoration*. By E. A. Freeman, Esq.," *Ecclesiologist* 7 (1847): 161–68, 163; my emphasis.

38. G. A. Bremner and Jonathan Conlin, "History as Form: Architecture and Liberal Anglican Thought in the Writings of E. A. Freeman," *Modern Intellectual Thought* 8, no. 2 (2011): 299–326, 320; J. Conlin, "'Development or Destruction': E. A. Freeman and the Debate on Church Restoration, 1839–51," *Oxoniensia* 77 (2012): 1–32.

39. [Samuel Wilberforce], "Essays and Reviews," *Quarterly Review* 109 (1861): 248–305, 293.

40. Ernest Renan, *Recollections of My Youth*, 3rd ed. (London: Chapman and Hall, 1897), 224.

41. William S. Petersen, "Mrs Humphry Ward on 'Robert Ellsmere': Six New Letters," *Bulletin of the New York Public Library* 74 (1970): 587–97, 591.

42. Petersen, "Mrs Humphrey Ward," 592

43. Goldhill, *Victorian Culture and Classical Antiquity*, 153–264.

44. See Friedrich August Wolf, *Prolegomena to Homer, 1795*, trans. with introduction and notes by A. Grafton, G. Most, and J. Zetzel (Princeton, NJ: Princeton University Press, 1985), 7.

45. Wolf, *Prolegomena to Homer*, 7. The same moves took place in the study of Sanskrit texts as it became clear that the dates for the written texts and their oral precursors were more and more unstable: see, for example, William Dwight Whitney, *Oriental and Linguistic Studies. First Series: The Veda; the Avesta; The Science of Language* (New York: Scribner, Armstrong, 1873) especially 73–99.

46. Wolf, *Prolegomena to Homer*, 46.
47. Wolf, *Prolegomena to Homer*, 196.
48. Wolf, *Prolegomena to Homer*, 226.
49. Thanks to Theo Dunkelgrün and Michael Ledger-Lomas, as well as to the members of the Leverhulme Project.

5

In the Beginning

Helen Brookman

In the beginning—of English literature—was Cædmon. Cædmon (pronounced /ˈkædmən/, "Cad-mun") was, says Bede, a seventh-century Northumbrian who was given the divine gift of poetry. According to Bede's Latin history, Cædmon was born around 650 at Whitby and lived a secular life until he was of advanced age. He was, by nature, distinctly *not* a poet. Bede tells how, at feasts, when the harp approached, and it was his turn to sing, Cædmon would leave the hall in shame. One night, he went out to the cattle shed, fell asleep, and was visited in a dream by a man who asked him to sing something. Cædmon protested (he did not know how, he said), but the man insisted. Cædmon then found he had been given the divine gift of song and sang a nine-line hymn about God the Creator.[1] He went on to transform the scriptures into vernacular poetry, singing of "the creation of the world and about the origin of mankind and all of the history of Genesis."[2]

Owing to Bede's account, itself an "aetiological myth that accounts for the origin of English poetry on devotional themes," Cædmon had in the nineteenth century—and to a large degree still has—a symbolic status as the first English poet.[3] He is the first English poet for whom we have both a name and an attributed work, and was the first to turn the secular, native verse form to religious subjects. In the eighteenth century, scholarly consensus held that the only poetry that could be reliably attributed to Cædmon were the nine lines recorded by Bede and known as Cædmon's *Hymn*, and modern scholars have returned to that cautious viewpoint.[4] However, in the nineteenth century, the potential recovery of this lost *oeuvre* of early English scriptural poetry strongly appealed, and many were keen to attribute some of the chief works in the known Old English canon to Cædmon. He was fêted as the "Saxon Homer," the "Milton of our Forefathers," and the "Father of English Song."[5] It became a matter of national pride that "the first of the Western European lands to break forth into singing was our own England."[6] Simon Goldhill argues in chapter 4 of this volume that the attempt to return *ad fontes* was "a constructive act of imaginative idealism"; this essay focuses on a similar series of attempts across the century to construct—earnestly and reverently rather than knowingly or playfully—a specific origin for English literature.

Philology, History, Literature: "Anglo-Saxon"
and the Formation of English

As a subject of formal study, English sprang up late in the life of the British university, and the story of the discipline's formation is one of instability and contestation. Establishing what "reading English" meant—what the curriculum should include and what students and practitioners of the discipline should *do* with its contents—was a part of the ceaseless nineteenth-century project to understand what "being English" meant. As the century moved to a close, in the context of the increasing professionalization and specialization of academic research, the boundaries of the emergent discipline—the methods and approaches it should use, the forms of knowledge it should produce, and how they should be organized—were repeatedly drawn, contested, and redrawn.[7]

Textual study was revolutionized by the influence of Germanic philology applied to English texts by editors and lexicographers. To one late-nineteenth-century writer, philology was "a master-science, whose duty is to present to us the whole of ancient life."[8] As has recently been observed, in the way it "brought a consciousness of the historical depth" of language, philology produced the "lexical-semantic equivalent of the dizzying sense of time" created by the advent of geology and, as it enabled new forms of access to the far past, it was "surrounded by the same glamour and excitement of the great archaeological investigations."[9] English also emerged from literary history and the developing practices of literary criticism. Following proposals for a new School of Literature at Oxford, debates raged about the relative significance of the linguistic, historical, and literary elements of textual study; of the medieval and the modern; and about the relationship between English and the study of the classics, history, and other modern European languages and literatures.[10] The successful establishment of English as a university subject did little to repair the faultlines between the diverse practices and subjects of study that are still visible in the modern disciplinary landscape.

"Anglo-Saxon" language and literature had a unique place in this story, one that is "deeply entangled in the politics of racial, national, and imperial identity."[11] First, from the 1820s on, as English emerged slowly as a taught subject in the metropolitan and civic colleges and universities, and flourished particularly in the university extension movement, it demanded a literary canon, beginning with the earliest material that was accessible in translation, to trace the historical development of the national literature.[12] The construction of this national canon and the teaching of it as the "English" curriculum was a key underpinning of the imperial project, which produced and relied upon increasingly racialized understandings of "Englishness." The fallacious and racist belief in a distinct, continuous "Anglo-Saxon" ethnicity (variously understood as Teutonic, Scandinavian, or Caucasian) was, from the mid-century on, bolstered by "scientific" developments in both the natural sciences and comparative philology. The idea of an "Anglo-Saxon race" was used frequently in providential, chauvinist narratives to account for the historical and cultural development of the nation, for the emergence of the national character, and for imperial expansion and the British domination of supposedly inferior "races."[13] As Chris Jones discusses, the placement of "Anglo-Saxon poetry" at the fountainhead of the national canon performs a backward "projection of the myth of cultural homogeneity," an historic "sense of literary English," to support this ethnic-nationalist ideology.[14] Second, as the prime fodder for "scientific," philological study of language, "Anglo-Saxon" played a crucial role in providing the robust "intellectual capital" to legitimize English as a worthy professional discipline rather than, in E. A. Freeman's infamous phrase, "mere chatter about Shelley."[15]

In 1887, in printed debates over the proposed Oxford School of Literature, the figure of Cædmon was deployed metonymically to stand for language study and for early English as the intellectual rigor and cultural and social value of English as a discipline were contested. A prominent advocate of a School of Literature (which he closely allied with classics), extension lecturer John Churton Collins sought to dissect the literary from the philological (in its newly narrow, late-nineteenth-century sense of "linguistically oriented erudition with a bias toward Middle and Old English," which Churton Collins felt "contributes nothing to the cultivation of taste").[16] He argued that, although it would "no doubt be very desirable that . . . the student . . . should be able to discuss the relative merits of *Beowulf* and of Cædmon's *Paraphrase* in the original, and that he should possess a competent knowledge of philology and . . . of history . . . to expect all this to be accomplished in the time now allotted to education is obviously absurd."[17] A fellow objector, Thomas Case, warned that in "nourishing our own language . . . from the savagery of the . . . Anglo-Saxons," "We are about to reverse the Renaissance."[18]

Regius Professor of Modern History, E. A. Freeman, opposing Collins and cleaving to the older, broader understanding of philology as the love of language *and* literature, supported the proposed School of Modern Languages. He sought to refute suggestions that the study of "Anglo-Saxon" was outside "literature," illustrating his argument by way of Cædmon and Milton: "Milton's *Paradise Lost* is confessedly 'literature'; to study it is a literary business. It would seem to be ruled that, if we bring in any reference to Cædmon, the whole business ceases to be 'literary'; it becomes the forbidden study of 'language.'"[19] Freeman believed his philological understanding of English—the way that "naturally takes [Cædmon] in," as distinct from Churton Collins's antiphilological approach, the way that "shuts [him] out"—was the only way "fit to take its place in a University course."[20] The questions that are here applied to Cædmon—Is his poetry "literature"? Is it "English"? Does it have a rightful place in the literary canon and in the university curriculum? If so, how should it be read, interpreted, and appreciated?—are indicative of fundamental issues underpinning the troubled formation of the discipline as a whole.[21] But how had this figure—who decades earlier had been little known—come to represent entire areas of knowledge in this heated rhetoric?

The essay explores how the Victorian Cædmon was brought into being through the processes of attribution and translation and then commemorated as a historical icon. Despite his celebrated status as an oral singer, the medium of the nineteenth-century encounter with Cædmon was chiefly textual and material, enacted via the printed book and the stone monument, as readers sought to produce something knowable and appreciable from a body of material that was often seen as distant, unstable, or obscure, in order to engender a "sense of time, place, and self."[22]

The search for the "Anglo-Saxon past" is often understood to emerge from a "desire for origins"—a desire in which, as Matthew X. Vernon reminds us, "dangers lurk."[23] Scholars in nineteenth-century studies have described the increasing sense the Victorians had of the "remoteness and unreclaimableness of origins."[24] The figures in this essay—for the most part writing and reading in the later decades of the nineteenth century—were engaged in an authenticating form of fabrication, holding in contradictory accord their awareness of the impossibility of a total recovery of the origins of English literature and their earnest desire to see it realized. They include figures who were of major significance in the development of the discipline, such as J. J. Conybeare (1779–1824) or Benjamin Thorpe (1781/2–1870), and others who were not, such as Robert Spence Watson (1837–1911) or the Reverend Stephen Humphreys Gurteen (1836–1898). The essay does not distinguish and consider only scholarship that was formative of accepted knowledge; rather, it seeks to capture a range of voices, elite and popular, which sought to perform various forms of ideological work. Including "lesser" scholarship—some of which contains rather wild contentions—and popular publications provides more overt evidence of a common desire to recover and secure the existence and value of the poet and his work, which is also often present beneath the scholarly caution of the more influential work. The breadth and variety of such response shows that, although he was less frequently remembered than Alfred the Great, Cædmon was a similarly appealing and malleable figure within nineteenth-century cultural memory of the "Anglo-Saxon past."

"*CÆDMON wrote them all*": Creating Cædmon and His Corpus

In 1832, the Society of Antiquaries published a "Prospectus of a series of publications of Anglo-Saxon and Early English literary remains."[25] The prospectus lamented that "for the small portion of Anglo-Saxon learning already rendered accessible to the student, we are in some measure indebted to foreign scholars" and deemed the fact that many important works were still unpublished "a subject of national reproach." It emphasized that the works would "be accompanied, in every case, with an English translation." A new urgency propelled such translation from the oldest form of the national language into the modern vernacular for the use of increasingly diverse readerships (which would ultimately include working class and female readers). The first work the society selected for publication in 1832 was an edition and translation of the Old English poem now called *Genesis*, then known as Cædmon's *Metrical Paraphrase* of the scriptures (or, later, his *Fall of Man*), a work that came to subsume and overshadow the *Hymn* in the nineteenth-century "Cædmonian" corpus.

While readers were faced with the rediscovery of lost early medieval material and the new availability of known material (of the major poetic codices, the Vercelli Book was discovered in Italy in 1822 and the Exeter Book edited by Thorpe in 1842), there was a conspicuous lack of "Anglo-Saxon authors." Other literary traditions possessed such foundational figures: most significantly, as Simon Goldhill has described in chapter 4 of this volume, the "genius Homer at the fountainhead of Western civilization." Although literary culture in the early Middle Ages—a period of scribal transmission and composite authorship—was largely unconcerned with the notion of a singular author, it became a crucial element of the paradigms used to understand and categorize literature, and to frame interpretation, in the nineteenth century. The biography and moral character of the real, historical person were crucial to the work's reception; Victorian readers "conspicuously read the author out of his work" in order to interpret the emerging corpus of otherwise troublingly anonymous early medieval poetry.[26]

In the eighteenth century, the "Anglo-Saxon poet" had emerged as an oral bard, in the model of Homer, Ossian, and the Norse skalds.[27] In the nineteenth century, although antiquary and archivist Sir Francis Palgrave (1788–1861) felt that Cædmon's historical person was unimportant (seeing him as a myth "floating upon the breath of tradition"), few others were content to reside in "the obscurity attending the origin" of the poems.[28] Rather, that obscurity provided space for the construction of Cædmon as author and as man. With little evidence to go on, everything recorded textually—and some things that were not—became significant. Cædmon's name was subjected to philological scrutiny. Palgrave observed that "that very same name is the initial word of the Book of Genesis in the Chaldee paraphrase . . . b'Cadmin or b'Cadmon' . . . a literal translation of . . . the initial word of the original Hebrew text."[29] The very name of the poet is here read as meaning "in the beginning." He was considered morally exemplary, especially given the global influence of English as the language of the empire. The author of *Cædmon: The First English Poet* (1873), Robert Tate Gaskin, felt that "it is a pleasing reflection that this first singer in a language destined to become that of a world-wide race was a good man."[30] To Henry Wadsworth Longfellow, he was "a pious, prayerful monk" with "all the simplicity of a child."[31]

For Cædmon to be established as the foundational national poet, he needed a corpus to rival Homer's or Milton's. National pride rested particularly on the establishment of the aesthetic qualities of Cædmon's work. Although the moment of Cædmon's divine inspiration provided a subject for contemporary poetry and art, the brevity and repetitiveness of the *Hymn* itself were challenging to readers looking for the origins of a subsequent poetic tradition of a very different character.[32] This was part of a broader ambivalence about the aesthetic value of Old English poetry. Geologist, antiquary, and professor of (first) Anglo-Saxon and (later) Poetry at Oxford, John Josias Conybeare admitted that "[the *Hymn*] will scarcely be thought to merit the praises bestowed upon it by [Bede]."[33] In 1832, Benjamin Thorpe defended his study of Cædmon's "mangled remains."[34] Yet Francis Palgrave, reading Thorpe's translation, felt that in "poetical imagery and feeling they excel all the other remains of the North."[35]

Although there was never a consensus about what might have constituted the Cædmonian corpus in the nineteenth century, there was a common desire, even among the most cautious scholars, to speak of Cædmon as the author of multiple poems. In 1828, Conybeare admitted that the *Hymn* was "the only well authenticated remain" of Cædmon.[36] He nonetheless considered seriously the attribution of *Genesis* to Cædmon—first suggested by editor Franciscus Junius in 1655 but rejected by George Hickes in the late seventeenth century—noting the use of the "same poetical ornaments and form" as in the *Hymn*. Cædmon's supposed authorship of *Genesis* was often extended to the other biblical poems found in the manuscript now known as Bodleian MS Junius 11 but then often termed the "Cædmon Manuscript" (*Exodus; Daniel; Christ and Satan*).[37] In scholarly and popularizing works, the careful voicing of uncertainty was underpinned by an evident desire for certainty. As Robert Spence Watson phrased it: "I have no wish to push matters too far, . . . but it is difficult not to believe . . ."[38]

Since Cædmon had been established as a historical author, his work could be understood in the language of aesthetic feeling; his poetry was "the production of the poet's brain in all its beauty of design and perfection of finish."[39] Rejecting notions that composing scriptural verse made Cædmon a derivative copyist, Gurteen asserted that Cædmon was "no mere paraphrast, but a man endowed with the soul and fire of the born poet."[40] Watson described how "the vigour, the picturesque power, the living energy of his language have not been surpassed by those upon whom the mantle of the oldest English poet, the true father of English poetry, has fallen," stating that the *Genesis* was "worthy to be admitted into the long list of glorious poems to which our country's bards have given birth."[41]

Although there was also frequent skepticism about "the large collection of very ancient poems now passing under his name," for many, this process of attribution and aesthetic appreciation was self-perpetuating.[42] Rediscovered in 1822, the vision-poem now named *The Dream of the Rood* was claimed as Cædmon's work by philologist and runologist George Stephens (1813–1895) in 1866.[43] Stephens concluded that "we have fragments of a religious poem of very high character, and that there was but one man living in England at the time worthy [of it]."[44] Stephens developed "unexpected confirmation" of this attribution, claiming that he had discovered in an illegible section of runic inscription on the eighth-century Ruthwell Cross, among the lines very similar to *The Dream of the Rood*, the words "CADMON ME FAWED *(made)*."[45] With this "discovery" (a fantasy that no other scholar has observed on casts, rubbings, photographs or in the original), Stephens felt that the "daring assumption" that Cædmon also wrote *The Dream of the Rood* had been "approved by the very stone itself." This was evidence of incomparable solidity, surpassing textual judgements. Stephens went on to attribute the long narrative poem *Judith* to Cædmon, on the basis of a shared "rhythmical peculiarity" and the same "color" and "Miltonic sublimity," declaring "*CÆDMON wrote them all* [emphasis in original]"[46] However, for modern readerships, it was not enough for scholars to make attributions based on meter or even stone. It was the process of translation that recreated these obscure and difficult poems as sublime and epic works worthy of their place in canon and curriculum.

"Anglo-Saxon poetry in the cadences of Milton": Translation and Epic

The chief motive behind efforts to translate Old English literature was that described by the Society of Antiquaries in their 1832 prospectus: to make the material accessible to students of the period. In his effort to present "Cædmon's Paraphrase, with all its beauties and all its faults . . . for the first time, before the public in an English garb," Thorpe aimed only to "[exhibit] a faithful text."[47] This desire to make the sense of the work accessible is evident in other translations, often accompanied by the intent to replicate the "Anglo-Saxon qualities" of the work: the "form of the original" (Bosanquet); "Anglo-Saxon modes of thought and of expression" (Gurteen); "the force of the grand old mother tongue" (Watson).[48] Watson gives one contemporary impression of "Anglo-Saxon verse" as "strong, manly alliterative verse, often no doubt bald and tame enough, always serious and somewhat tinged with sadness, but with a rough, fresh beauty and power which is all its own."[49] The increasing appreciation of the aesthetics of Old English poetry, following advancements made in philology, did not occur uniformly or without contradiction (with, on occasion, the same writer expressing "both admiration and distaste . . . simultaneously").[50] The actuality of Old English verse was difficult to assimilate into the Victorian understanding of poetics; a translator had to balance the inherent qualities of the text with contemporary conventions and tastes. While its antiquity, "manliness," and Englishness were celebrated, many translators sought to align early medieval literature with, and read it through the lenses of, more familiar and acceptable literary styles, making use of features and forms associated primarily with other periods of literature entirely. This produced the "pseudo-continuity" that Chris Jones observes in some contemporary Anglo-Saxonist poetry, in which "the ideals of later English poetry are written back over the imagined literature of the Anglo-Saxons," creating "desired forms." [51] Later critics have often dismissed such approaches to translation as egregious misreadings that fail to appreciate the inherent poetic qualities of Old English verse on their own terms. But this maneuver provides a useful insight into how the Victorians experienced the "Anglo-Saxon past," being motivated by a sincere desire to capture and revivify a literary quality that had been obscured by time and loss of understanding.

The Cædmonian *Genesis* was retrospectively understood as the first English epic, a primitive analogue to classical epic.[52] Other literary traditions had their national epics—like France's *Chanson de Roland*—and this was one of gaps in the cultural-historical landscape that Old English literature was deployed to fill. The recovery of lost epics and the writing of new ones became a way of "expressing the national spirit."[53] The British-born American charity organizer Stephen Humphreys Gurteen, who wrote *The Epic of the Fall of Man: A Comparative Study of Cædmon, Dante, and Milton* in 1896, argued that Cædmon's *Genesis* could not be denied "the lofty title of epic." As well as the "Saxon Homer," Cædmon was the "Milton of our Forefathers." Milton was positioned as the leading English literary example of religious epic; Cædmon was then imagined as the initiator of the genre. Cædmon was closely associated with Milton not only because they both composed long poems based on the Book of Genesis. That Franciscus Junius and John Milton knew one another at the time Junius was editing the *Cædmon Manuscript* and Milton was writing *Paradise Lost* provided fertile ground for speculation about Cædmon's influence on Milton.[54]

To some writers, Cædmon's artistry was inferior to Milton's. Isaac D'Israeli, literary historian and father of Benjamin, (1766–1848) cautioned that one must not "mistake nature in her first poverty, bare, meagre, squalid, for the moulded nudity of the graces."[55] To others, determining the quality and originality of these works was essential for establishing their value as objects of study; to William F. H. Bosanquet, translator of *The Fall of Man; or, the Paradise Lost of Cædmon* (1860)—later translations used loftier titles than "Paraphrase"—Cædmon's poem "resembles so strongly the sacred writings that one approaches the subject with a feeling of reverence."[56] To those suspicious of Milton's Puritanism, Cædmon's poem was preferable, as it was "not tainted with theological error or polemic banter, with which Milton's is much defaced."[57]

This close association between Milton and Cædmon was a crucial element of the conception of the Cædmonian *Genesis* as the first English epic, as described most emphatically by Gurteen: "It is an epic of the whole of creation; of infinitude in space and time; and, in common with *Paradise Lost*, is charming, illogical, and at times incomprehensible."[58] The asserted status of Cædmon's poem within this tradition was verified by the mimicry and amplification of "epic" styles and forms in its translation.

One of the most evident differences between the Old English *Genesis* and Milton's *Paradise Lost* is their poetic form, particularly their meter.[59] Understanding of the Old English poetic line—what Isaac D'Israeli termed its "elastic rhythms"—was newly emerging in the nineteenth century, and was not uniform among these writers.[60] The interpretation furthest from modern understanding is displayed by Bosanquet, who states that the "metre of Cædmon's first poem I believe to be in the Heroic measure of five feet, making ten or eleven syllables, . . . the same as Chaucer's and Shakespeare's." Bosanquet achieves this improbable scansion by some creative lineation and the identification of a large amount of "elision, or blending of syllables," however unlikely ("*Sweglbosmas*, a disyllable").[61] He admits that "the difficulty of dividing the poem into lines of five feet increases as one proceeds; but I believe . . . may be overcome." Underlying this belief was a deep investment in the longevity of classical English verse forms: "To Caedmon, then, I believe English poetry is indebted for the heroic line; or rather to that inspiration which . . . Caedmon received in the stable."[62] If Bosanquet could name Cædmon (rather than Chaucer) as the inventor of heroic verse, then he could claim it as a "native" form—not influenced by French—and even trace its invention directly to God.[63]

Other translators did not go quite so far, but they did choose epic meters for their modern translations. D'Israeli deemed Conybeare's translations "unrivalled" in 1841, but commented on the extent to which his use of pseudo-Miltonic blank verse altered the work: "If the rude outlines are to be retouched, and a brilliant colouring is to be borrowed, we are receiving Anglo-Saxon poetry in the cadences of Milton and the 'orient hues' of Gray."[64] Yet despite this perception of anachronism, Conybeare's style continued to be influential. Gurteen, too, chooses blank verse for his translation. His *Athenaeum* reviewer—who felt Gurteen was "essentially a caterer for the 'public'"—noted that the translator "feels himself a genuine pleasure in these antique metres, and we shall be glad indeed if he can inspire others with it," accepting it as a sufficient vehicle for popularizing the poetry.[65] Gaskin uses heroic verse in cantos to reveal the hidden epic qualities of the text. This was deemed by one reviewer to be "superlatively unfit to convey any idea of Cædmon's rugged directness," who felt that "poetic quality—nay, common literary accomplishment—is totally absent."[66] These metrical markers allow the translators to associate Cædmon—whose work appeared so alien and primitive at first encounter—with not only Milton, but Homer, Virgil, Chaucer, and Shakespeare.

By aligning these works with the greatest works of literature of other periods, these translators aimed to "affirm and legitimate their place at the fountainhead of the English tradition."[67] These creative artifacts were not knowing satires. Bosanquet described the reverential mode of scholarship that produced them: "When speaking of the Poems of the Father of English song, and perhaps the first Christian author . . . of modern Europe, one assumes instinctively a calm and subdued air." It is not only Cædmon's originary role that inspires this veneration, but the fact that these translations act as proxies for both the original "sacred writings" and their "Anglo-Saxon" adaptations. Although they are evidently creations, their authors were concerned primarily and seriously with seeking to produce texts that more authentically delivered to their readers the "truth" of the original version.

"Time-fretted ruins": *Local Memorials and Poetic Lineages*

On September 22, 1898, the *Times* (London) reported, "Yesterday, in beautiful weather, an interesting ceremony was performed in the ancient parish churchyard at Whitby on the occasion of the unveiling of a monument to Cædmon, the first English Christian poet."[68] A committee led by Canon Hardwicke Rawnsley, one of the founders of the National Trust, had raised subscriptions to erect a Northumbrian sandstone cross near the top of the 199 steps. The dedication reads, "To the glory of God and in memory of his servant Cædmon, who fell asleep hard by, A.D. 680." The stone cross was fashioned after the Ruthwell Cross, a replica of which had been purchased for the casts gallery at the Victoria and Albert Museum in 1849.[69] As well as relief carvings of Cædmon, Hilda, Christ, and David, the modern cross displays the text of Cædmon's *Hymn* in runic characters, minuscule script, and modern English translation. Although there are no known runic versions of Cædmon's *Hymn*, a replica cross with runic inscriptions sought to speak of Cædmon's antiquity.[70]

The monument was unveiled by the poet laureate, Alfred Austin (1835–1919). The gathered company included literary scholar Israel Gollancz, geologist G. A. Lebour, and the Bishops of Bristol and Hull. Among those who, "unable to be present, had signified their deep sympathy with the work and the occasion" were architect Alfred Waterhouse, Anglo-Saxonists J. A. Earle, A. S. Napier, and W. W. Skeat (who provided the translation for the cross), Robert Spence Watson (author of *Cædmon: The First English Poet*), and a collection of bishops (both Anglican and Roman Catholic), canons, deans, and the Archbishop of York. It was, the Bishop of Bristol felt, the "most beautiful cross in the North of England, and that means the world."[71]

Austin was asked to give an address, reported in full in the *Times*, "partly no doubt, by reason of the ancient office I have the honour to fill, and partly, I believe, because of my own northern blood." The new poet laureate reminded his audience that "there was recently unveiled another cross to another English poet, Cædmon's latest descendant in the tomb, the never-to-be-forgotten author of 'In Memoriam.'" The Tennyson monument, a granite stone cross at Freshwater on the Isle of Wight, had been erected in 1897. Austin had recently written about another Alfred closely associated with English national identity, King Alfred the Great, in his verse-drama *England's Darling* (1896).[72] Austin used this occasion to trace a venerable lineage of poets laureate, from Whitby poet Cædmon ("Hilda's Laureate") to himself, a Yorkshireman: "He is the half-inarticulate father of poets yet to be and it is, as I understand it, not only to the lisping ancestor, but to his full-voiced descendants in this island throughout all time, that this memorial cross has been erected." Even as he cast Cædmon's work as primitive and incoherent, Austin used the opportunity presented by the object and the occasion to self-authorize, conferring the monumental qualities of Cædmon and Tennyson on the office of laureate and thus on himself.

The memorializing function of the Cædmon cross served the construction of locally as well as nationally constituted identities. The placement of the physical form of the stone cross in the dramatic Whitby landscape was emphasized by many accounts. The location, at the top of the cliffs, in the shadow of the abbey, had been made famous by the publication of *Dracula* the previous year. In his address, Austin drew on the natural landscape:

> "Think of what your billow-beaten bluffs, and wind-swept, beck-furrowed moorlands must have been in the time of Cædmon! Then Nature wielded an unchallenged sceptre; and it was from her mystic voice and majestic presence that Cædmon learned to shape his verse. (Cheers)."[73]

His speech is rhetorically crafted for his audience and the occasion, drawing on local dialectal terms ("beck"), incorporating Anglo-Saxonizing alliterative patterns ("billow-beaten bluffs"), and painting the surrounding landscape as a source of poetic inspiration.

In the preface to the first edition of *Cædmon: the First English Poet* (1873), which he termed "humble," "popular," and of "no more than local value," Whitby writer Robert Tate Gaskin stated that the work was "printed with the idea that there may be many visitors to this romantic district who will be glad to be reminded that they are looking upon scenes once familiar to the earliest English poet."[74] In the second edition (1902), he noted the "growing interest" in Cædmon, claiming that it was "now not difficult for the reader to learn the story of the homely bard who, thirty years ago, was almost an unknown character." Gaskin's touristic intent is evident in some of the more romantic passages of his work—"The sun rises early upon the time-fretted ruins of Streonshalh [Whitby]"—and he places Cædmon as a figure of timeless significance: "Learning and artistic ability and deft hands have united to give to this age a pillar of stone, destined, we may reasonably believe, to teach its lessons far into the distant future."[75] A reader could come to Whitby, read Gaskin's book, view inspirational nature, behold the cross, and experience the feelings that proximity to England's first poet ought to inspire.

The cross had a particular significance for local readers who could remember the nearby Abbey Plain cross, an "ancient, cemetery cross, a fluted column, with its deeply-worn rows of steps," which had long been "the target of irreverent lads, who have helped on the too tardy efforts of time and weather to destroy its graceful rood." In celebrating the new cross, Gaskin regrets "the heedless merriment with which we have dealt havoc upon a beauty we did not comprehend." Printing a lithograph of the Abbey Plain cross and a photograph of the Cædmon memorial, he highlights a curious disparity: while the authentic historical (albeit non-medieval) monument was neglected—unacknowledged in the local landscape and damaged by unwitting destructiveness—the modern recreation was unveiled by the poet laureate and celebrated in the *Times*.

Cædmon was cast as an accessible figure to local readerships as well as Gothic tourists. To the members of the Whitby Literary and Philosophical Society "who wish to inform themselves on this subject," W. W. Skeat recommended a "little book on Anglo-Saxon Literature . . . published in a cheap form by the Society for Promoting Christian Knowledge."[76] Robert Spence Watson found Cædmon a useful figure in teaching "an English Language and Literature Class of Men and Women . . . at the Literary and Philosophical Society in Newcastle-upon-Tyne," to many of whom "the idea of an English Literature prior to Chaucer" was new.[77] Watson emphasized that Cædmon was, in his words, a "rude, uncultivated boatman" (Watson inventing this occupation); that Cædmon was of simple birth and "unlettered" was important to advocates of working men's education such as Watson and Rawnsley. The latter expanded on the connections between Cædmon's example and the modern "World of Labour" in a poem written to commemorate the unveiling in 1897.[78]

Ahead of the unveiling, *The Academy* printed a piece introducing its readers to Cædmon. It recounts Cædmon's life, praises his "Paradise Lost" as "the foundation-stone of English literature," and imagines the gathered "scholars and poets . . . in the manner of latter-day Englishmen, hold[ing] a *gebeorscipe* [a feast] of their own, and instead of passing the harp, call[ing] on one another to make speeches and give toasts." It urged Austin to "celebrate the occasion in verse," suggesting (echoing Milton's description of Shakespeare) that there could be no "finer theme than that of the first English singer "warbling his wood-notes wild" to those stark ancestors of ours who little dreamed what glories were to be achieved by their race in the future."[79] Other admirers of Cædmon were similarly invested in such explicitly racialized Anglo-Saxonism: Gaskin describes Cædmon's work as "some of the earliest writings of our race."[80] In the cultural moment surrounding the unveiling of the Whitby cross, Cædmon's admirers are constructed as a community of latter-day "Anglo-Saxons" celebrating their ancestral inheritance; Cædmon's artistic achievements are understood to anticipate, teleologically, their own. Chris Jones identifies this as "a form of *translatio imperii*," which "creat[es] the historiographical patterns by which the institutions, values and achievements associated with the Anglo-Saxons are transferred into nineteenth-century Britain."[81] This memorializing activity can be seen to emerge from the same impulse that fuels the practices of attribution and translation of Cædmon's work: the desire to create and celebrate a figure who could offer connection and continuity with the earliest English past.

Conclusion

A confident counterpart to Macaulay's New Zealander, the "Anglo-Saxon" Cædmon stood "almost alone in Europe amid the ruins of the literature of the old world, presaging the greatness of the future"; looking back at him, one of his modern admirers felt, "we become impressed with the large share our country has had in building up the literature of the Christian world, and with the great antiquity of our own."[82]

To these readers, Caedmon was the poet of origins. He was the first poet, the originator of English literature. In his *Genesis*, he sang about the origins of the world and humanity. Even his name was read to mean "In the Beginning." The emergence of a corpus of pre-Conquest poetry and the need to assimilate it within the national canon posed crucial and unsettling questions about the nature and form of English poetry, especially in the absence of authorial figures through which to frame it. The existence of a named poet, recorded by as reliable a source as Bede, who was said to have composed poetry on the very subjects of the anonymous poems, was too tempting an opportunity to resist.

In the Victorian encounter with Cædmon, this desire for origins engendered—fittingly, for the poet of *Genesis*—a process of creation. The historical reality of Cædmon was doubtful, and facts were scarce; he could, as Isaac D'Israeli writes, "vanish in the wind of two Chaldaic syllables."[83] Nonetheless, he was brought into being so that enthusiasts could claim much of the emerging Old English poetic corpus as his work. Although only nine lines of poetry are directly associated with his name, the poet laureate could call him to "serve and stand as the type of the English poet."

Across these scholarly attributions, popularizing translations, and public commemorations, Cædmon was implicated in the multiple strands of disciplinary "English" that were winding, inconclusively, around one another at this period. First, his construction was the fruit of a deeply philological imagination: from linguistic and textual remains, he was brought into being by language. Second, he was deployed for literary-historical ends, as canon and curriculum were formed and contested. Finally, he was increasingly appreciated and commemorated for the perceived literary quality and influence of his work. For many of

Cædmon's readers, poetic value—however they conceived and reimagined it—was an integral element of the attraction and utility of the literature of the past.[84] Longfellow's view of 1838, that the Caedmonian *Genesis* is "very remarkable, both in a philological and in a poetical point of view," was long sustained.[85]

Where it was lacking, the necessary evidence was constructed: the inventive decipherment of an illegible inscription; the willful elision of a few inconvenient syllables. Indeed, alongside creation and construction, we can perceive a broader interpretative process of elision, in both cautious and more reckless variants. Encountering Caedmon, the reader elides, omits, blends—in the spirit of "truth" rather than deceit—the elements that cannot otherwise be assimilated into the idealized vision of the first English poet. The project of translation was crucial in making these poems comprehensible to modern readers and ensuring that their literary clothing was congruent with contemporary notions of what canonical poetry should sound and look like. Once translated, they were available for readers to experience as proxies for the "originals" and thus feel the appropriate reverence and awe as they witnessed the birth of English poetry.

In this narrative of English literature, Cædmon is not merely understood as the primitive progenitor of English poets. His Victorian creators made a series of cross-period connections that enabled them to found multiple social and cultural identities and perform ideological work with localizing, nationalizing, and imperializing aims. He was figured variously as a vernacular English poet; a Christian poet; a Miltonic epic poet; a Romantic nature poet; a poet laureate; an ethnic cultural ancestor; a local Whitby bard. In the pages of modern books—and at locations like Whitby Abbey—Cædmon's admirers found a range of temporally mobile resonances, in which elements of past and present were themselves elided.

Notes

1. In 1832, Benjamin Thorpe translated the *Hymn* as follows: "Now must we praise / the Guardian of heaven's kingdom, / the Creator's might, / and his mind's thought; / glorious Father of men! / as of every wonder he, / Lord eternal, / formed the beginning. / He first framed / for the children of earth / the heaven as a roof; / holy Creator! / then mid-earth, / the Guardian of mankind, / the eternal Lord, / afterwards produced; / the earth for men, / Lord Almighty!" Thorpe follows the antiquarian practice of setting what are today conceived of as nine long lines (each with two half-lines) as eighteen individual short lines. See *Cædmon's Metrical Paraphrase of Parts of the Holy Scriptures*

in Anglo-Saxon, ed. Benjamin Thorpe (Published for the Society of Antiquaries of London, 1832), xxii–xxiii.

2. Bertram Colgrave and R. A. B. Mynors, eds, *Bede's Ecclesiastical History of the English People* (Oxford: Clarendon Press, 1969), xxii–xxiii.

3. John D. Niles, "The Myth of the Anglo-Saxon Oral Poet," *Western Folklore* 62, no. 1 (2003): 16. Since 1962, the *Hymn* has been the first text in the *Norton Anthology of English Literature*, performing the "awesome, traditional duty of inaugurating all of English literature." See Kevin S. Kiernan, "Reading Cædmon's 'Hymn' with Someone Else's Glosses," *Representations* 32 (1990): 9.

4. If, indeed, that much could be concluded. The possibility that the Old English poem as we have it (recorded by Bede's translators and annotators) is a "retranslation" of Bede's Latin paraphrase rather than an "original effusion" was contemplated by scholars throughout the nineteenth century, John Josias Conybeare, *Illustrations of Anglo-Saxon Poetry* (London: Printed for Harding and Lepard, 1826), 7. The idea is reconsidered by Kiernan, "Reading Cædmon's Hymn with Someone Else's Glosses," 162–64, but rejected by scholars such as Daniel Paul O'Donnell, "Bede's Strategy in Paraphrasing Cædmon's *Hymn*," *Journal of English and Germanic Philology* 103, no. 4 (2004): 417–32 (423).

5. Conybeare, *Illustrations of Anglo-Saxon Poetry*, 7; Society of Antiquaries, "Prospectus of a Series of Publications of Anglo-Saxon and Early English Literary Remains Under the Superintendence of a Committee of the Society of Antiquaries of London," in Thorpe, ed., *Cædmon's Metrical Paraphrase*, v, 5.

6. Robert Spence Watson, *Cædmon, the First English Poet* (London: Longmans, Green, 1875), 98.

7. Recent work has understood disciplinary formation as a "fractious and muddled affair" that occurred in the context of professionalization and specialization. See Josephine M. Guy, "Specialisation and Social Utility: Disciplining English Studies," in *The Organisation of Knowledge in Victorian Britain*, ed. Martin Daunton (Oxford: Published for the British Academy by Oxford University Press, 2005), 200, 202.

8. "Familiar Studies in Homer," *Athenæum*, June 25, 1892, 816.

9. Matthew Townend, "Victorian Medievalisms," in *The Oxford Handbook of Victorian Poetry*, ed. Matthew Bevis (Oxford: Oxford University Press, 2013), 175; Chris Jones, "Anglo-Saxonism in Nineteenth-Century Poetry," *Literature Compass* 7, (2010): 365.

10. The "varied practices" of "linguistic scholarship, editing, literary history, evaluative criticism" that made up the new discipline of English stemmed from the "confluence of two ancient historically related fields of knowledge—philology and rhetoric" (James Turner, *Philology: The Forgotten Origins of the Modern Humanities* [Princeton, NJ: Princeton University Press, 2014], 272, 254).

11. David Clark and Nicholas Perkins, *Anglo-Saxon Culture and the Modern Imagination* (Woodbridge: D. S. Brewer, 2010), 4. At the time of writing, the continued widespread usage of "Anglo-Saxon" to denote the period of English history c.450–1066, and its literature, language, and culture, is under particular scrutiny within medieval studies. Thanks largely to decolo-

nization and antiracist work by medievalists of color and scholars of critical race theory, and in the context of its recent (re)appropriation by resurgent white supremacists, "Anglo-Saxon" is being recognized by many as an inaccurate term that is rooted in racialized ideologies of whiteness and which carries a long history of colonialist and racist utilization in both Britain and the United States. Given this context, I have used quotation marks throughout this piece to mark the nineteenth-century usages of "Anglo-Saxon" and the terms it modifies (past, poetry, history, "race") as historically situated discursive constructions and to foreground their ideological functioning. See Adam Miyashiro, "Decolonizing Anglo-Saxon Studies: A Response to ISAS in Honolulu," published July 29, 2017, online at http://www.inthemedievalmiddle.com/2017/07/decolonizing-anglo-saxon-studies.html; see also Medievalists of Color, "On Race and Medieval Studies," published August 1, 2017, online at http://medievalistsofcolor.com/statements/on-race-and-medieval-studies; M. Rambaran-Olm, "Anglo-Saxon Studies, Academia and White Supremacy," published June 27, 2018, online at https://medium.com/@mrambaranolm/anglo-saxon-studies-academia-and-white-supremacy-17c87b36obf3; David Wilton, "What Do We Mean by Anglo-Saxon? Pre-Conquest to the Present," pre-published September 13, 2019, online at http://wordorigins.org/documents/Wilton_JEGP_AS_Paper_Pre-Publication_Draft.pdf.

12. Alexandra Lawrie, *The Beginnings of University English: Extramural Study, 1885–1910* (Basingstoke: Palgrave Macmillan, 2014), 23.

13. See Reginald Horsman, *Race and Manifest Destiny: The Origins of American Racial Anglo-Saxonism* (Harvard University Press, 1981), "Chapter 4: Racial Anglo-Saxonism in England," 62–77. On racism and "Anglo-Saxon" in university curricula in the United States, see Mary-Dockray Miller, *Public Medievalists, Racism, and Suffrage in the American Women's College* (Basingstoke: Palgrave Macmillan, 2017).

14. Chris Jones, *Fossil Poetry: Anglo-Saxon and Linguistic Nativism in Nineteenth-Century Poetry* (Oxford: Oxford University Press, 2018), 22.

15. Carol Atherton, "The Organisation of Literary Knowledge: The Study of English in the Late Nineteenth Century," in Daunton, ed., *The Organisation of Knowledge*, 229.

16. Turner, *Philology*, 272.

17. John Churton Collins, "Review of the Petition Addressed to Oxford Hebdomadal Council for the Foundation of a School of Modern Literature at Oxford," in *The Nineteenth-Century History of English Studies*, ed. Alan Bacon (Aldershot, UK: Ashgate, 1998), 229; orig. pub. 1887.

18. Thomas Case, *An Appeal to the University of Oxford against the Proposed Final School of Modern Languages* (Oxford: Parker & Co., 1887), 5.

19. Edward Freeman, "Literature and Language," in Bacon, ed., *Nineteenth-Century History of English Studies*, 286–87; orig. pub. 1887.

20. Although his "Anglo-Saxonist" discourse is not explicitly racialized in this particular context, Freeman's broader understanding of "Anglo-Saxon history" was embedded in white supremacist thinking; his "firm convictions on the ethnic superiority and providential calling of the modern Anglo-Saxons led him at times to the extreme limits of racism," (Hugh A. MacDougall, *Racial*

Myth in English History: Trojans, Teutons, and Anglo-Saxons (Montreal: Harvest House; Hanover, NH: University Press of New England, 1982), 101.

21. A Final Honour School of English language and literature was established at Oxford in 1894; see Haruko Momma, *From Philology to English Studies: Language and Culture in the Nineteenth Century* (Cambridge: Cambridge University Press, 2013), 164–70.

22. Clark and Perkins, *Anglo-Saxon Culture*, 10.

23. Matthew X. Vernon, *The Black Middle Ages: Race and the Construction of the Middle Ages* (Basingstoke: Palgrave Macmillan, 2018).

24. Gillian Beer, "Origins and Oblivion in Victorian Narrative," in *Arguing with the Past: Essays in Narrative from Woolf to Sidney* (London: Routledge, 1989), 12.

25. Society of Antiquaries, "Prospectus of a Series of Publications," iv.

26. A similar attempt to "assimilate Anglo-Saxon poetry within the author-centred canon of English literature" made use of the ninth-century Cynewulf; see Jacqueline A. Stodnick, "Cynewulf as Author: Medieval Reality or Modern Myth?," *Bulletin of the John Rylands University Library of Manchester* 79, no. 3 (1997): 25–39.

27. Roberta Frank, *The Search for the Anglo-Saxon Oral Poet*, 2nd ed. (Manchester: Manchester Centre for Anglo-Saxon Studies, 1993).

28. Sir Francis Palgrave, "XII. Observations on the History of Caedmon," *Archaeologia; or, Miscellaneous Tracts Relating to Antiquity* 24 (1832): 341, 343.

29. Palgrave, "History of Cædmon," xxiv.

30. Robert Tate Gaskin, *Cædmon: The First English Poet*, 2nd ed. (London: Society for Promoting Christian Knowledge, E. & J. B. Young, 1902), 68.

31. Henry Wadsworth Longfellow, "Anglo-Saxon Literature," *North American Review* 47 (July 1838): 110.

32. Eleanora L. Hervey, "The Vision of Cædmon," *Illustrated Periodicals* 51 (1866); Sarson C. J. Ingham, "Cædmon's Vision," *Leisure Hour*, Sept. 18, 1880, 597–99.

33. Conybeare, *Illustrations of Anglo-Saxon Poetry*, 5.

34. Thorpe, *Cædmon's Metrical Paraphrase*, v, xvi.

35. Palgrave, "History of Cædmon," 343.

36. Conybeare, *Illustrations of Anglo-Saxon Poetry*, 7.

37. Stephen Humphreys Villiers Gurteen, *The Epic of the Fall of Man* (New York: G. P. Putnam, 1896), 90; Thorpe, *Cædmon's Metrical Paraphrase*, viii.

38. Watson, *Cædmon, the First English Poet*, 38.

39. Gurteen, *Fall of Man*, 386.

40. Gurteen, *Fall of Man*, 93–94.

41. Watson, *Cædmon, the First English Poet*, 119.

42. John Earle, quoted in Gaskin, *Cædmon: The First English Poet*, 12. The most notable skeptic was W. W. Skeat, who anticipated the modern position.

43. George Stephens, *The Old-Northern Runic Monuments of Scandinavia and England* (London: J. R. Smith), 411.

44. Stephens, *Old-Northern Runic Monuments*, 411.

45. Challenged by Watson, *Cædmon, the First English Poet*, 98; accepted by Gaskin, *Cædmon: The First English Poet*, 61.

46. Stephens, *Old-Northern Runic Monuments*, 420.

47. Thorpe, *Cædmon's Metrical Paraphrase*, xiv, xvi.

48. William H. F. Bosanquet, *The Fall of Man; or, Paradise Lost of Cædmon, Translated in Verse from the Anglo-Saxon, with a New Metrical Arrangement of the Lines of Part of the Original Text, and An Introduction on the Versification of Cædmon* (London: Longman, 1860), v; Gurteen, *Fall of Man*, v; Watson, *Cædmon, the First English Poet*, 45.

49. Watson, *Cædmon, the First English Poet*, 98.

50. Jones, "Anglo-Saxonism in Nineteenth-Century Poetry," 360.

51. Jones, "Anglo-Saxonism in Nineteenth-Century Poetry," 360, 362–63.

52. Gurteen, *Fall of Man*, 95. María José Mora has shown how the Old English genre we term *elegy* was similarly fabricated in the nineteenth century; see "The Invention of the Old English Elegy," *English Studies* 76, no. 2 (1995): 139.

53. Simon Dentith, *Epic and Empire in Nineteenth-Century Britain* (Cambridge: Cambridge University Press, 2006), 67.

54. Gurteen felt it was a "moral certainty" that Milton had read Cædmon, (*Epic of the Fall of Man*, 131). Cf. Gaskin, *Cædmon: The First English Poet*, 72; Bosanquet, *Fall of Man*, xxxv; Watson, *Cædmon, the First English Poet*, 38–39.

55. Isaac D'Israeli, *Amenities of Literature, Consisting of Sketches and Characters of English Literature* (Paris: A. and W. Galignani), 43–44.

56. Bosanquet, *Fall of Man*, vii–viii.

57. Bosanquet, *Fall of Man*, viii.

58. Gurteen, *Epic of the Fall of Man*, 95–97.

59. The Old English alliterative meter comprises two half-lines, each with two stressed syllables and a varying number of unstressed syllables, and an intricate rule determining the alliteration of the stressed syllables. Milton wrote in blank verse (unrhymed iambic pentameter).

60. D'Israeli, *Amenities of Literature*, 42.

61. Bosanquet, *Fall of Man*, x, xxiv.

62. Bosanquet, *Fall of Man*, xxxiii.

63. On "this manoeuvre of . . . extending the roots of . . . contemporary forms and fashions further back into history," see Jones, "Anglo-Saxonism in Nineteenth-Century Poetry," 364

64. D'Israeli, *Amenities of Literature*, 57.

65. "The Epic of the Fall of Man: A Comparative Study of Cædmon, Dante, and Milton," *Athenaeum* 3625 (1897): 449–500.

66. "Cædmon: The First English Poet. By Robert Tate Gaskin," *The Academy and Literature* 62 (1902): 528.

67. Cary H. Plotkin, *The Tenth Muse: Victorian Philology and the Genesis of the Poetic Language of Gerard Manley Hopkins* (Carbondale: Southern Illinois University Press, 1989), 150.

68. "The Cædmon Memorial at Whitby," the *Times* (London), September 22, 1898, 10.

69. Clark and Perkins, *Anglo-Saxon Culture*, 4, n. 5.

70. The 1898 Cædmon cross at Whitby has recently been used to "exemplif[y] many of the key features of nineteenth-century medievalism" (Townend, "Victorian Medievalisms," 167).

71. "The Cædmon Cross at Whitby," *Artist: An Illustrated Monthly Record of Arts, Crafts and Industries* 25, no. 233 (1899): 32–33.

72. On the association of Tennyson with Alfred the Great, see Joanne Parker, *England's Darling: The Victorian Cult of Alfred the Great* (Manchester: Manchester University Press, 2007), 193.

73. "The Cædmon Memorial at Whitby," 10.

74. Gaskin, *Cædmon: The First English Poet*, n.p.

75. Gaskin, *Cædmon: The First English Poet*, 64–65.

76. Gaskin, *Cædmon: The First English Poet*, 9–10. On the importance of the local "Lit and Phil" to the emerging discipline, see Lawrie, *Beginnings of University English*, 19–21.

77. Watson, *Cædmon, the First English Poet*, v.

78. Wendy Joy Darby, *Landscape and Identity: Geographies of Nation and Class in England* (Oxford: Berg, 2000), 160.

79. "The First English Poet," *The Academy*, no. 1376 (1898): 276.

80. Gaskin, *Cædmon: The First English Poet* (1902) 3rd ed. n.p. Similarly, as Chris Jones notes, the French historian H. A. Taine believed Cædmon's and Milton's work were similar because "both have their model in the race" (*Fossil Poetry*, 104).

81. Jones, *Fossil Poetry*, 24.

82. Bosanquet, *The Fall of Man*, ix.

83. D'Israeli, *Amenities of Literature*, 35.

84. In 1936, J. R. R. Tolkien derided the aesthetic obliviousness of Old English scholarship, noting that the eighteenth-century appreciation of *Beowulf* as *Poesis* had been overwritten by the later interventions of *Historia* and *Philologia*. The different "lines of scholarship" have recently been characterized as "firmly entrenched" in the late nineteenth century, entailing "a curious blindness" to non-philological concerns, but Cædmon's example shows that, for many, philological, poetic, and historical modes of thought were productively enmeshed; see his "Beowulf: The Monsters and the Critics," *Proceedings of the British Academy*, 22 (1936): 4; and Niles, *Anglo-Saxon England*, 252–53.

85. Longfellow, "Anglo-Saxon Literature," 106.

6

Under False Pretenses

Astrid Swenson

The Multiplication of Fakes

It was the fate of every great curator to buy at least one forgery, the *Times* (London) chimed in 1937 on the occasion of Sydney Cockerell's retirement as director of the Fitzwilliam Museum in Cambridge. Only the man himself seemed "never to have tripped."[1] Alas, unbeknownst to the admiring columnist, this most Victorian of twentieth-century curators proved no exception to the rule. Although ever concerned with the authenticity of acquisitions, his desire to turn the Fitzwilliam from a "pigsty into a palace" by adorning it with the finest works of art, enticed him in 1926 to buy a little statuette extolled as a masterpiece of the recently discovered Minoan civilization and the "earliest Greek marble statuette in existence."[2] Having paid an astronomical £2,750 to make her a Cambridge resident, Cockerell was in for a nasty surprise when sailing soon after in her ancestral lands: on a visit to the museum at Candia, he learned that his

---

Warmest thanks are due to the Cambridge Victorian Studies Group and in particular to Adelene Buckland and Sadiah Qureshi for their insightful comments.

statue was a forgery. Cockerell's goddess (like several of her "sisters," which also appeared on the market), had not been made by the ancients, but by two modern Greeks, who worked as restorers in the employ of the British archaeologist Sir Arthur Evans. Apparently, the discoverer of Knossos had tutored them well in the arts of "restoration," for they duped him too. It was Evans who authenticated the marble in the first place. Had it not been for the price, he might have been tempted to purchase her.[3]

The story of the little goddess is typical for "the mercenary motives of the forger and the duping of the hapless expert" as significant features of forgery.[4] It also reveals how long a shadow the Victorian discoveries of "new" pasts cast over the intertwined development of forgery and expertise. As pasts started to multiply in the Victorian age, so did forgeries. The democratization of collecting and the diversification of collectables made the nineteenth century the golden age of forgery.[5] Access to the past ceased to be the privilege of a few aristocratic collectors. New industrial classes, public museums, and educational institutions emerged as buyers, and forgers eagerly supplied them with objects. The production of forged antiquities and Old Master paintings, popular since the Renaissance, soon extended to fakes from other periods, paralleling, and partaking in, the development of new disciplines discussed in this book. Invented medieval epics and fake marginal notes on Shakespearean texts accompanied the growth of literary history; forged Jackal-headed Egyptian deities and cuneiform tablets from Mesopotamia appeared on the heel of archaeological discoveries, while counterfeited flint stones were produced as part of the search for a more distant human past. Once scholars moved from European and near Eastern history further afield, fakers followed, supplying Mughal paintings, Chinese oracle bones, pre-Hispanic Peruvian mummy masks, and Maori family heirlooms.[6] Counterfeiters also exploited the new scientific uncertainties about the authenticity of the Bible and the age of the world. They used the constant unearthing of "new" old civilizations to play to dreams that even more pasts submerged underground and under the sea were just on the cusp of discovery. Biblical scholars who unsuccessfully stomped the Dead Sea looking for the stones of the damned cities of Sodom and Gomorrah to prove the impact of God's wrath on the wretched were thus consoled with chips from a pillar of salt pretended to be Lot's wife.[7] Archaeologists trying to substantiate the theory of evolution fell fool to artificially aged monkey jaws in their desire to establish the great chain of being, while faking cosmogonists played with dreams of uncovering the even more distant origin of the universe.[8]

Telling what was real and what was false was not made easier by the fact that the developing historical disciplines had to deal with old fakes in addition to new ones. Ever since Lorenzo Valla had unmasked the *Donation of Constantine* in the fifteenth century, antiquarians had been weeding authentic charters from forgeries through linguistic and textual criticism. But as archaeologists dug deeper into the past, they had to make sense of even earlier "pious frauds" from the Neo-Babylonian period or "transcriptions" of worm-eaten documents from eighth-century Memphis.[9] Fakes did not have to be ancient, or even dead, however. In the early nineteenth century, false hair from Marie Antoinette was as popular as sham Greek sculptures. Like the many imposters impersonating French aristocrats on the streets of post-revolutionary London, fake artifacts from the recent past fed the craze to experience the terrors of contemporary history up close.[10]

This proliferation of fakes created anxieties not only among art experts and scholars but haunted the Victorian imagination more largely.[11] Forgeries enhanced fears about the trustworthiness of appearance in a society transformed by progress, and destabilized beliefs in the validity of origins already shaken by true scientific discoveries. They seemed to laugh in the face of all the earnest attempts to establish knowledge about human origins through research or fictionalization discussed by Helen Brookman and Simon Goldhill in chapters 5 and 4 in this volume. The anxieties about being duped resulted in a "rich rhetoric on forgery," offering the forger, the expert, and the dealer as archetypes through which selfhood was defined in relation to gender, class, and race.[12] Although financial forgery ceased to be punished by hanging in 1830,[13] artistic forgery was increasingly vilified. Crucially, what was condemned as forgery also changed, as the late Victorian years fundamentally redefined notions of authenticity. With a virulence hardly seen outside Britain, all kinds of hitherto accepted artistic practices from copying to restoration were now reviled as "lifeless forgeries."[14] Arguably this hatred of the "fake" is one of the strongest aesthetic legacies the Victorians bestowed on the twentieth century.

Yet, in the Victorian imagination, the forger appears not only as a villain but also as a hero admired for his craft. Some of the rich literature on nineteenth-century forgery has pointed to the ambiguous feelings the Victorians expressed about faking.[15] However, the analyses focus predominantly on the anxieties provoked. Not only are tales of forgery used to understand fears about the "violation of genealogy" and "threats to the purity of nations,"[16] but more broadly, the story of the fake in the nineteenth century has been interpreted as "one of subjects who depend on objects that betray them, of humans who define themselves through things that lie." Here a forgery is seen as a transgressive autonomous object because it lacks proper authorship, and is therefore bound neither to its attributed nor its actual creator. Its autonomy rendered it somehow "*both* an object *and* a person" that never allowed humans to forget this independence.[17] For many, *The Picture of Dorian Gray* best captures the fears of an object developing a life of its own, horrifying and threatening the creator by reversing power relations and laws of time.[18]

One could, however, also turn this interpretation on its head: If anything, the forger rather than the object became threatening as he made himself the new master of time, collapsing the distance between the past and the present, viscerally experiencing the sensations of another age by recreating its material culture: time traveling. Perhaps the opposition to fakery was so strong not because of the objects' sudden autonomy but because of their loss of sovereignty. By fabricating ageless objects, man put himself in the stead of God as the creator of history. After the dethroning of biblical creation through the elongation of geological time, forgeries tempering with the historical record threatened to destroy all viable indicators for determining the origins of species and the descent of man by replacing them with other origins.

While the attacks on the fabrication of the past in the late Victorian years necessitate explanation, it is useful to remember that throughout the long nineteenth century attitudes to faking and forging remained more complex and varied. This essay is therefore less about anxieties than about enthusiasm. Rather than treating forgeries as the dark underbelly of the expanding art market, it is worth revisiting them as a central feature of Victorian historical culture. Forgeries show in a pertinent way how this historical culture constantly connected the preoccupation with different pasts now divided into disciplinary silos; how it mixed the textual and the material, the popular and the elitist; and how it was shaped by global connections through the circulation of objects, people, and narratives. A number of exhibitions, such as the British Museum's *Fake? The Art of Deception* have brought forgeries out of the vaults and showed the diversity of fakes and fakers, but mostly the significance of particular fakes remains discussed only by the discipline whose history was shaped by them. Building on this substantial scholarship, and reconnecting debates across different areas of Victorian art, science, and culture, this essay considers not only how "false pretenses" challenged the veracity of the past, but how faking offered new opportunities for understanding it, as forgeries were used as tools to discover the processes behind the making of historical artifacts, often turning the scientist into a faker. Forgery was thus often just another facet of the search for origins, rather than the opposite. Relatedly, the essay draws greater attention to the fact that forgery was used not only to build rhetorical barriers and consolidate a white, male word of expertise, but also to create a space for those excluded from this world by class, gender, or race to include their versions of the past and subvert the usual power relations. Thus I argue that telling lies about the past was not merely misleading. Victorian audiences reveled in fakes and forgeries, which brought them into visceral contact with the feelings and forms of past worlds; they were willing participants in frauds that offered them new kinds of connection with the past; and they threw themselves with energy into creating neo-Gothic and neoclassical buildings, lining their streets with historical fakes. Where then was the line between fakery and inspiration?

Forgeries in Historical Culture

With hindsight, many forgeries are easily identifiable because of their Victorian features. A group of supposedly "Moabite" sculptures, discovered at Moses Shapira's shop in Jerusalem in 1878, for instance, looked just a tiny bit too much like Napoleon III, Lord Kitchener, and a notorious Christian from a Greek orthodox church in Jerusalem to have been made in the biblical era. Yet, at the time, the resemblance to his very own likeness did not stop Kitchener himself from acquiring some of the sculptures for the Palestine Exploration Fund in London.[19] Despite the increasingly sophisticated methods developed by scholars to catch forgers, often decades elapsed before fakes were unmasked.[20] In the meantime many forgeries had the opportunity to shape academic and popular histories lastingly. A brief glance at famous forgeries in the possession of the British Museum suffices to reveal that hardly any period remained unaffected. Research on artistic development of the late seventeenth and eighteenth dynasties in Egypt, for instance, was much shaped by a forged limestone statue inscribed with the name of Queen Tetisheri between its acquisition in 1890 and its discreditation four years later. Other forgeries had a much longer shelf life. Prehistoric scholarship was dominated for almost half a century by "Piltdown Man"—supposedly the missing link between ape and men—which was only unmasked as a modern fabrication in the late 1940s.[21] What had been the most widely reproduced portrait of Julius Caesar, a bust acquired by the British Museum from the collector James Millingen in 1818, was identified even later as an early nineteenth-century forgery from Rome. Together with false Renaissance sculptures, especially those by the famous forger Bastianini, it shaped generations of art historians' and amateurs' perception of classical beauty. It was their very contemporaneity, which now seems saccharine, that rendered Renaissance sculpture so popular in the nineteenth century and shaped taste well into the twentieth. The art historian Kenneth Clark, for instance, recalled how his generation found the Renaissance via Victorian translations:

> I was taught drawing at school in a dismal room that contained about a dozen casts, which I was required to draw in pencil. As I drew them every week for four years I got to know them fairly well. Four of them were by Bastianini, one was the British Museum Caesar . . . and one was the bust of a female saint which . . . is undoubtedly nineteenth century. There was not a single cast of an authentic work by Mino da Fiesole, Rossolino, still less Donatello, simply because they would all have contained an element of style that would have upset my drawing master. The same I am sure, was true of art schools all over the country.[22]

Not only undiscovered forgeries shaped the culture of the past. Although many forgeries were banished to the crypts as soon as they were identified as such, a considerable number of objects of doubtful provenance remained in the public sphere. An "Etruscan" terracotta sarcophagus for instance, which the British Museum bought from the dealer Alessandro Castellani in 1871 stayed on display until 1935 despite numerous indications of its inauthenticity: Soon after the acquisition, it transpired that the sarcophagus' inscription came from a brooch in the Louvre. In 1875, Enrico Penelli, brother of the dealer Pietro Penelli from whom Castellani had purchased the sarcophagus, claimed that he had fabricated it. While Pietro denied this and threatened legal action against his brother, other signs led many outside the museum to suspect a forgery: the reliefs on the chest seemed based upon Athenian black-figure vase painting of an earlier period than the figures at the top. As to these figures, not only was the nudity of the man unparalleled in other work, but the dress of the female had an uncanny resemblance to nineteenth century underwear. As exhibition by the British Museum gave it authority, however, these doubts were immaterial for most viewers and the sarcophagus was reproduced in countless popular books on the Etruscans.

Other forgeries, whose authenticity had long been questioned, also continued to influence historical culture because of their importance for national mythmaking. The faithfulness of Geoffrey of Monmouth *Historia Regum Brittaniae*, for instance, had been debated for centuries. Yet whether the very ancient Welsh book, upon which it was supposedly based, had been real or invented, mattered little for the development of popular history. Via Malory's *Morte d'Arthur*, it made it into Henrietta Elizabeth Marshall's children's book *Our Island Story* (1905) which through many reeditions shaped historical consciousness in Britain lastingly.[23] Other works enshrined in *Our Island Story* were of equally questionable provenance, such as Richard Cirencester's map of Roman Britain, fabricated by Charles Bertram (1723–65), the son of a silk dyer who had immigrated to Copenhagen. He offered the English Antiquarian William Stukeley, one of the founders of the Society of Antiquaries, a "copy" of a manuscript history of Roman Britain with a map. Karl Wex showed in 1845 that a misquotation of Tacitus stemmed from an edition of 1497 (thus much later than the assumed fourteenth-century origin), and B. R. Woodward and J. E. B. Mayor demonstrated through textual criticism that the text was a skillful pastiche of Caesar, Tacitus, Solinus, Camden, and others. Yet by then it had penetrated the fabric of historical consciousness. It had not only led to a renewed interest in the mapping of Roman Britain (a field largely abandoned thence) but also inspired the Ordnance Survey to put Bertram's invented Roman names, such as the Pennines, on the maps of modern Britain.[24]

Forgeries might not appear to have been quite as important for nation-building in Britain as in other parts of Europe.[25] Yet it was the famous eighteenth-century British forgeries that gave the inspiration to many of the nineteenth-century Continental ones. In turn Continental artists contributed to shaping the image of the British Isles by playing with the motives invented by forgers: by the time an 1807 edition of James Macpherson's Ossianic poems showed that they did not stem from a rediscovered lost manuscript but were augmented versions of traditional Gaelic originals, Macpherson's fiction had achieved topographic reality: "Fingal's Cave" had been given its name and was soon to be immortalized by Felix Mendelssohn. Welsh revivalism remained equally shaped by known forgeries as the Scottish one: the poems that Iolo Morganwg, the pseudonym of Edward Williams, added in 1789 to those of Dfydd ap Gwilym to rekindle interest in Welsh literature were widely read throughout the nineteenth century—with the eighteenth-century poems somewhat more popular than the originals. While the authenticity of another famous eighteenth-century forgery, Thomas Chatterton's poems, was still debated, and only finally put to rest by Walter Skeat's edition of 1871, Wordsworth, Keats, Wallis and Ruskin had incorporated him into the canon.[26] Other forgeries shaped ideas about the nation in more indirect ways too: torture chairs and chastity belts were fabricated in the eighteenth and early nineteenth century for the mushrooming neogothic castles. As no tangible sources of the second could be found, and evidence of the first had mostly disappeared with the abolition of torture, they needed to be faked to invoke the horrors of a violent Catholic past and to remind British Protestants of their modernity and moderation.[27]

Yet another group of forgeries was knowingly integrated into the canon because of their beauty rather than their usefulness for a particular historic narrative. In the process, the forger was often allowed to rejoin the ranks of the artists. The Florentine forger Giovanni Bastianini for instance, who so shaped Victorian Renaissance tastes, "came out" after the dealer M. de Novilos did not share the profit for a bust bought by the Louvre in 1865. His revelations were received graciously. The artist Giovanni Costa, when finding out that a bust of Savonarola, which he had bought to prevent it from leaving the country, was a fake, expressed his pleasure that such a talented artist was among the living. The Victoria and Albert Museum subsequently bought a bust of Lucrezia Donati as a "Bastianini" for a price normally paid for Renaissance work.[28] Although "authenticity" and provenance were becoming ever more important as the century progressed, collectors to some degree remained indifferent as long as an object's beauty was sufficient. Alexandre Dumas fils famously shook the hand of the Corot forger Paul Desire Trouillebert when he discovered how well he had been duped.[29] The "Tiara of Saitaphernes," supposedly a headdress offered to the Scythian King Saitaphernes by the city of Olbia in the Crimea not only made its creator's career as an artist in his own right but started an international bidding war once it was identified as a forgery. After the Louvre purchased it in 1896, a lively debate arose about its authenticity in the periodicals, until an artist from Montmartre claimed it as his own. This in turn, prompted a Russian goldsmith to write a letter to a French paper arguing that this claim was fraudulent and that it was his friend Israel Rouchomovsky, a goldsmith from Odessa, who was the creator. Rouchomovsky was called to Paris, where he remade part of the object for an independent expert. A successful career as an artist followed. In 1903 Rouchomovsky even won a medal at the Paris Salon for a number of his "archaeological-style objects," some of which had appeared as proper antiques on the market a few years earlier.[30]

Finally, some fakes became icons of popular culture out of sheer notoriety. The impresario Phineas T. Barnum, who had made a business of trying to bring the world's national treasures, from Shakespeare's house to Mayan temples, to the United States of America,[31] wanted to acquire not only the Tiara of Saitaphernes once it had been identified as a fake,[32] but also the giant statue of a "pre-historic man" that the American George Hull had carved and buried on his brother's farm until it acquired patina, and which he then exhibited to visitors for a fee. After Hull rejected the offer, Barnum made a copy. The judge from whom Hull sought an injunction ruled, "that it was hardly a crime to exhibit a 'fake.'"[33]

Fakes in the Service of Science

Meanwhile, the ubiquity of forgeries made scholars hone their instruments to detect frauds and a good part of the textual and iconographic methodologies of the developing disciplines of literary criticism, history, art history, and theology were developed to deal with forgeries old and new, as discussed in chapters 4 and 5 of this volume. Scholars also increasingly used the methods established in one discipline to detect forgeries in another. Of all historical subdisciplines, numismatics had one of the longest traditions in reflecting on forgery—as not only modern forgeries, made for collectors, were circulating, but also old ones, forged as currency.[34] Numismatic methods for detection played therefore a crucial role for discovering frauds in other areas. The antiquary John Evans, in particular, established the principles for the documentation, authentication, and interpretation of flint stones by establishing analogies. With regard to flints found in the Somme valley, he argued, for instance, that having "a coating of carbonate of lime forming an adherent incrustation upon them" was "for those weapons what the patina is for bronze coins and statues, a proof of their antiquity." He, like many of his contemporaries, also used the analogy with monetary forgery to argue that in the end, "the best advice to give to collectors is that given to the public by the Bank of England: . . . look not only at the note itself, but also at those from whom you receive it."[35]

While his own standing as a businessman contributed much to his authority, personal and institutional status allowed forgers elsewhere to hide in plain sight. A good number of the most notorious forgers operated from within the international scholarly establishment. Carl Wilhelm Becker (1772–1830)—a friend of Johann Wolfgang von Goethe, an intimate of Prince Carl von Isenberg, and one of the greatest numismatic forgers—is said to have started his career as a forger to take revenge on another collector who had first sold him a fake and then ridiculed him for his ignorance. Eventually Becker was able to deceive not only his deceiver but also major private and public collections in Europe. Rivalry among colleagues might also have turned the British Museum into a "forger's den."[36] One of the best known literary forgeries—marginalia by an "old corrector" on a 1632 Folio of Shakespeare—was made in 1852 by either John Payne Collier, a leading editor of Elizabethan and Jacobean literature, director of the Shakespeare Society, vice president of the Society of Antiquaries of London, and secretary of the Royal Commission on the British Museum, in an attempt to gain authority and shape the interpretation of Shakespeare through his comments on the text, or by his colleague Sir Frederick Madden, who, out of jealously, supposedly fabricated the folio and pinned it on Collier.[37]

The forgery that perhaps caused the greatest scientific sensation, the prehistoric "Piltdown Man," was also at heart a scholarly scandal. Discovered in 1912, it enthused those looking for the missing link between ape and men.[38] A range of earlier forgeries had staked different locality's claims to the origins of man. In 1863 the lower half of a human jaw was found in Moulin Quignon in France. Its authenticity was soon challenged (in particular, in England, by Evans) as fake flint stones had also been found at the site. The New World tried next. In 1866, a Californian gold miner claimed to have found a skull inside a mine in the Table Mountains some time before George Hull carved the giant statue of a man that attracted Barnum's attention.[39] But the search for the missing link continued, fueled by the publication of Charles Darwin's *The Descent of Man* (1871), the discovery of Java Man in 1891, and Heidelberg Man in 1907. When "Piltdown Man" was discovered in 1912, its human skull and apelike jaw promised to locate the origins of the human race in England. It was only when later finds from China, Java, and Africa made "Piltdown Man" appear atypical—unlike him, other prehistoric men had an apelike skulls, and manlike jaws—that doubts arose. Fluorine dating challenged the skull's age in 1949, and in 1953 J. S. Weiner discovered that the wear of the teeth was the only feature that differentiated Piltdown's jaw from that of an ape. Since then there has been much speculation regarding its creator. The key figure seems to be the solicitor Charles Dawson, who regularly sought the limelight through the discovery of marvelous objects. In the 1890s he gave his name to *Plagiaulax dawsoni*, a mammal with reptilian ancestors; found three new species of iguanodon; and found the oldest cast iron statue.

He also discovered a stag's-horn hammer from the submarine forest at Bulverhythe, a "transitional boat—half-coracle and half-canoe," a "Neolithic stone weapon with a wood shaft," a "Norman prick spur," a "transitional" horseshoe, and a new race of man with thirteen dorsal vertebra. As Mark Jones has argued, it is unlikely that he was at the heart of one of the biggest forgeries purely by chance. In March 1909, Dawson wrote to Smith Woodward, one of the other discovers, that he was "waiting for the big discovery that never seems to come."[40]

Yet scholarly involvement in forgeries transcended polarized versions of deceiving or unmasking. They also involved using the act of forgery as an analytical tool in creative ways. Despite all his efforts to uncover fakes, for John Evans in particular, forgeries were not to be dismissed: not only would paying attention to them prevent being duped in the future, but forgeries could acquire value through their rarity. Moreover, and perhaps more importantly, they could also provide "a unique heuristic experience, reflecting backward, as it were from fraud to fabrication."[41] Many numismatists believed that frauds could provide insight into period skills, but for flints they offered even more useful insights. "Their very forgery," so he argued, "could refute those who considered candidate exemplars as mere 'plays of nature.'"[42] Evans's experimental flint-making helped to secure his status as an expert. One of the most notorious forgers of the period, Edward Simpson, also known as "Flint Jack" (although he forged objects from many other periods too), was apparently curious to see Evans's work and predicted that Evans "was likely to attain an equal degree of eminence" as a forger as Flint Jack himself. For Evans, this appreciation by a man who had tricked many collectors and curators, "only confirmed the heuristic value of the authenticating experiment."[43]

A similar logic underpinned the "restoration" of historic buildings—according to the French architect Eugène Emmanuel Viollet-le-Duc, who transformed medieval buildings—and ideas about restoration like no other in Europe. While some of his contemporaries, and many of those born later, condemned the restoration of buildings as a falsification akin to forgery,[44] for Viollet-le-Duc, "restoration" was above all a scientific act that contributed in essential ways to the knowledge of the past. In his *Dictionnaire Raisonné de l'Architecture* Viollet-le-Duc argued that restoration did for architectural history what Cuvier's comparative anatomy (with his wizard-like reconstructions of entire ancient species from fragments of bone or "functional correlation") did for the advancement of natural history, or Champollion's deciphering of hieroglyphs did for archaeology. Only by learning the techniques of the medieval master could the architect restore, and only by restoring—that is by de-and re-assembling the structure of a building, could he learn about the past.[45] Yet while Viollet-le-Duc's famous definition of restoration began with the words "to restore, is not to remake," as he went along, the line between old and new, fake and genuine was less clear for him than for the numismatist Evans. Although for Viollet-le-Duc, restoration defined the nineteenth century as uniquely modern, because it accepted consciously its difference from the past, the act of restoration also allowed to overcome the temporal distance. By bringing a building "back to life," the architect could be possessed by the creator's ghost and leave a building in which the viewer would find it hard to distinguish past and present.

Mixing analytical with creative methods, developing academic disciplines such as archaeology and art history thus appropriated consciously or unconsciously the creative processes already employed by the literary and historical writings of the late eighteenth century—when the development of literary history, literary forgery, and the historical novel had been closely intertwined. Debating the forgeries of Macpherson and Chatterton not only helped to develop more sophisticated principles for literary history, but also inspired new literary creations, such as Walpole's *Castle of Otranto*, which, like Scott's *Antiquary*, borrowed the trope of the rediscovered manuscript from literary forgery.[46] That the trope was becoming rapidly overused is indicated by Walpole's parody in his *Hieroglyphic Tales* ("the Hieroglyphic Tales were undoubtedly written a little before the creation of the worlds, and have ever since been preserved by oral tradition, in the mountains of Crampcraggiri, an uninhabited island, not yet discovered").[47] This did not stop its popularity throughout the nineteenth century, however, as Jocelyn Betts shows in his discussion of William Henry Smith's meditations on the future in *Thorndale*.[48] Forgeries and the historical novel had a common aim in "the literary making of the past."[49] All invoked the imagination as a necessary supplement to scientific methods. Not only did the forger exhibit "a carnivalesque irreverence towards the sanctity of various conventions designed to limit what is permissible in literary production,"[50] but the use of similar methods, and the self- identification of novelists with forgers, "suggest the pervasiveness and popular dissemination of sceptical attitudes towards historical certainty."[51]

Authentic Forgeries

While the lines between academic discourse and fiction continued to be blurred in popular historic novels during the following century,[52] historical certainty was becoming rather more important in other fields. Forgery was ending the life of entire groups of objects as collectables or objects of study: medieval ivories and prehistoric painted pebbles for instance were so frequently forged that their reputation was permanently damaged.[53] This coincided with a growing denigration of forgers as effeminate, and dealers as antinational, often Jewish, figures. From the middle of the century, objection rose against imitation as a legitimate means to advance knowledge or diffuse art. What had long differentiated the forgery from the copy, the restoration, the replica, or even the parody was the intention to deceive—but that differentiation disappeared increasingly. Too creative a re-use of existing texts became a matter for copyright lawyers.[54] According to Briefel, the copyist was seen as even more deviant than the forger—being female rather than male (and often a prostitute or spinster). Many copies and casts that were once the pride of museums were moved into storage or destroyed with a sledgehammer (an episode also known also as the "Battle of the Casts").[55] But it was restoration that earned the biggest scorn. Ruskin provided generations of future campaigners with fiery quotes in the *Seven Lamps of Architecture*. Chastising the tampering with the historic and aesthetic record, Ruskin branded restoration a "lie from start to end," equating it to the annihilation of the traces of age on a beloved mother's face, and likening the once vaunted summoning of the spirit of the original creator to a sinister act of necromancy. William Morris rather more concisely brought it to the point in the expression "lifeless forgery" and started to campaign internationally against the evils of restoration—a fight in which he was soon joined by a new breed of "conservationists" elsewhere.[56]

While often observed, the turn against forgery and assorted practices is less often properly explained. Literary texts ranging from *Trilby* to *The Marble Faun*, written by male writers from England, France, and the United States, betray a number of fears in a largely defensive lexicon. Anxieties about the loss of economic control and national prestige are expressed alongside fears of homoerotic bonds, of Jews gaining power in the (art) world, and of the loss of racial purity more largely.[57] Many of the manifestos against restoration or historicism also display fears about alienation through commodification in an industrialized society, and about the inability to leave an art proper to the "nineteenth century."[58]

On closer inspection, the literary critics of the later nineteenth century, however, continued to find pleasure in playing with artistic convention, much like their eighteenth-century predecessors. With regard to the artists who verbalized many of the pleas for greater "authenticity," the shift could be explained not only by anxieties but also by the search for new aesthetic and creative experiences. Historicism, once so exiting and new, had by the end of the century become the art of the fathers, and the new generation that was arguing to clearly separate old and new styles was clearly searching for a style of its own: most of the prominent artists who argued against restorations, from William Morris in Britain, to Victor Horta, Auguste Rodin, or Hermann Muthesius on the Continent, were also involved in various artistic and life-reform movements.[59]

Moreover, the verbosity of critics who imposed the "you-must-not-fake" mantra upon later generations should not necessarily be mistaken for the attitude of the majority,[60] or even seen as absolutes for the writers of said opinions. When we turn from rhetoric to practice, quite a different picture emerges, and anxieties sat very close to enthusiasm. Many of the authenticity gurus immersed themselves in fabricated experiences just like the rest of the Victorians. A literature of appreciation existed alongside the literature of disdain: on viewing Viollet-le-Duc's first major restoration, the church of the Madeleine in Vézelay, Ruskin, for instance, "went round the aisles, disdainful at first but, gradually warming to the intelligence and skill of the great modern architect, he confessed that if restoration might be done at all, it could not have been done better."[61] A brief glance at the magnificently restored dining hall in Queen's College, Cambridge, also suffices to realize that many of the interventions on historic buildings performed by Morris were as inventive as those of the reviled restorers. The pleasure of recreating an imagined past also outlasted the shift in public opinion that the anti-restorers had themselves achieved. While their principles became enshrined in international charters through the League of Nations in the interwar years, an anti-restorationist of the early hour, the Arts and Crafts designer C. R. Ashbee, restored Jerusalem's town walls in Mandatory Palestine in a manner that would have made Viollet-le-Duc proud.[62] Similar anachronisms of the appreciation of the "fake" can be seen elsewhere: While many museums destroyed their casts, others continued to commission new ones, and calls to create a "National Cast Museum" in London were at their loudest in the 1930s.[63] What is more, even when it came to these "original casts" (a wonderful oxymoron reserved for casts taken directly from the original), things were not simple: casts in an "inauthentic" pure and white, neoclassical aesthetic were still preferred to an "authentic" recreation of the polychrome original.[64]

Although the language of authenticity is one of the strongest legacies the Victorians gave to later generations, the line between fakery and inspiration was thus never clearly drawn. Until the very end of the period, telling lies about the past was not seen as necessarily misleading. Fakes and forgeries show rather how much the Victorians sought the visceral contact with the feelings and forms of past worlds. They also reveal the simultaneously exclusionary and participatory nature of access to the past. While many criticisms of forgery were used to consolidate a white and male world of expertise and economic power, other tales of forgery challenged this order. For example, in *Ruth*, Elizabeth Gaskell reveals the gendering of punishment poignantly by juxtaposing the story of a woman who is ostracized for forging an identity for herself and her illegitimate child as a widow to the fate of Richard Bradshaw, a gentleman's son, who hardly suffers for the forgery of signatures on insurance shares.[65] Charlotte Brontë also used the language of forgery to challenge normative views of femininity—indeed, she herself "*forges* a subject marked by the misbehavior of Victorian fictions of sex and self" through the figure of Lucy as "a self in translation as well as in transition" in *Villette*.[66] While it might go too far to see forgery always as an act of resistance (when often it was born of desperation), the subversive and democratizing qualities of forgery certainly deserve further investigation.[67] Forgery not only allowed more people access to and ownership of the past, but also provided a space for those excluded from the world of art and expertise to contribute to its production and con those in position of power with ingenuity. Two illiterate workmen, William Smith and Charles Eaton, for instance, created fake medieval badges and claimed to have found them by the Thames.[68] From a more global perspective, closer attention to the provenance of many fakes and forgeries helps us not only to better understand the transnational nature of historical culture, but also to rethink agency in an age of imperial collecting. Noticing Western demand, Persian potters, for instance, created fake Chinese blue and white pottery from the seventeenth century onward.[69] The forgeries by countless Palestinian, Greek, and Chinese "hands," restorers, and dealers' assistants during the nineteenth century can be seen—if not as acts of resistance—at least as successful subversions of the market forces and antiquities legislation bent on the spoliation of ancient cultures.

Forgeries thus unravel many of the paradoxes at the heart of (Victorian) historical culture. They helped to fabricate national myth but also exposed the transnational nature of objects and narratives. They fostered disciplinarization, yet consistently challenged the new boundaries between academic disciplines. They were used to establish an exclusionary world of expertise but also offered unique means to subvert power relations. They created anxieties as much as they provoked enthusiasm. They show that, for some, the past was no laughing matter; for others, it was a burlesque to be staged with gusto. They not only destabilized existing origins but also created comfort by inventing new ones. In the end, they should perhaps be best understood simply as one of many faces of the broader Victorian quest to understand and enjoy the past discussed in this book. They show just how many, often contradictory, yet ultimately complementary ways coexisted to understand "authenticity" in the quest for knowledge through a range of creative practices. The inquiry into Victorian attitudes toward forgery thus complements Simon Goldhill's accounts of *fiction* in the return to original sources, and Helen Brookman's analysis of the *creation* of an "original" set of sources (chaps. 4 and 5 in this volume), by revealing the importance of *inspiration* through the act of faking sources. While Goldhill suggested that the search for original sources prompted more and more fictionalizations, but not a relinquishing of the idea that one can get to the truth, Brookman shows how, despite a long-standing critical tradition claiming that the Victorians increasingly abandoned the search for "origins," in some emerging fields, such as philology and English Literature, the search was perceived as ever more urgent but required a series of genuinely pursued (and yet ultimately inaccurate and fictional) tactics. The fake, on the other hand, has no authentic originary moment, belongs to no original place, but makes the forger the originary point of the object. While this grasp of creative (and, by extension, divine), agency at times troubled contemporaries, it was not always understood as a false pretense. In many senses, for the Victorians, it really often did not matter where a thing came from or if it was authentic or "original": fakes were *required* to replace "lost" or only imaginary originals and to evoke the required feelings about the past; the stories they acquired as they moved through their lives were just as important as their provenance and were valued precisely for their inauthenticity.

Notes

1. *Times* (London), July 13, 1937.
2. K. Butcher and D. W. J. Gill, "The Director, the Dealer, the Goddess and Her Champions: The Acquisition of the Fitzwilliam Goddess," *American Journal of Archaeology* 97 (1993): 383–401.
3. Stella Panayotova, *I Turned It into a Palace: Sydney Cockerell and the Fitzwilliam Museum* (Cambridge: Fitzwilliam Museum, 2008), 89.

4. Ian Haywood, *Faking It: Art and the Politics of Forgery* (New York: St. Martin's Press, 1987), 2.

5. Aviva Briefel, *The Deceivers: Art Forgery and Identity in the Nineteenth Century* (Ithaca, NY: Cornell University Press, 2006), 1–18.

6. Mark Jones, ed., *Fake? The Art of Deception* (London: British Museum Publications, 1990).

7. Astrid Swenson, "Sodom," in *Cities of God: The Bible and Archaeology in Nineteenth-Century Britain*, ed. David Gange and Michael Ledger-Lomas (Cambridge: Cambridge University Press, 2013), 197–227.

8. Adelene Buckland, "Losing the Plot: The Geological Anti-Narrative," *19: Interdisciplinary Studies in the Long Nineteenth Century* 11 (2010), https://doi.org/10.16995/ntn.578.

9. Jones, *Fake?*, 60.

10. Peter Fritzsche, *Stranded in the Present: Modern Time and the Melancholy of History* (Cambridge, MA: Harvard University Press, 2004); Billie Melman, *The Culture of History: English Uses of the Past, 1800–1953* (Oxford: Oxford University Press, 2006); Tom Stammers, "Collecting Cultures, Historical Consciousness and Artefacts of the Old Regime in Nineteenth-Century Paris" (PhD diss. University of Cambridge, 2010).

11. Briefel, *Deceivers*; Sara Malton, *Forgery in Nineteenth Century Literature and Culture: Fictions of Finance from Dickens to Wilde* (New York: Palgrave MacMillan, 2009); Scott Carpenter, *Aesthetics of Fraudulence in Nineteenth Century France* (Aldershot, UK: Ashgate 2013).

12. Briefel, *Deceivers*, 2.

13. Randall McGowen, "From Pillory to Gallows: The Punishment for Forgery in the Age of the Financial Revolution," *Past and Present* 165 (1999): 107–40.

14. Society for the Protection of Ancient Buildings, *Manifesto* (London, 1877).

15. Briefel, *Deceivers*.

16. Malton, *Forgery*, 40.

17. Briefel, *Deceivers*, 18, 16.

18. Briefel, *Deceivers*, 2; Malton, *Forgery*, 134, 136.

19. "The Moabite Pottery," *Palestine Exploration Fund Quarterly Statement* (1878), 88–102; Charlotte Trümpler, "Die Moabitischen Fälschungen," in *Das Große Spiel: Archäologie und Politik zur Zeit des Kolonialismus (1860–1940)*, ed. Charlotte Trümpler (Cologne: DuMont, 2008).

20. For other problems of determining what was real, see the discussion on the Antediluvian Child by Qureshi, on biblical manuscripts by Goldhill, on the search for Atlantis by Pettit, and on geological specimens by Buckland in this volume.

21. Jones, *Fake?*, 162, 93.

22. K. Clark, "Forgeries," *History Today*, November 11, 1979, 724–33, cited in Jones, *Fake?*, 198.

23. Jones, *Fake?*, 64–66; Henrietta Elizabeth Marshall, *Our Island Story: A Child's History of England* (London: T. C. & E. C. Jack, 1905).

24. Rosemary Sweet, *Antiquaries: The Discovery of the Past in Eighteenth-Century Britain* (London: Hambledon and London, 2004), 175–81; O. G. S.

Crawford, "Archaeology and the Ordnance Survey," *Geographical Journal* 59, no. 4 (1922): 245–53.

25. See János M. Bak, Patrick J. Geary, and Gábor Klaniczay, eds., *Manufacturing a Past for the Present: Forgery and Authenticity in Medievalist Texts and Objects in Nineteenth-Century Europe* (Leiden: Brill 2014).

26. Jones, *Fake*, 67, 154.

27. Jones, *Fake*, 59, 70.

28. Jones, *Fake*, 197.

29. Briefel, *Deceivers*, 4.

30. A. Vayson de Pradenne, *Les Fraudes en Archéologie Prehistorique* (Paris: n.p., 1932), 519–73; Jones, *Fake?*, 33.

31. See Donna Yates, "Publication or Preservation at a Remote Maya Site in the Early Twentieth Century," and Melanie Hall, "Plunder or Preservation? Negotiating an Anglo-American Heritage in the Later Nineteenth Century in the Old World and the New: Shakespeare's Birthplace, Niagara Falls, and Carlyle's House," both in *From Plunder to Preservation: Britain and the Heritage of Empire, c. 1800–1940*, ed. Astrid Swenson and Peter Mandler (Oxford: Oxford University Press, 2013), 27, 220, 241–63.

32. Briefel, *Deceivers*, 7.

33. Haywood, *Faking It*, 95.

34. Jeffrey Spier and Jonathan Kagan, "Sir Charles Frederick and the Forgery of Ancient Coins in Eighteenth-Century Rome," *Journal of the History of Collections* 12, no. 1 (2000): 35–90.

35. Quoted in Nathan Schlanger, "Coins to Flint: John Evans and the Numismatic Moment in the History of Archaeology, "*European Journal of Archaeology* 14 (2011): 469.

36. Haywood, *Faking it*, 71.

37. G. F. Hill, *Becker, the Counterfeiter*, 2 vols. (London: Spink, 1925); Jones, *Fake?*, 144–46; Haywood, *Faking It*, 71–88.

38. On this search, see Qureshi (chap. 1) and Mandler (chap. 2) in this volume.

39. Haywood, *Faking It*, 95.

40. Jones, *Fake?*, 95.

41. Schlanger, "Coins to Flint," 469.

42. Schlanger, "Coins to Flint," 469.

43. Schlanger, "Coins to Flint," 470; R. A. Munroe, *Archaeology and False Antiquities* (London: Methuen, 1905), 117.

44. On restoration, see also Goldhill's essay (chap. 4) in this volume.

45. Eugène-Emmanuel Viollet-le-Duc, "Restoration," in Viollet-le Duc, *The Foundations of Architecture: Selections from the Dictionnaire raisonné*, by Viollet-le Duc, ed. Berry Bergdoll, trans. Kenneth D. Whitehead (New York: George Braziller, 1990), 197–98. On his reception in Britain, see Astrid Swenson, *The Rise of Heritage: Preserving the Past in France, Germany and England, 1789–1914* (Cambridge: Cambridge University Press, 2013).

46. Anne H. Stevens, "Forging Literary Histories: Historical Fiction and Literary Forgery in Eighteenth-Century Britain," *Studies in Eighteenth-Century Culture* 37 (2008): 217–32, at 219; Anthony Grafton, *Forgers and Critics*:

Creativity and Duplicity in Western Scholarship (Princeton, NJ: Princeton University Press, 1990), 5.

47. Cited in Stevens "Forging," 222.

48. Jocelyn Betts, chap. 11 in this volume.

49. Haywood, *Faking it*.

50. Kenneth Knowles Ruthven, *Faking Literature* (Cambridge: Cambridge University Press, 2001), 3–4.

51. Stevens, "Forging," 229.

52. Simon Goldhill, *Victorian Culture and Classical Antiquity: Art, Opera, Fiction, and the Proclamation of Modernity* (Princeton, NJ: Princeton University Press, 2011).

53. Jones, *Fake?*, 92–93.

54. See Clare Pettitt, *Patent Inventions: Intellectual Property and the Victorian Novel* (Oxford: Oxford University Press, 2004).

55. Mary Beard, "Cast: Between Art and Science," in *Les Moulages de sculptures antiques et l'histoire de l'archeologie*, ed. Henri Lavagne and Francois Queyrel (Geneva: Droz, 200), 157–66.

56. Swenson, *Rise of Heritage*.

57. Briefel, *Deceivers*, 126; Malton, *Forgery*, 20, 40.

58. Swenson, *Rise of Heritage*, 105.

59. Swenson, *Rise of Heritage*, chap. 2.

60. See Peter Mandler, *The Fall and Rise of the Stately Home* (New Haven, CT: Yale University Press, 1997).

61. William Gershorn Collingwood, *Ruskin Relics* (London: Isbister, 1903), 50.

62. Simon Goldhill, "The Cotswolds in Jerusalem: Restoration and Empire," in Swenson and Mandler, eds., *Plunder to Preservation*; Astrid Swenson, "Crusader Heritages and Imperial Preservation," *Past & Present* 226, supp.10 (2015): 27–56.

63. Astrid Swenson, "Musées de moulages et politiques patrimoniales: Regards croisés sur la France, l'Allemagne, l'Angleterre au XIXe siècle," in *Les Musées de la Nation: Créations, transpositions, renouveau. Europe XIXe–XXe siècles*, ed. A-S. Rolland and H. Murauskaya (Paris: Editions L'Harmattan 2009), 205–19.

64. Roberta Panzanelli, Eike D. Schmidt, and Kenneth Lapatin, eds., *The Color of Life: Polychromy in Sculpture from Antiquity to the Present* (Los Angeles, CA: The Getty Research Institute, J. Paul Getty Museum, 2008).

65. Malton, *Forgery*, 59.

66. Laura E. Ciolkowski, "Charlotte Bronte's *Villette*: Forgeries of Sex and Self," *Studies in the Novel* 26, no. 3 (1994): 218–19.

67. Leeann D. Hunter, review of Sara Malton, *Forgery in Nineteenth-Century Literature and Culture: Fictions of Finance from Dickens to Wilde* (New York: Palgrave MacMillan, 2009), *Review 19*, online at http://www.review19.org/Reviews_PDF/0039_Malton.pdf. On the democraticizing qualities of fabrications of the past, see also chap. 7 in this volume.

68. Briefel, *Deceivers*, 4.

69. Jones, *Fake?*, 109.

7

Through the Proscenium Arch

Rachel Bryant Davies

Weapons poised and horses mid-charge, the horseback warriors from *The Siege of Troy* (fig. 7.1) are frozen in time at the height of their mythical duel. The clash of their historical bricolage of armor is almost audible. The immediacy of this engraving, one of a series of "favorite horse combats," draws in viewers as imagined participants in the epic Trojan War. At the same time, sketched scenery on the horizon, glimpsed below and beyond the horses, evokes the upward gaze of ringside spectators watching live circus reenactments against the raised theatrical stage backdrop beyond.

This double perspective encapsulates the phenomenon of the theatrical souvenir: vibrantly melodramatic, yet captured on flimsy paper; memento of a specific performance, yet recognizably similar to combats depicted from other plays. Ancient and modern conflicts alike (see fig. 7.9) were recreated at venues such as Astley's Amphitheatre in London, where, as we shall see, theatrical reviewers jokingly weighed the relative authenticity of such reenactments alongside the sorts of familiarity with canonical narratives gained through traditional schooling.[1]

This is the story of these irreverently playful interactions with classical mythology as marketed for children.

FIGURE 7.1. "Skelt's Favorite Horse Combats, No. 3": lower half of sheet showing "In the Siege of Troy." Source: Sheet published "by M. Skelt II, Swan St. Minories, London. Price Halfpenny" [1835–37] (author's collection).

Burlesque theater—(melo) dramatic parodies that used anachronisms, puns, and mock seriousness to uproariously send up or comically defer to their ostensible sources—is only beginning to be taken seriously as a form of engagement with the classical past, while toy theaters are just starting to emerge as vital links between children's literature and other consumer culture; however, this widespread, enjoyable aspect of Victorian encounters with the past reveals a less constrained culture than has yet been properly understood.[2] The absolute fusion of ancient and modern is the hallmark of these recreational, commercial entertainments and their souvenirs, which inextricably entangled playful and scholarly modes for their consumers, across class, gender, and age divides.

Toy-theater sheets and miniature portraits were a craze that seems to have begun at the start of the nineteenth century; by 1812, these tiny keepsakes were sold in sufficiently detailed sets that owners could use them to reenact plays in model theaters. Historical, legendary narratives, such as this version of the Trojan War, were extremely successful as entertainments, so these miniatures afford a ringside view of the sorts of immersive, and often irreverent, encounters with these pasts that they both recorded and enabled. In other regions of the nineteenth-century cultural landscape (as Mandler's chapter 2 in this volume explains), history might seem like a portrait gallery, or travelers' destinations like a laboratory. Theatrical performances and souvenirs, however, offered an immersive past: a dressing-up box where participants could play out historical, legendary, and mythical events. The miniaturized perspective of the proscenium arch held a special appeal. G. K. Chesterton celebrated his experience that "by reducing the scale of events it [his toy theater] can introduce much larger events" since, as he argued, "You can only represent very big ideas in very small spaces."[3] It is this transformation, first for life-size and then for table-top stages, that enabled virtual time travelers to experience vast swaths of history "through the proscenium arch."

London's spectacular population explosion throughout the nineteenth century propelled a frenzied expansion of the entertainment scene. The popularity of historical and legendary tales meant that different pasts were constantly reenacted in all sorts of competing dramatic genres and performance media. Since 1737, only two theaters in London had been allowed to stage spoken dramas, but many more venues popped up to entertain the rapidly expanding population. Their theatrical performances were all censored by the Lord Chamberlain, and it was not until the mid-century that censorship was restricted to moral objections by the Theatres Act (1843). Such venues used increasingly inventive ways to dodge the ban on spoken drama, incorporating pantomime, music, dance, and stunning visuals, as well as borrowing elements from both canonical and current culture.[4] This variety made for a vibrant performance culture that delighted in bringing the past back to life for audiences who not only enjoyed spectating, but also immersed themselves in collecting mementos and recreating stage settings.

Theatrical souvenirs bridge these hybrid evolutions in performance culture with the rapid growth of print and childhood cultures which, over the course of the nineteenth century, would transform the past into an accessible, imaginative play-space; a "foreign country" that was perceived as remarkably—and even, at times, dangerously—familiar. As urban expansion drove rapid increases in the number of leisure venues and types of entertainment on offer, new forms of popular consumption raised their visibility. The increased speed and reduced cost of printing prompted the spread of brightly colored playbill advertisements, provided a platform—in newspapers and magazines—for lengthy dramatic reviews, and enabled the production of intricate theatrical souvenirs, which further exploited the commercial potential of imaginative time-travel. Miniature portraits, along with printed toy-theater backdrop and character sheets intended for recreating staged performances at home in table-top model theaters, enable us to enter those imagined landscapes.[5] They not only illustrate how the past was imagined visually, but also reveal how theater-goers and consumers at home became virtual time travelers.[6] It is striking how often reviews judged theatrical plots by the criterion of what "school-boys, little and large" would recognize;[7] at the same time critics emphasized, both in disparagement and approval, the socially mixed audiences to which many venues catered. The pasts recreated at these venues and in such souvenirs were most often those memorized in the schoolroom, analyzed in scholarly tomes, and debated by learned societies.

Classical antiquity was an especially vibrant inspiration. Not only did the classics underpin educational curricula, social hierarchies, and much political discourse at the time, but, as many elite travelers would note, childhood experiences of reading classical literature prompted an emotional response to the physical landscapes they visited abroad.[8] In particular, Homer's epic tale of the Trojan War shows just how much the past animated different branches of the theatrical entertainment market: the myth had long been a staple of fairground entertainment (as depicted in William Hogath's painting of *Southwark Fair* in 1732). The myth was also familiar to different audiences from operas, ballets, and ballads. As controversies over the authenticity of the epic narratives and the existence of Troy intensified from the late eighteenth century, however, they converged fruitfully with a new, parodic form: burlesque dramas. Between 1819 and 1893, numerous comic adaptations spoofed Homer's *Iliad* and Virgil's *Aeneid*.

One of the most exciting was an equestrian reenactment of *The Siege of Troy* at Astley's Amphitheatre in 1833, from which the combat scene in figure 7.1 was drawn.[9] This famous circus-cum-theater was one of London's most popular and socially inclusive venues; it was described that year as "favourite summer resort alike of the infant and the adult."[10] Its re-creations of historical, legendary narratives were box-office hits that, as we shall see, spawned some of the most remarkable miniature collectibles.

Through analyzing how this popular version of *The Siege of Troy* was miniaturized for home performance and preservation, I engage in exactly the kind of interpretative act examined previously in this part: many toy-theater sheets were so comprehensive and so minutely detailed that, even if they don't quite live up to their claims to be accurate representations of their "original" theatrical productions, they nonetheless give us the most comprehensive insights we can possibly hope to achieve. Some sets, such as Hodgson's Scenes and Characters in the *Giant Horse*, also drawn from *The Siege of Troy* at Astley's in 1833, even specified how far forward backdrops should be placed and included pairs of wings and moveable parts and props down to fencing;

scripts issued with these sets even listed which colors to use when painting plain, rather than ready-colored, sheets.[11]

As the other essays in part 2 of this volume amply demonstrate, the idea of reaching for an "original" is always a creative act of interpretation. Here, I show that originality and playfulness in comic performances—even, or especially, when fabricated, faked, and fictionalized—trumped original, classical, canonical narratives. Conversely, interactive toy versions aspired, in overtly adapted miniature form, to recreate these performances in table-top size while retaining in exact detail the "original" burlesque experience. At the same time, such fresh, playful perspectives were often perceived as more "authentic" than scholarly engagement: burlesques could exhibit more classical taste than faithful adaptations, while children and others attending popular theater could be described as more capable readers, spectators, or producers than classically educated adults and scholars. Marketed to consumers young and old, these miniaturized myths brought ancient characters forward in time just as much as they enabled their owners to visit the classical past both in the theater and, through theatrical souvenirs, at home.

"Extraordinary Feats"

The favorite horse combat in *The Siege of Troy* (fig. 7.1) encapsulates how spectacular recreations of classical antiquity circulated far beyond the original venues and audiences. The dramatic joust commemorated in this portrait epitomizes the carnivalesque spectacles with which Astley's Amphitheatre purported to recreate the past; in addition, the popularity of the comic stagings of the Trojan War myths throughout the century reveals the coexistence of a range of ludic and scholarly encounters with the past, within the same cultural imagination. The very existence of these theatrical portraits demonstrates the extent of popular exposure that such historical performances achieved and the number of consumer spin-offs they inspired. It is clear that both children and adults, male and female, across a wider range of social classes than might be assumed, experienced classical antiquity and other pasts "through the proscenium arch."

For only a halfpenny—half the usual price for a sheet of characters—the proud owner of "Favorite horse combats, No. 3" gained two sets of jousts, with some scenery thrown in: *"The Siege of Troy"* depicts the Greek army's tents, the walls of Troy, and Mount Ida in the distance. The upper half of the sheet shows a similarly composed combat from *Wallace, the Hero of Scotland*. Both productions were performed at Astley's Amphitheatre in 1815 and 1833 respectively, and had also inspired full-scale toy-theater sets. This sheet was published by Martin Skelt, one of the dynasty that traded in various combinations between 1827 and 1872. Their family business specialized in producing cheaper sheets, and often seems to have bought up or copied other publishers' stock.

Sheets for performance in what was affectionately known as "the Juvenile Drama" were usually sold for "a penny plain and twopence coloured," a well-known phrase that gave Robert Louis Stevenson and A. E. Wilson the respective titles for their essay and book. Stevenson's celebration of "Skeltery" and "Skeltdom" vividly evokes "the giddy joy" of "breathlessly devour[ing] those pages of gesticulating villains, epileptic combats, bosky forests, palaces and war-ships, frowning fortresses and prison vaults."[12] Full-scale sets usually included both characters and scenery, which could comprise backdrops, wings, and props. Since characters were often provided in different positions and groups, a full set could number as many as twenty or more sheets, with a script for sixpence. Once the cost of a model theater is added in too, this pastime seems reserved for middle- and upper- class consumers. However, in addition to cheaper versions, such as those by "the immortal Skelt," which combined figures with scenery,[13] it is important to take into account Stevenson's recollections of gazing through his local shop-window at the "theatre in working order" and "linger[ing] there with empty pockets," as well as his admission that several of his own sets were incomplete, or "imperfect." Clearly their imaginative impact was just as powerful even when experienced at a remove, in part, or without a complete performance.

The extent of the impact of such unconventional presentations of the ancient world as experienced at Astley's Amphitheatre is demonstrated by the sheer scale of the toy-theater sheets that attempted to reproduce them. One of the most beautiful, extensive (and expensive) complete sets is a miniature production that claims to reproduce the same spectacle which inspired Skelt's portrait (fig. 7.1): *The Siege of Troy* at Astley's in 1833. This set, in contrast to Skelt's later portraits, was engraved with the same year as this production by Orlando Hodgson, one of the leading toy-theater publishers between 1822 and 1832. Considered "one of the most illustrious figures of the palmy days of the Juvenile Drama," he is now celebrated for his "splendid exuberance." *Giant Horse*, his last publication, has been praised as one of his "most striking efforts."[14]

Importantly, these sheets are not only theatrical portraits, but also interactive toys: Hodgson pioneered plays "more capable of being performed," since they now included "enough sheets for every character

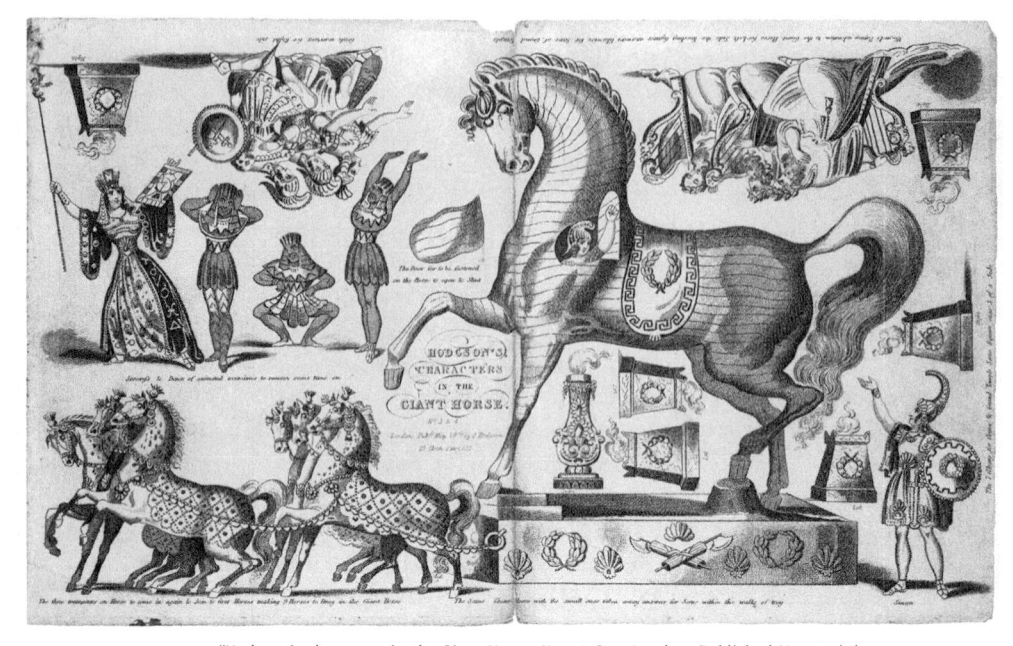

FIGURE 7.2. "Hodgson's characters in the Giant Horse, Nos. 3 & 4. London, Published May 29th by O. Hodgson, 10 Cloth Fair, 1833." Double-size sheet (24cm X 38cm) showing (clockwise from top Left) "Greek warriors for Right side"; "Minstrels paying homage to the Giant Horse for Left side, the kneeling figures answers [sic] likewise for Some of Grand Temple"; "The 7 altars for arena & Grand Temple Scene Square ones 3 of a Side"; Sinon; "The three trumpeters on Horse to come in again & join first Horses, making 9 Horses to Drag in the Giant Horse—The Same Giant Horse with the small ones taken away answers for Scene within the walls of troy" [sic]; "Sorceress & Dance of animated mummies to remain some time on"; "The Door for to be fastened on the Horse to open and shut." Source: Marcus Stone Collection, Victoria and Albert Museum: Theatre Performance Collections archives: HC 277. Photo: V&A Images, Victoria and Albert Museum.

to appear in all the positions required" as well as "all the scenes and wings necessary."[15] This set, entitled "Hodgson's Scenes and Characters in the *Giant Horse*" was published piecemeal between at least May and December 1833. In addition to a generous selection of character and scenery sheets at the usual prices, it included a double-sized sheet that was, fairly enough, priced at double the usual rate (fig. 7.2). This sheet included the eponymous Giant Horse, mounted on a pedestal drawn by six horses. As we shall see, many additional characters enabled owners to recreate the numerous grand processions that punctuated this spectacle. Carefully slotted in the gaps between these larger figures are smaller parts: seven altars for use in other scenes, and a flap to cover the gap in the horse's neck, from which the hidden army will emerge. These are just some of the intricate details, finely realized in beautiful copper engravings, with which Hodgson ensured that his miniature version of *Giant Horse* could recreate the spectacular effect of the original equestrian burlesque.

These details would have been complicated to manage. Their sheer fiddliness has often been pointed out in support of the idea that toy-theater sheets were aimed at young men. It is certainly the case that adults did enjoy staging miniature performances: George Speaight is an example of an adult practitioner who not only conjured up spectacular performances but undertook invaluable scholarly research into the genre. It is also likely that Stevenson sums up the experience of many in his disbelief that "any child [could] twice court the tedium, the worry, and the long-drawn disenchantment of an actual performance." However, we must not overlook the fact that he counts Juvenile Dramas as "evidences of a happy childhood," and Chesterton analyses toy theaters as a pastime more suited to children than adults, while describing his adult attempt to get up a performance of *St. George and the Dragon*. Moreover, the first script to be simplified for toy-theater production in 1831 by William West, "The greatest publisher of the Juvenile Drama," was described in its subtitle as "suited to the capabilities of his little patrons, the managers and real minor performers."[16] This reflects the age-range of the original show at Astley's Amphitheatre, whose the Easter pantomime—in 1833, *The Siege of Troy*—was "[t]he great event to which childhood begins about Christmas to look forward, with an impatience not to be checked by mince-pies, and with a curiosity not to be satiated with snap-dragon."[17]

These miniaturizations were certainly also enjoyed by adults, and overlapped with other (more adult) souvenirs and pastimes, including theatrical portraits. But, as images such as John Leech's caricatures of a schoolboy's holiday pastime emphasize (figs. 7.3 and 7.5), the Juvenile Drama was, as its name suggests, certainly marketed to, and consumed by, children. Leech's series of three scenes depicting the progress of a toy-theater production illustrated a comic picture-book entitled *Young Troublesome; or, Master Jacky's Holidays*. The book itself draws on dramatic conventions, including a list of *dramatis personae*, and is structured around a series of scenes captioned by verses. Master Jacky, the eldest sibling, calls on his brothers and sisters to help him prepare and perform a toy-theater production of *The Miller and His Men* in the model of Drury Lane he has been given by his uncle. The second image of the series (fig. 7.3) depicts the family painting, cutting, and gluing the characters. Leech's illustration, originally published in London in 1849, presents this activity as a familiar process in this popular pastime. This in an important corrective to much of the preserved evidence, which is most often uncut sheets, or Stevenson's infamous assertion that "to cut

FIGURE 7.3. "Further preparation of the Miller and his Men." Source: John Leech, *Young Troublesome; or, Master Jacky's Holidays*, wood engraving by William Jay after John Leech's etchings (orig. pub. London: Bradbury and Evans, December 1849). Source: This image was stereotyped and printed by George Charles and published by Willis Pope Hazard (Philadelphia, PA: 1860), p. 9, then painted by an owner (author's collection).

the figures out was simply sacrilege." After describing in loving detail the job of choosing watercolors, he does admit, "I cannot deny that joy attended the illumination," but nonetheless maintains that, "when all was painted, it is needless to deny it, all was spoiled."[18]

The powerful resonance of the stories that inspired Stevenson is all the more telling since the sheets were left largely unperformed. An example of a different treatment of these sheets, of which Stevenson would have approved, also shows the popularity of coloring as an activity in its own right (fig. 7.4). This is a theatrical portrait of Andrew Ducrow holding a horseback pose that he used as part of *The Siege of Troy*, the

Roman Gladiator. He was renowned for "his extraordinary feats," and their inclusion in this spectacle was noted as particularly "to the delight of his juvenile visitors."[19] This portrait, most likely published by Skelt, has been brightly colored and cut out, but then mounted onto thick album card.[20] We cannot know whether it was first used for performance but, like figure 7.1, this sheet was not sold as part of a toy-theater set. Rather, it is one of many portraits that depict Ducrow in various roles. Such portraits were often extravagantly decorated and adorned by "tinselling" with various materials to create special effects, such as armor:[21] Hodgson's description of Menelaus and Ulysses's armor and Penthesilea's gown as "richly spangled" suggests this technique.[22] At the other extreme from Stevenson's reverence for the complex toy-theater sheets he purchased, Chesterton confides, in *Tremendous Trifles*, that, in his adult attempts, "I drew and coloured the figures and scenes myself. Hence I was free from the degrading obligation of having to pay either

FIGURE 7.4. "Mr. Ducrow as the Roman Gladiator." Source: Portrait [by Skelt], colored, cut out, and mounted on an album leaf (author's collection).

a penny or twopence; I only had to pay a shilling a sheet for good cardboard and a shilling a box for bad water colours."[23]

Master Jacky's family, in Leech's depiction of their theatrical preparations, have adopted Chesterton's approach. Leech foregrounds their table strewn with paint palettes and brushes, while the prominence of the new model theater highlights the family's goal: performance. Figure 7.3 affords an interesting insight into nineteenth-century coloring practices. It is taken not from a first edition of *Young Troublesome*, which featured hand-tinted engravings in handsome, vellum-bound covers, but from a later, cheaper, pamphlet version, whose owner has partially colored the illustrations after purchase. It is also striking that this stereotyped edition, which included cheaper wood engravings after Leech's original etchings, was produced in Philadelphia, Pennsylvania, in 1860—evidence that the Juvenile Drama circulated widely across geographic, as well as social, boundaries, for family fun. It is fitting that figure 7.6 is a photograph of a plate from this edition, which has itself been cut out and stuck on cardboard to become, not a toy-theater figurine, but a greeting card.

Note that Master Jacky's sisters are involved in the preparation (fig. 7.3), although they will be relegated to the audience for his performance (fig. 7.5): a portrait of Madame Ducrow (fig. 7.6), who also performed at Astley's, emphasizes that female participation onstage (both in female and more risqué "breeches" roles) was also reflected in both character sheets and solo portraits.[24] In her relatively respectable role "as Queen of the Amazons," Madame Ducrow (probably Louisa Woolford, a famous equestrienne in her own right, who became Ducrow's second wife in 1838) most likely elaborates this role from the 1833 *Giant Horse*.[25]

"Cheap Knowledge"

Leech's illustrations and verses rely for their comedy on shared, or at least recognizable experiences of interaction with toy theaters, while Skelt's sheets demonstrate a wider and longer circulation of depictions from a particular performance far beyond the audience that might have experienced the original show. Newsprint also ensured the wider circulation of these fun versions of antiquity, enabling readers who could not reach Astley's for the life-size version of the show to recreate the performance in their mind's eye. It is quite possible that reviews further whetted consumer appetites for theatrical souvenirs. The sheer number of reviews and spread of their distribution across different titles, as well

FIGURE 7.5. "Grand evening rehearsal of the Miller and his men, and terrific explosion in the housekeeper's room: Christmas card, undated." [John Leech, from *Young Troublesome; or, Master Jacky's Holidays*]. MS Thr 944 (62a), Houghton Library, Harvard University.

as links between the contents of the 1833 *Siege of Troy* and contemporary articles, expose the coexistence of playful and serious recreations of the past within the same cultural discourse.

Only a fortnight before the premiere, the title page of a penny paper entitled *The Tourist* featured a large woodcut of "Virgil's Tomb," accompanied by a lengthy article that quoted various classical scholars in support of the structure's identity. Since Virgil's burial was supposed to have been at Mount Posilippo, near Naples in Italy, this article may well have inspired the otherwise inexplicable renaming of Mount Ida by Troy (often called Ilium) as "MOUNT PAUSELLEPPO & ILIUM in the Distance." This notable landmark in both classical epic narratives of the Trojan War—Homer's *Iliad* and Virgil's *Aeneid*—features in Hodgson's lovely opening backdrop with the same title. Similar cross-fertilization from other penny papers may not only have inspired the production itself, but also supplied spectators with the requisite familiarity with the spectacle's classical allusions.

Penny publications aimed to supply "wholesome reading material" for the newly literate, with a focus on "the elements of cultural literacy that would enable working-class readers to take part in the intellectual life of the nation and at the same time furnish a model of behaviour."[26] Six months earlier, the *Penny Magazine of the Society for the Diffusion of Useful Knowledge* had run an illustrated series with detailed synopses and analyses of various scenes from the *Iliad* and *Aeneid*.[27] The very month after *Giant Horse* opened, *Chambers' Edinburgh Journal*—a penny paper only founded a year earlier—also summarized the *Iliad* for its series entitled "Popular Information on Literature."[28] Most notably, another of Ducrow's horseback poses, "The Dying Gladiator," based on the statue now known as "The Dying Gaul," would have been familiar to the estimated one million readers of *The Penny Magazine* from the cover page of January 1833. In light of the number of arena combats featured in *The Siege of Troy*, the statue was accompanied by an explanation of gladiatorial shows in ancient Rome.[29] Patricia Anderson's description of how such articles forged links between high culture and working people's experience can fruitfully be applied to spectators at Astley's, especially in light of the interactive souvenirs, since Ducrow's performance onstage would literally have rendered the sculpture "no longer just a figure of antiquity," but also "a role-model for the contemporary reader-viewer."[30]

Such articles represent precisely the sort of "cheap knowledge" at which the *Times* (London) sneered in its review of the "very gorgeous" piece at Astley's. Though the critic praised the fact that it had "more than an ordinary share of processions, civil, military and religious, and an almost uninterrupted series of combats," as well as "terrific engagements of whole battalions, and even of armies," he also claimed that it was not necessary "[t]o go into any detail of the events on which the story is founded . . . in the present age of cheap knowledge."[31] In fact, as we shall see, the exact plot of this particular version of *The Giant Horse, or The Siege of Troy* was far from well known. As the *Athenaeum* would point out, "As this theatre is meant mainly for the entertainment of school-boys, little and large, it is lucky that the victors *are* known to them, for really it would be difficult to decide which side has the best of it at Astley's."[32] Nonetheless, wider familiarity with the received epic narratives would have helped a greater number of the socially varied spectators common at the circus to follow the outline of the action and to appreciate at least some of the comic innovations. Such familiarity enabled the repackaging, as socially inclusive entertainment, of aspects of classical culture that are otherwise most visible from this period in debates among scholarly, literary circles. "Cheap knowledge," as embodied in these articles, helped bridge some of the chasm between working-class readers and spectators, and the aristocrats who would have been familiar with the *Iliad* and *Aeneid*, probably in the original Greek and Latin, and might well have visited the Troad, which was often a stop *en route* to the Holy Land.

Previous essays in this volume have underlined the links between journeys around the world and through time: prior to the development of economically accessible modes of transport in the later-nineteenth century, the Grand Tour traditionally linked journeying across the world with the experience of classical antiquity. Even as horizons expanded for a wider range of travelers, the range and extent of theatrical recreations of historical tales and events gave "armchair traveling" a whole new dimension. It was not until 1836 that John Murray's first travel guide appeared (*A Hand-book for Travellers on the Continent*) and, in 1855, Thomas Cook first guided his tourists to Europe (although they did not venture as far as Rome until 1864). Between these two developments, the Great Exhibition at Hyde Park and its reincarnation as the Crystal Palace at Sydenham had enabled many more to travel through time and space as they journeyed through the parkland, with its "antediluvian monsters," and the courts inside, with their chronological layout of the fine arts and technology.[33] At the same time, many classical sites were being mapped and excavated, leading to bitter controversies and the announcement of many exciting discoveries. It is no accident that, in 1874, an American journalist, writing in a London literary journal, described as "theatrical" Schliemann's announcement of digging up "King Priam's Treasure"—which, for many, proved Hisarlik's identity as the site of Homer's Troy.[34] Excavations of mythological sites were the newest examples of the "startling survivals of antiquity, the dusty revivals of mythic men" that were already familiar from theatrical re-creations.

Audiences at Astley's Amphitheatre included "young and old, fashionable and unfashionable" spectators, among whom, as reviews and the lengthy run of more than one hundred performances show, *The Siege of Troy* was "an undiminished favourite."[35] At the other end of the newspaper spectrum, *The Age*, a cheap weekly with a slightly dubious reputation, agreed with the *Times* (London) and the *Athenaeum* that "[t]his splendid establishment has been the great attraction of the holiday week."[36] It was precisely this popularity that rendered the "tale of Troy—nothing less than the *Giant Horse* himself!" an especially suitable topic for spectacular recreation.[37] As the *Times* had pointed out, such prior familiarity also meant that playbills, advertisements, and

FIGURE 7.6. "Madame Ducrow as Queen of the Amazons" [possibly in *The Siege of Troy*]. Source: MS Thr 942 (136), Houghton Library, Harvard University.

reviews could focus on highlighting the comic, spectacular elements of the show. Some focused on the staging, such as the *Morning Chronicle*, which praised "[t]he intricate stage arrangements of nearly every scene" and approved of the impression that "complicated scenery, costumes, and stage properties . . . appear to have been got up with an eye to their effect and propriety only."[38] After detailing "the galleys and barges with their splendid sails and decorations floating in a painted ocean, spread over the saw-dust of the equestrian circle!," another review claims the redundancy of "attempt[ing] to describe what each will go to see for himself." This enthusiastic commentator provides a vivid glimpse of the eager anticipation of performances at Astley's, especially by children, and the particular excitement surrounding *Giant Horse*. It was this mixture of "steeds of magical capacity" with theatrical excitement that Skelt's portrait captures and Hodgson's toy-theater set, somewhat ambitiously, attempted to re-create.[39]

Leech's depiction (7.5) emphasizes the comic potential inherent in the production of miniature performances when the explosion in "The Miller and His Men" destroys the model of Drury Lane—and demolishes much of the housekeeper's parlor. Ironically, this is an accurate reflection of the dangers inherent in contemporary theatrical production techniques: it was not uncommon for theaters to burn down, and Ducrow would never recover from the third major fire at Astleys in 1841. After reports of an early morning fire backstage among props during *The Siege of Troy*, then, it is understandable that the *Morning Chronicle* was moved to reassure readers with a prominent notice in its business column, that, although one set had been damaged, "[t]he accident will not interrupt the continuance of the regular performances."[40] This also demonstrates the level of consumer saturation: readers of all sorts of newspapers would have been unlikely not to have heard about *The Siege of Troy*, while anyone moving through London would have seen the large, bright playbills, of which figure 7.7 is the most magnificent example.

In much the same way as Skelt's "Favorite Combats" series revisited productions that other toy-theater publishers had earlier miniaturized in full, reviews show that the original props for *The Siege of Troy* that survived that fire were subsequently repurposed in other spectacles, as well as reproduced in miniature. Notices in at least two very different papers revealed that repainted chariots from the last act were re-used in a spectacle entitled "St. George and the Dragon": seemingly rather a flop, its main claim to fame was only that "all the fine cars exhibited were transplanted from Astley's, having been well battered in the *Siege of Troy*."[41] The revival of *The Siege of Troy* in July and August 1840, when the circus was going through a rough patch due to Ducrow's illness and absence, was noted by the *Penny Satirist*, and the show was further recalled when the venue revisited the topic with a completely new plot in 1854.[42] "[T]he historical accuracy of the house has improved since the old days," pronounced the *Times* (despite not reporting any confusion back in 1833), "when Helen was made a Trojan, and Paris a Greek."[43] It is to this unconstrained, inauthentic irreverence, which reshaped the epic plot as a circus entertainment, that we now turn.

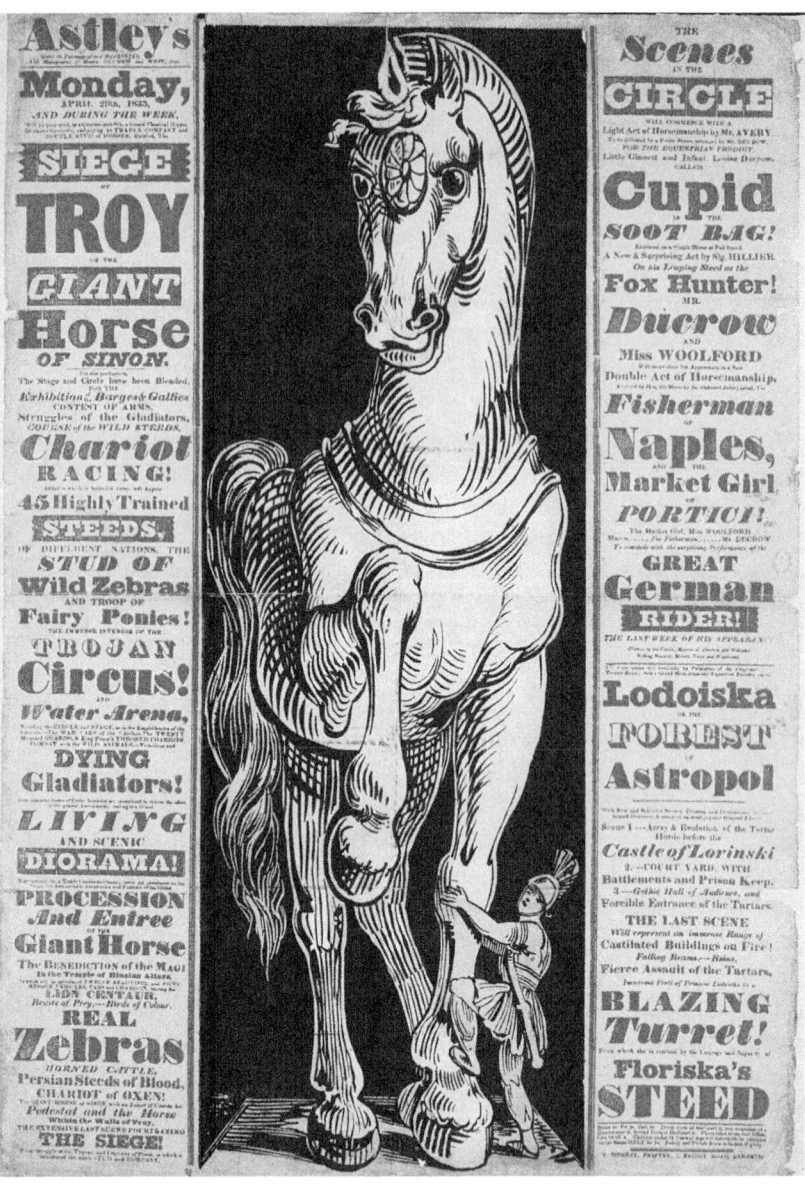

FIGURE 7.7. Playbill for *The Siege of Troy; or, The Giant Horse of Sinon,* Astley's Amphitheatre, April 29, 1833. Source: Printed by T. Romney. Woodcut and letterpress, London. Photo: V&A Images, Victoria and Albert Museum (V&A S.2–1983).

"A Brilliant Jumble"

"[A]ny detail of the events on which the story is founded" would be "unnecessary," claimed a review in the *Times* (London) of *The Giant Horse, or the Siege of Troy* at Astley's in 1833.[44] The *Standard* (London) also expected that "[t]he title will give to our readers full information as to the story of the piece."[45] These reviews entirely missed the "brilliant jumble" of plot and characters which was, as playbills such as the one shown in figure 7.7 emphasize, such an attraction and that the *Literary Gazette*, at eightpence, attempted to clarify: "*Helen* is *Hecuba's* own daughter; the horse is a *Trojan* device; *Paris* besieges *Troy*; the *Greeks* are its inhabitants, and become the victims of its besiegers: lastly, Bacchus sports a pair of spurs, and sundry of the heroes wear the jack-boots of modern horse-guards."[46] Unfortunately, this synopsis only adds to the "jumble." It is interesting that Bacchus's incongruous appearance is not mentioned elsewhere (or depicted in the toy-theater sheets): such variations suggest that details of the performance fluctuated over its lengthy run. More importantly, the toy-theater evidence is entirely clear that the Horse was still full of Greek soldiers, and it was only Menelaus, Paris, and Helen (as yet unmarried to Menelaus) whose nationalities were swapped. Skelt does not name his warriors (7.1), but they are probably Menelaus, usually known as King of Sparta, and Paris, Prince of Troy, fighting over Helen.

These innovations in the mythical love triangle, unmentioned in Skelt's portrait but perpetuated by Hodgson's toy-theater sheets, seem intended as deliberate jokes rather than mistakes. The *dramatis personae* of the script, as abridged for toy-theater performance, listed Paris as "Grecian" and Menelaus as "Trojan." This change was emphasized throughout the play, especially in their clearly identified first entries, when Paris is hailed as "mighty Grecian conqueror," and Menelaus describes himself as the son of Priam and Hecuba, King and Queen of Troy.[47] The *Morning Chronicle*'s reviewer did not entirely approve, describing himself as "on this particular point decidedly conservative," but milked the situation for a joke of his own, "ask[ing] Mr. Amherst . . . whose Classical Dictionary he has consulted in regard to the proper names, we not having hitherto ever heard either of Menelaus of Troy, or Paris of Greece."[48]

It is striking that this review nonetheless categorized the show as a "duly classical" Easter entertainment. The popularity of this enactment of *The Siege of Troy* and the souvenirs it spawned inevitably raises questions as to what sort of classics spectators were learning, recreating, or rejecting. A similar, and apparently also contradictory, opinion is espoused by another reviewer, who praised Ducrow's "classical taste" in the same breath as he excused the show's departure from classical mythology: "If Homer ha[s] not been invariably kept in sight—what wonder!" He describes, approvingly, how the show is "not all strictly Greek and Trojan."[49] This label, however, does not necessarily refer to the exchange of characters noted by other reviewers, but may mean the intermingling, evident from the playbills, of different pasts in this recreation of *The Siege of Troy*. In addition to Ducrow's horseback *pose plastique* as the Roman Gladiator (fig. 7.4), there were many other Roman elements. Most obviously, the warriors' costumes, in this era prior to the excavation of Bronze Age sites such as Mycenae and Troy, draw on both Roman and contemporary ideals. Hodgson's characters are more finely detailed, especially in their varied helmets and in the scaling of their armor, which recalls ancient Near Eastern iconography. Skelt's portrait does differentiate between the medieval Scottish/English combat and the ancient Greek/Trojan duel, but achieves this only through relatively minor changes to the details of weaponry and the scenery, rather than through the overall composition.

What Skelt's flattening of differences between ancient and more modern pasts demonstrates, along with the selection of characters represented on Hodgson's sheets, is a kaleidoscopic view of history, in which different sorts of antiquity were jumbled together with contemporary culture for maximum comic effect. This greedy assemblage is evident in the "Sorceress & Dance of animated mummies" in figure 7.2, which Hodgson clearly considered essential for a full reenactment of the Procession of the Giant Horse. It is notable that the Sorceress's costume, despite the embroidered border of Greek-inspired letters, is closer to Helen's Regency-style dress in figure 7.8 than those of the "animated mummies," who wear purportedly "authentic" costumes of the sort that would have been familiar from another entertainment industry, the display of native peoples. The Egyptian mummies in figure 7.2 seem incongruous next to the Virgilian character of Sinon in that figure, as do the modern characters such as Diorus and the Drunken Guard also reproduced by Hodgson in figure 7.8. A Greek warrior and Trojan captive, Sinon's testimony that the wooden horse is not dangerous is instrumental in the ultimate downfall of Troy, as narrated by Aeneas to Dido. His presence on the sheets is guaranteed by this important role, despite the fact that he speaks no lines in Hodgson's Juvenile Drama script, which was "Printed from the acting copy with . . . the whole of the stage business as performed at Astley's Amphitheatre."

FIGURE 7.8. Plate 10 of Hodgson's characters in the *Giant Horse*. Top row, L–R: "Drunken Guard; Diorus & Boy; Soldiers with Torches"; Bottom row, L–R: "Menelaus; Paris in Chains; Helen; Trophy & Banner for 1st Arena, temple & Procession"; up the RH side: "The whole except banner-bearers for Act 2nd, Scene 1st. July 17 1833."

Although the character of Napoleon is not in the script or the theatrical souvenirs, nevertheless he did appear on stage in the performance of *The Siege of Troy*, and there are many portraits of him in other spectacles (fig. 7.9). The introduction of other modern, anachronistic characters proves essential to the plot: Diorus, for example, ignores Menelaus's bribe and refuses to "condemn the Grecian Paris," while the Amazons send a vital message leading to Paris's escape from a melodramatic ruined temple.[50] Napoleon's appearance, however, ruffled plenty of critical feathers. *The Figaro in London* was disgusted at this "horrible state of absurdity." Its critic did not appear to object to the anachronism of this "conqueror of kingdoms" being "lugged neck and shoulders into the same arena with Paris and Menelaus":[51] since anachronism and incongruity were the stock-in-trade of burlesque, an objection on anachronistic grounds would have missed the point of the entertainment. Rather,

FIGURE 7.9. "Mr. Gomersal as Napoleon Buonaparte [Bonaparte] [in The Battle on Waterloo]." Billy Rose Theater Division, New York Public Library. New York Public Library Digital Collections. http://digitalcollections.nypl.org/items/510d47dd-ecc2-a3d9-e040-e00a18064a99.

he claimed to resent the irrelevance of the intrusion "merely because they have got a biped on the establishment who is said to be a good representative of the Emperor Napoleon" (he especially deplored the actor's habit of taking snuff and "following up the filthy proceeding by drawing his fingers down the sides of his inexpressibles").[52]

The "biped" in question was Gomersal, who played Menelaus, and in 1824 had "appeared for many nights as Napoleon Bonaparte in the Hippodramatic Spectacle of the *Battle of Waterloo*," a role that *Punch* would later call his "pet part."[53] Since Napoleon's appearance at Troy is otherwise unrecorded, it is possible that Gomersal's Menelaus (the villain of the piece) shared mannerisms with his Napoleon. Either way, the link reveals the fantastically kaleidoscopic, quasi-classical antiquity of the staging, far removed from the ongoing topographic attempts to locate the mythical action of the Homeric and Virgilian in the physical landscape of the Troad. It also demonstrates how classical mythology in *The Siege of Troy* was seen as part of a wider historical narrative, in which Egyptian mummies, Roman gladiators, Amazon warriors, and modern characters (including, perhaps, Napoleon), could share in the epic action.

The spectacles, with their clever alterations (or, indeed, incongruous accuracy), could then be appreciated all the more by those in the know, while looking "through the proscenium arch" could blur boundaries between adults and children. The breakdown of this divide, further blurred by children performing table-top recreations to adult audiences, as in figure 7.5, was enjoyed by Leigh Hunt, former editor of the *Examiner*, who "confess[ed]" that, at another classical burlesque two years earlier, "we were children enough (nay old enough rather)" to show "emotion."[54] Such evidence reveals, not a linear model of elite reception diffused passively downward through society, but a messy web of classical interactions between and among both elite and less-elite participants, who each transferred and adapted, appropriated and appreciated different sorts of classical knowledge. These vibrant burlesques and their miniaturization of classical mythology for private reperformance illuminate one of the many instances of culturally embedded classics in nineteenth-century London, as well as the enjoyment evident in these classical encounters. As the *Graphic* explained, "The old Greek and Roman mythology is, after all, the best of all subjects for stage burlesques."[55]

We see a vital, dynamic approach to playing with the past in which comedy precedes accuracy. Playful interactions with epic are also evident in later periodicals (in stories, puzzles and described re-enactments) but these later versions are more obviously, ostensibly didactic. The huge success of the daring, democratizing blend of burlesques and toy-theaters is important, precisely because they could be appreciated on different levels. Ceding center-stage to their cultural impact, rather than the creative process, reveals not only a greater social dispersal of knowledge about, but also a complex model of interaction with, classical mythology. This small sample of keepsakes inspired by *The Siege of Troy* shows that we cannot draw distinctions between original and imitation or elite and popular as clearly or as easily as previously thought. These heavily mediated forms of classical knowledge trade on a desire for an unmediated encounter with antiquity through more visually visceral, direct experiences "in the spirit" of the past—encounters which might be more, rather than less, authentic than educated, textual engagement. In the theater, originality trumped the original classical text; publishers competed to market the most "authentic" version of this new "original" for home recreation by the children whose playful interaction and adaptation of this latest "original" souvenir could be perceived as faithful to the ultimate classical source. Some critics and reviewers muttered, but most embraced the retellings and recognized that ignorance, incomprehension or confusion was not at stake in this more liberated classical culture: the majority of participants playing with, plundering and poking fun at Britain's classical heritage did not care about getting it right: originality was more important than the original.

Notes

1. See e.g. Jacky Bratton, "Theatre of War: the Crimea on the London Stage 1854–5," in *Performance and Politics in Popular Drama: Aspects of Popular Entertainment in Theatre, Film and Television, 1800–1976*, ed. David Bradby, Louis James, and Bernard Sharratt (Cambridge: Cambridge University Press, 1981), 119–38; Barbara Gribling, *The Image of Edward the Black Prince in Georgian and Victorian England: Negotiating the Late Medieval Past* (London: Boydell & Brewer, 2017), 125–8; Rachel Bryant Davies, *Troy, Carthage and the Victorians: The Drama of Classical Ruins in the Nineteenth-Century Imagination* (Cambridge: Cambridge University Press, 2018), 131–145; 187–194.

2. See the pioneering inclusion of burlesque in Edith Hall and Fiona Macintosh, *Greek Tragedy and the British Theatre, 1660–1914* (Oxford: Oxford University Press, 2005); more recent studies include Edmund Richardson, *Classical Victorians: Scholars, Scoundrels and Generals in Pursuit of Antiquity*; Edmund Richardson, "The Harmless Impudence of a Revolutionary: Radical Classics in 1850s London," in *Greek and Roman Classics in the British Struggle for Social Reform*, ed. Henry Stead and Edith Hall (London: Bloomsbury, 2015); Laura Monrós-Gaspar, *Cassandra, the Fortune-Teller: Prophets, Gipsies and Victorian Burlesque* (Bari, Italy: Levante, 2011); Laura Monrós-Gaspar, *Victorian Classical Burlesques: A Critical Anthology* (London: Bloomsbury, 2015). Bryant Davies, *Troy, Carthage and the Victorians*, examines all available burlesques that retell Homeric and Virgilian epics; Rachel Bryant Davies, *Victorian Epic Burlesques: A Critical Anthology of Nineteenth-Century Theatrical Entertainments after Homer* (London: Bloomsbury, 2018) provides editions and commentary on four burlesques of the *Iliad* and *Odyssey*.

3. G. K. Chesterton, "The Toy Theatre," in *Tremendous Trifles*, chap. 23 (Mineola, N.Y.: Dover, 2007), 120.

4. On popular theater audiences and entertainments, see Peter Bailey, *Popular Culture and Performance in the Victorian City* (Cambridge: Cambridge University Press, 1998); Susan Bennett, *Theatre Audiences: A Theory of Production and Reception* (London: Routledge, 1997); Michael Booth, *Theatre in the Victorian Age* (Cambridge: Cambridge University Press, 1991); Jacky S. Bratton *New Readings in Theatre History* (Cambridge: Cambridge University Press, 2003); Jim Davis and Victor Emeljanow, *Reflecting the Audience: London Theatregoing, 1840–1880* (Iowa City: University of Iowa Press, 2001); Jane Moody, *Illegitimate Theatre in London, 1770–1840* (Cambridge: Cambridge University Press, 2000); Richard W. Schoch, *Queen Victoria and the Theatre of Her Age* (Basingstoke: Palgrave Macmillan, 2004).

5. For discussion of toy theater's role in the imaginative play of subsequently famous consumers, see Lizz Farr, "Paper Dreams and Romantic Projections: The Nineteenth Century Toy-Theatre, Boyhood and Aesthetic Play," in *The Nineteenth-Century Child and Consumer Culture*, ed. Dennis Denisoff (Farnham: Ashgate, 2008), 43–62. For links between juvenile dramas and literature, see Suzanna Rahn, *Rediscoveries in Children's Literature* (Abingdon: Routledge, 2013), 23–37, and Andrew O'Malley, *Children's Literature, Popular Culture, and Robinson Crusoe* (Basingstoke: Palgrave Macmillan, 2011), 131–53. George Speaight, *The Juvenile Drama: A Union Catalogue* (London: Society for Theatre Research, 1999) lists ten surviving toy-theater sets on classical subjects.

6. On home theatricals, see Lynn M. Voskuil, *Acting Naturally: Victorian Theatricality and Authenticity* (Charlottesville, VA: University of Virginia Press, 2004); Sara Hudston, *Victorian Theatricals: From Menageries to Melodramas* (London: Bloomsbury, 2000).

7. "Astley's," *Athenaeum*, July 6, 1833, 444.

8. For recent discussion of Victorian uses of the classical past, see Simon Goldhill, *Victorian Culture and Classical Antiquity: Art, Opera, Fiction and the Proclamation of Modernity* (Princeton, NJ: Princeton University Press, 2011); Richardson, *Classical Victorians*; Christopher Stray, *Classics Transformed: Schools, Universities and Society in England, 1830–1960* (Oxford: Clarendon Press, 1998); Christopher Stray, ed., *Remaking the Classics* (London: Bloomsbury, 2013).

9. For detailed analysis, including discussion of the 1854 rewrite as *The Siege of Troy; or, The Giant Horse and Miss-Judgment of Paris*, see Bryant Davies, *Troy, Carthage and the Victorians*, 125–202.

10. "The Theatres," the *Standard*, April 9, 1833, 1. On Astley's as circus, see Brenda Assael, *The Circus and Victorian Society* (Charlottesville, VA: University of Virginia Press, 2005); Maurice Wilson Disher, *Greatest Show on Earth: As Performed for Over a Century at Astley's* (London: G. Bell & Sons, (1937); Marius Kwint, "The Legitimization of the Circus in Late Georgian England," *Past and Present* 174, no. 1 (2002): 72–115; Arthur Hartley Saxon, "The Circus as Theatre: Astley's and Its Actors in the Age of Romanticism," *Educational Theatre Journal* 27, no. 3 (1975): 3, 299–312; Helen Stoddart, *Rings Of Desire: Circus History and Representation* (Manchester: Manchester University Press, 2000).

11. See Bryant Davies, *Troy, Carthage and the Victorians*, 148–204, for almost the complete set of Hodgson's *Giant Horse*.

12. Robert Louis Stevenson, "A Penny Plain and Twopence Coloured," *Magazine of Art* 7 (1884): 227–32; reproduced with an introduction in Glenda Norquay, *R. L. Stevenson on Fiction: An Anthology of Literary and Critical Essays* (Edinburgh: Edinburgh University Press, 1999), 75–76.

13. A. E. Wilson, *Penny Plain, Twopence Colored: A History of the Juvenile Drama* (London: Harrap, 1932), 51.

14. Wilson, *Penny Plain*, 51.

15. George Speaight, *Juvenile Drama: The History of the English Toy-Theatre* (London: Macdonald, 1946), 58.

16. William West, *Olympic Revels; or, Prometheus and Pandora* (London: West, 1831); and Speaight, *Juvenile Drama*, 214, 122. I therefore disagree with the claim, in an otherwise fascinating study, that "dramas produced for toy-theatres were not intended for children"; see Deborah Philips, *Fairground Attractions: A Genealogy of the Pleasure Ground* (London: A. & C. Black, 2012), 49.

17. Unattributed clipping, British Library scrapbook: "Astley's Amphitheatre," [June 3, 1833].

18. Stevenson, quoted in Norquay, *Stevenson on Fiction*, 75.

19. "Theatricals," *The Age*, May 5, 1833, 141.

20. A very similar, uncoloured and entire, sheet is held at the Victoria and Albert Museum, Theatre and Performance Collections archive, London: THM/DIS/2016/MS/1.

21. Some spectacular examples have been digitized by the Museum of London.

22. Orlando Hodgson, *The Siege of Troy . . . as Performed at Astley's Amphitheatre* (London: O. Hodgson, 1833), 1 (held at the Victoria and Albert Museum: Theatre and Performance Collections archive, London: THM/234/1/8/6).

23. G. K. Chesterton, "The Toy Theatre," 123.

24. For contrasting discussion of Madame Vestris on toy-theater sheets, see Rachel Bryant Davies, "Olympic Revels for 'Little Patrons': Classical Mythology in Nineteenth-Century Children's Toy Theatres, *Journal of Victorian Culture*, forthcoming.

25. "Queen of the Amazons" appears on plate 2 of Hodgson's characters, on foot in a very similar pose and costume: see figure 3.1 in Bryant Davies, *Troy, Carthage and the Victorians*, 166.

26. Kathryn Prince, *Shakespeare in the Victorian Periodicals* (New York: Routledge, 2008), 25, 19–20.

27. See, for example, "Parting of Hector and Andromache," and "The Laocoon," both in *Penny Magazine of the Society for the Diffusion of Useful Knowledge*, November 3, 1832, 306; and November 10, 1832, title page, respectively.

28. "Popular Information on Literature," *Chambers' Edinburgh Journal*, May 25, 1833, 130. The journal was founded February 4, 1832.

29. "The Dying Gaul," *Penny Magazine*, January 12, 1833, title page.

30. Patricia Anderson, "Pictures for the People: *Knight's Penny Magazine*, an Early Venture into Popular Art Education," *Studies in Art Education* 28, no. 3 (1987): 133, 136, 138.

31. "A Very Gorgeous Piece: Astley's Amphitheatre," *Times* (London), April 9, 1833, 3.

32. "Gorgeous Piece."

33. See Kate Nichols, *Greece and Rome at the Crystal Palace: Classical Sculpture and Modern Britain, 1854–1936* (Oxford: Oxford University Press, 2015).

34. W. J. Stillman, "Homer's Troy, and Schliemann's," *Cornhill Magazine*, July 1874, 663. On Homeric archaeology, see David Gange and Rachel Bryant Davies, "Troy," in *Cities of God: The Bible and Archaeology in Nineteenth-Century Britain*, ed. David Gange and Michael Ledger-Lomas (Cambridge: Cambridge University Press, 2013), 39–70; and Bryant Davies, *Troy, Carthage and the Victorians*, 47–124.

35. Unattributed clipping, [June 3, 1833].

36. "Theatricals." *The Age*, June 2, 1833, 173.

37. Unattributed clipping, British Library scrapbook: "Astley's Amphitheatre," August 14, 1833.

38. "Astley's Amphitheatre," *Morning Chronicle*, May 18, 1833, 3.

39. Unattributed clipping, August 14, 1833.

40. "Accident at Astley's," *Morning Chronicle*, July 25, 1833, 4.

41. "The Drama," *The World of Fashion and Continental Feuilletons*, February 1, 1834, 30; "The Theatres," *The Satirist, and the Censor of the Time*, January 5, 1834, 6.

42. "Theatricals," *Penny Satirist*, July 25, 1840, 2.

43. "Astley's Amphitheatre," *Times* (London), September 8, 1854, 7.

44. "Gorgeous Piece," 3.

45. "Astley's," *Standard* (London), April 9, 1833, 1.

46. "Unrehearsed Stage Effects,." *Literary Gazette*, June 8, 1833, 364.

47. Hodgson, *Siege of Troy*, 2, 9.

48. "Astley's Amphitheatre," *Morning Chronicle*, April 9, 1833, 3.

49. Unattributed review of *Giant Horse*, August 14, 1833: British Library scrapbook, "Astley's clippings from newspapers, vol. 3, 1806–1856."

50. Hodgson, *Siege of Troy*, 12.

51. "Theatricals," *Figaro in London*, June 22, 1833, 100.

52. "Theatricals," *Figaro in London*.

53. "The English Napoleon," *Punch*, January/June, 1844, 56. Gomersal as Napoleon is the subject of many theatrical portraits, including four in the Jonathan King collection at the Museum of London, which can be viewed online (ID 99.132/417; 99.132/256b; 99.132/523a; 99.132/523b).

54. Leigh Hunt, "Olympic Pavilion," *Examiner*, January 4, 1831, 3.

55. "Jesting, Murder and Burlesques," *Graphic*, February 5, 1870, 219.

Part Three: Time in Transit

8

On Pilgrimage

Michael Ledger-Lomas

On June 10, 1874, the people of Bedford assembled on St. Peter's Green to watch Lady Augusta Stanley unveil a statue of John Bunyan (fig. 8.1). Funded by the Duke of Bedford and sculpted by the queen's favorite, Joseph Edgar Boehm, it attracted thousands of visitors, who could also inspect "relics" of Bunyan. Anglican and Dissenting school children were given illustrated copies of *The Pilgrim's Progress*, before sitting down to a ton and a quarter of cake and one hundred gallons of tea, their joint feasting interpreted as a mark of Christian unity. Later that afternoon, a crowd in the new Corn Exchange—another gift from the duke, built in an appropriately progressive and mercantile Renaissance style—heard Lady Augusta's husband, Arthur Penrhyn Stanley, the dean of Westminster, deliver a speech entitled "The Character of John Bunyan, Local, Ecclesiastical, and Universal." It was a meditation on how a pilgrimage to Bedford, the "certain place where was a den where he laid down and dreamt a dream," might decode but also extend the meaning of *The Pilgrim's Progress*.[1] Stanley read the book not only as an allegory of Bunyan's surroundings but also of England's subsequent progress in Christian charity. That had been in short supply during Bunyan's lifetime, when, as a Dis-

FIGURE 8.1. Unveiling the statue of John Bunyan, at Bedford. The *Illustrated London News*, June 20, 1874.

senter, he had been locked up for unlicensed preaching. Yet since that day, the "giant" of "Old Intolerance" had been slain. The Toleration Act of 1689 had sprung open "the gates of the prison-house of prejudice and intolerance" and begun to reintegrate Dissenters into the nation, a process Stanley now considered virtually complete.[2] As "one of the few books which act as a religious bond to the whole of English Christendom" *The Pilgrim's Progress* was prophetic of but also instrumental in this rapprochement of Church and Dissent, rivaled only by the Bible in supplying the "one uniting element" to otherwise divided Protestants.[3]

This essay presents pilgrimage as a venerable way in which Victorians invested the passage of time with religious significance. Many essays in this volume suggest that the experiences of traveling through time afforded by such new and developing disciplines as archaeology, anthropology, and geology involved their practitioners in traveling through space and vice versa.[4] As literary scholars have recognized, Victorian Protestants—a necessarily crude collective noun—inherited by virtue of their religion the pilgrim's journey as a rite and a metaphor that plotted their individual and national present onto the most important of timelines: the salvation history of Christianity.[5] Pilgrims moved both forward and back in time. For Stanley, pilgrimage was an advance into the future: from bondage to freedom, from ignorance to wisdom, from materiality to the Celestial City. Yet, as Simon Goldhill's chapter in this volume suggests, Protestants often wished to travel backward: to the origins of Christianity or of their churches.[6] For such pilgrims, travel did not so much sanctify as abolish time, just as in Clare Pettitt's essay, the "weary pilgrimage" of Childe Harold leads him beyond the seriality of enlightened historical consciousness.[7] While Protestants condemned the Roman Catholic cult of the saints, with its notion that dead bodies were reservoirs of salvific power, they accepted that it might be a duty and a pleasure to visit the graves of their religious ancestors or the places they had inhabited.[8] Such visits powerfully realized their remoteness in time while briefly bridging it. That urge dribbled from religion to literature, as pilgrims flocked to the rooms in which Shakespeare was born and Keats died, or erected the Caedmon cross (1898) to mark the spot at Whitby where the poet supposedly drew inspiration for his hymns.[9] As William Godwin wrote in his *Essay on Sepulchres* (1803), a founding text of necro-tourism, "I never understood the annals of chivalry so well, as when I walked amid the ruins of Kenilworth Castle."[10] Canonical texts were increasingly read as precise evocations of places and vanished times: visitors to Stoke Poges believed that Thomas Gray had not so much composed neoclassical reflections on *a* country churchyard as described minutely *the* churchyard of Stoke Poges.[11]

This essay proposes that the venerable urge to be a pilgrim shaped how Victorians thought about the passage of time and the scholarly investigation of the past. It continually draws attention to the contradictions embedded in pilgrimage, which though hidden in Stanley's patter were exacerbated by the very scholarly disciplines that promised to guide pilgrims more accurately to their destinations. The first part uses Holy Land pilgrimage to suggest that Victorian Protestants inherited a tension in Christian thought between a definition of pilgrimage as physical contact with holy bodies or sacred ground and one which insisted that it was a quest to purify faith from local, transitory attachments. In following in the footsteps of Christ, Protestants revived Holy Land pilgrimage as a mass phenomenon, even though Reformers had destroyed its rationale by suggesting that journeys to scriptural sites brought no spiritual blessings.[12] The steamship made it easier to visit Bethlehem and Jerusalem—the birthplace and the tomb of Christ—while the efforts of Protestant theologians to delineate a historical Jesus stoked the desire to visit sites associated with him. Yet the knowledge about

Palestine amassed by travelers, biblical critics, and topographers could undermine pilgrimage by turning up evidence against the authenticity of traditional sites. The result was a determination to substitute new sites for the old, or a more radical insistence that it was better to understand pilgrimages not as journeys to places, however venerable, but as an allegory for the Christian life. Scripture supported that attitude, defining believers as "strangers and pilgrims on the earth" who "seek a country" (Hebrews 11:13–14). Pilgrimage was not an event in the Christian's life; it *was* the Christian's life, defined as a quest for a heavenly homeland.[13] Christianity distinguished its followers from those of other religions, many Protestants imagined, by propelling them forward to salvation rather than locking into the past. The lives of Christians were not bounded by ritual devotion to their dead, but defined by their future reunion with Christ.[14] In its most radical form, this emphasis on the shift from repetitive to progressive conceptions of religion led freethinkers to imagine that the truest Protestants were those who cast off the historic forms of Christianity altogether, precisely because they accepted its soteriology.

The second part of this essay uses the writing and reception of national and ecclesiastical histories to suggest another way in which Christian pilgrimage entailed a more problematic relationship to the past than the polite genuflections prescribed by literary tourism. The study of modern pilgrimage by contemporary anthropologists tends to present it as a fractious exercise, in which pilgrims jostle with one another, church hierarchies, and local cultures. If authoritative texts such as the Bible provide shared points of reference, individuals and groups can display "ritual creativity," fashioning rival itineraries that bypass established locations or prize overlooked ones.[15] The understanding of pilgrimage—how to pay obeisance, where and to whom—was similarly fractious in Victorian Britain and divided along denominational and partisan lines, especially at home. Before the Reformation, pilgrimage had largely involved not journeys to the Holy Land but shorter trips to the bodies of local or national saints.[16] Protestants had scrapped this cult of the saints as unscriptural, but the tendency to envision a national past as a fixed constellation of luminaries died hard. Yet denominations differed on whom to include in their little pantheons. Their rival maps of Britain were thick with wayside shrines to heroes and martyrs that often meant little to outsiders.[17] As a liberal dean and favorite of Queen Victoria, Stanley itched to claim that British or at least English history was tending to the creation of an inclusive civic identity.[18] Yet the mooted pluralism of that history favored the continued existence of religious communities whose pasts were too humble or sectarian to be easily incorporated into a national story.

: : :

On the Sunday before Ascension Day in 1853, the young Arthur Stanley preached in the English Church at Jerusalem on pilgrimage to the Holy Land. He had chosen as his text John 16:28, "I came forth from the Father, and am come into the world: again, I leave the world, and go to the Father." Christianity, Stanley argued, initially resided in the memory of apostles, whose recollections of Jesus "would linger yet, for days and years, after the Sun of Righteousness itself had set." Yet the religion of Christ surpassed the earthly "relations" of the historical Jesus. Christ was now with the Father and, as such, "He was to be to them what He was to all mankind, to all mankind what He was to them—the Light, the Life of all; no longer in the flesh but in the spirit."[19] The "historical and local character of Christ" authorized, even impelled, later ages to visit his homeland. It was impossible not to wax sentimental over the "footsteps of the greatest and holiest Presence (to use no higher word) that this earth has ever seen."[20] Yet the "blessing" of pilgrimage could be a snare, for "local associations, questions of sacred topography, scenes of sacred events, however solemn, however interesting, however edifying, are not, and never can be, religion itself." "It is not because He *was here* [in Jerusalem], but because He *is risen*, that we are Christians." Geographical distinctions had no meaning for a universal religion. Stanley's hearers had journeyed from the periphery to the emotional center of Christianity only to recognize that "our communion with Him will be far deeper and truer in the every-day intercourse of common life on our return to our wonted duties, than in the very midst of the scenes of His labours and His teaching."[21]

Stanley echoed a Christian critique of pilgrimage that went back to the Church Fathers.[22] Yet his message was at odds with its setting: a church which, as Francis Frith's widely reproduced photograph of it demonstrates, gave chunky expression to the Church of England's and by extension the British state's investment in the Holy Land. It was the seat of a bishopric founded to discourage penetration of the region by French Roman Catholics and—for some of its millenarian supporters—to hasten the conversion of the Jews to Christianity.[23] Stanley did not share such eschatological fantasies, but he was nonetheless preaching at the end of a journey during which he had intensively studied the topography of the Bible. When published as *Sinai and Palestine* (1856), his findings became a second Bible for Anglophone pilgrims. Stanley was not then discouraging his contemporaries from going to the Holy Land—just refining their interest in it. His investigation of scriptural topography distanced Protestants from the improper localization of piety. Pilgrimage was legitimate only if it was textual—that is, if its itineraries were policed by a critical investigation of the Scriptures. When collated with Scripture, Sinai and Palestine became a Fifth Gospel: a commentary on the other four, but also revelation itself.[24]

Protestant pilgrimage in the Holy Land was thus enabled by the developing disciplines of topography, biblical criticism, and archaeology, but it also shaped what the practitioners of those disciplines thought significant and how their findings were used. The maps produced by the Palestine Exploration Fund guided pilgrims before they served troops, while if the pilgrim's gaze superficially resembled the tourist's, it differed from it in probing for truth.[25] The application of a Scriptural template to modern Palestine also meant that disappointment was intrinsic to Holy Land pilgrimage, because travelers who came equipped with scholarly guidebooks looked for places not as they currently were but as they had been in Scriptural times. Everything in modern Bethany, clucked Stanley's friend the Reverend Norman Macleod, "is squalid as in Skibereen, Connemara, or, alas! some villages in the Hebrides."[26] Worse still, Holy Land sites were carpeted with "monkish" chapels, their relics an improper materialization of Stanley's spiritual religion. Some visitors might feel that the price of engaging in pilgrimage at all was tolerance of these fibs as cues to devotion. Passing the supposed Garden of Gethsemane, the early Victorian artist William Henry Bartlett wondered if it was not too close to the main road to have been a "place of especial retirement. . . . Be this as it may, this group of eight olive-trees has upon it a certain stamp of sadness which cannot fail to affect those who are prepared to *feel* rather than disposed to cavil." Nobody, "however hardy his scepticism," could read the scriptural narrative "said to have occurred within these shades, and not feel himself strangely and powerfully influenced."[27]

Yet if the lure of pious frauds could be strong, Protestants in the Holy Land were generally loath to countenance the heuristic adventures with facsimiles or historical fictions described in Goldhill and Astrid Swenson's essays (chapters 4 and 6 in this volume). They shared the peculiar literalism of Victorian literary tourists: as one read texts to meet their writers, so in pursuing the latter it was important to avoid error. Godwin's *Essay* had argued that "feeling and scepticism in the same question 'cannot live together.' When I meet the name of a great man inscribed in a cemetery, I would have my whole soul awakened to honour his memory.... I must not be intruded on by any idle question, that this is perhaps but his idle grave."[28] Protestant pilgrimage exaggerated that insistence: unlike Byron's rambles, say, it did not represent an escape from but was tightly corseted by enlightenment commitments to measurement, accuracy, and chronology, as embodied in the developing discipline of archaeology. Because Roman Catholic and Orthodox Christians appeared to be careless of such considerations, archaeology in the Holy Land—or for that matter, in other sites of Scriptural significance and confessional rivalry, such as Rome—often packed a confessional sting.[29] Protestant scholars repelled by the ornate Church of the Holy Sepulcher in Jerusalem thus produced alternative suggestions for Christ's tomb, all based on the assumption that the New Testament was less a timeless address to the believer's condition than a grid of references to be superimposed on today's landscape. This method emboldened the architectural writer James Fergusson to locate the tomb beneath the Dome of the Rock and the soldier and evangelical visionary Charles Gordon to insist that he had located it in a garden near a skull-shaped hill—the original of Golgotha.[30] Fergusson inveigled his speculations into William Smith's authoritative *Dictionary of the Bible* (1860–63), while, despite public skepticism, Anglican clergy and grandees stumped up £4,000 for the freehold of Gordon's Garden Tomb.[31]

The abhorrence of pious frauds could then generate speculations as baroque as any saint's reliquary. A safer tactic adopted by Protestants in Bible lands was to relinquish the attempt to pinpoint exact locations and to settle for general evocations of Christ's surroundings. Their writings reveal a preoccupation with landscape, reflecting the premise that while cities crumble and monuments vanish, lakeshores and hillsides are unchangeable and thus timeless points of reference. Travelers could feel that, in moving through the vistas witnessed by Christ and his disciples, they trod in their "footsteps" and inhabited their gaze, a form of contact reproduced for a domestic public in mechanically reproduced images of ever greater sophistication and cheapness. The inconvenience, even danger, associated with these landscapes—ranging from hungry fleas to bandits—enabled travelers to share in the experiences and sufferings of Christ and the apostles. To travel by camel or to accept the ornate rhythms of eastern hospitality was to reproduce the tempo of the Bible. Speed and ease were the enemies of such identification. When the French writer Pierre Loti traveled to Jerusalem in the 1890s, he did so with a Bedouin caravan and was dismayed when they fell in with Thomas Cook's tourists in pith helmets. "Even more than our Oriental,

8. ON PILGRIMAGE

FIGURE 8.2. David Roberts, *Pilgrims Approaching Jerusalem* (1841), Royal Holloway Picture Gallery (accession no. THC0063), University of London.

our religious dream was shaken." Loti was jealous for the Holy Land precisely because he was, as a confirmed agnostic, embarrassed by his inability to respond to Jerusalem itself. His first trip to the Holy Sepulcher did nothing for him, nor did a nighttime trip to Gethsemane.[32]

For the lapsed Catholic Loti, more intent on an authentic performance of orientalism than on his destination, the brisk Protestants were the ignorant intruders. It reminds us that pilgrims, like tourists, always bumped into others whose reactions were an impediment or provocation to forming a proper response to a site.[33] For Protestants, encounters with the rich deposits of devotion at Christian shrines reminded them that Jerusalem and other sites were not just points of origin for the New Testament, but exhibitions of the lived history of Christian geopiety. In David Roberts's vast canvas, *Pilgrims Approaching Jerusalem* (fig. 8.2 [1841]), for instance, the eye is caught not so much by the dun-colored city but by the glittering river of pilgrims coursing toward its walls, their rich robes suggestive of "Oriental" exoticism and the crusader piety that created modern Jerusalem. The corporeal rituals of Orthodox pilgrims in modern Jerusalem challenged Victorians to reaffirm, and occasionally to regret, their persnickety, exegetical approach to the sacred.[34] The eye-popping ethnographic detail in William Holman Hunt's *The Miracle of the Holy Fire* (fig. 8.3)—painted in the nineties but based on earlier experiences—deliberately mocks the ritual it records.[35] Standing proxy for Hunt's repulsion, as an unconventional evangelical, at the mob's devotions is the lady at the bottom of the canvas, who shepherds two flaxen-haired children away from the melee. But eastern pilgrims could move as well as disgust. Loti expresses the ambivalence of his unmoored contemporaries when he describes a last visit to the Holy Sepulcher, "where it is as if the air has been made softly heavy by the prayers of centuries." Personally unmoved, he ogled a gnarled Oriental man who knelt there in fervent prayer and whose "exterior manifestations" were impossible to modern Europeans. Though Loti did in the end squeeze out a tear, he reassured his readers that they did not need to visit Jerusalem to seek Christ, who was everywhere. Yet this centerless religion does not seem superior to the pilgrim faith Loti witnessed in the Church, merely different.[36]

Outside Jerusalem, Protestant observers were more willing to concede that the rites and monuments of Christian pilgrimage were not offenses against historical truth but living forms. Visiting Loreto to see its Holy House, which supposedly flew there from Bethlehem in 1294,

FIGURE 8.3. William Holman Hunt, *The Miracle of the Holy Fire* (1892–99), Fogg Art Museum, Harvard Art Museums.

Stanley grasped that "in the mind of the Catholics it is not merely that they do not reject the fiction, but that it seems hardly to enter into their consideration of the place. . . . They look upon it simply as a part of Palestine become their own—the most holy spot on earth—the scene of the Incarnation." If "the best work for angels now, would be to lift up the House once again, & drown it in the depths of Adria," Loreto nonetheless embodied a doctrinal pillar of Christianity, the Incarnation, and had generated practices as engraved as the lines worn by the knees of its shuffling pilgrims in the basilica's floor.[37] Material survivals of medieval piety were not just burst chrysalises to be swept aside by intellectual advances, but should be read as evidence for Christianity's development. While "development" was popularized by John Henry Newman's apology for accepting the extra-biblical beliefs and practices of the Roman Catholic Church, it caught the imaginations of Protestant scholars who wished to embed Christianity's past within the natural history of culture. From the mid-nineteenth century, Christian universalism accommodated an emphasis on continuities in the character of races and nations, as church historians happily spoke of "Latin," "Teutonic" or "Eastern" Christianity or "Semitic" religion.[38] Both developments allowed for generous readings of embodied Christianity as expressions of national or racial character.

That lesson was applied at home, for England had once had its Loretos. If they could not be incorporated into a continuous narrative of English religion, then centuries of devotion preceding the Reformation must be surrendered to the Roman Catholics. Thanks to Chaucer, Canterbury Cathedral was the most celebrated of such pilgrimage sites. Stanley, who became a canon at the cathedral in 1851, assessed the cult of the murdered Thomas Becket in his *Historical Memorials of Canterbury Cathedral* (1855). He began by noting that in the middle ages an interest in the holy dead had degenerated into a "mania" for the "corporeal elements and particles (so to speak) of religious objects." Under the grip of this "medieval feeling," the people of Canterbury had behaved at times more like "a Neapolitan mob in disappointment at the slow liquefaction of the blood of St Januarius, than . . . the citizens of a quiet cathedral town in the county of Kent!" Stanley could not master his disgust at the "localisation of religion" involved in creeping to the body of Becket, and he regarded the Reformation's destruction of his bejeweled shrine as a blessing.[39] But what was important about the resulting "void" in the cathedral was how the Church of England positioned itself in relation to it. It should neither fill it by reviving Roman Catholic devotions, as Newmanite admirers of Becket wished to do, nor exacerbate it by abusing "old saints and pilgrims," as evangelicals did. Rather "in proportion to our thankfulness that ancient superstitions are destroyed, should be our anxiety that new light, and increased zeal, and more active goodness, should take their place. Our pilgrimage cannot be Geoffrey Chaucer's, but it may be John Bunyan's." Canterbury was no longer a destination, but "a stage in our advance," winning "in the generations which are to come after us, a glory more humble, but not less excellent, than when a hundred thousand worshippers lay prostrate before the shrine of its ancient hero." It was characteristic of the English that they had preserved the "broken pavement" where the shrine once stood to measure their advance in spiritual discernment. This magnanimity made Stanley's readers "richer than were our fathers" and allowed them to regret those necessary acts of iconoclasm, for "the further we recede from the past, the more eager seems now our craving to attach ourselves to it by every link that remains."[40]

Among the Victorians, preoccupation with the material remains of past piety was not then a hallmark of religious conservatism. The Tractarians, whose hagiographies of Becket Stanley disliked, undoubtedly hoped that Gothic architecture and liturgical scholarship could reconstitute religious communities and protect them against Dissent and worldliness.[41] Yet a progressive faith was just as compatible with piecing together broken fossils of faith. A vignette in Stanley's book on Canterbury recreated Erasmus and the Reformer John Colet's visit to the cathedral: Colet contemptuous of its mummery, Erasmus smiling in his skepticism. When they leave Canterbury and pass the nearby lazar house of Harbledown, an inmate offers them a relic—Becket's shoe leather—to kiss. Colet is disgusted, but Erasmus drops a penny into his box. Stanley's moral is the need to bear with those whose superstitions one is determined to abolish and to respect their material traces once their power is vanquished. It is a suggestive reading, given that Erasmus can also feature in Goldhill's essay (chapter 4 in this volume) as the patron of the cocksure search for pristine sources of faith. Stanley's account not only described Erasmian pilgrimage but encouraged it: going to Harbledown, "picturesque even now in its decay" and seeing its relics (such as the dole box, which survived) offered a chance to think about material memories that were bent but not broken by England's spiritual revolutions.[42]

Stanley's Erasmian courtesy involved sympathizing with earlier ages of Christianity while turning aside from their corruptions. His emphasis on the limits of iconoclasm reflected his position as an entitled custodian of national treasures, first at Canterbury and then at Westminster. The

Protestant *avant garde* went beyond his Burkean pose, arguing that the spiritualization of religion might entail the open abandonment not just of Christianity's traditional shrines but of some of its core beliefs. The increasing number of Protestant theologians trained in or touched by German idealism were preoccupied with demonstrating that Christianity was absolute religion, or at least the purest of all historic faiths. This required Christianity to be compatible with an interminable process of intellectual and moral improvement, a position that sat oddly with the conviction that the New Testament was a normative account of what a church and its moral life should be.[43] Was it Protestant to claim that Christians could not improve on Jesus Christ—a person who appeared to be ever more bounded by the limited horizons of his culture? Thinking British Protestants followed the struggles of German theologians with these problems and also confronted the most provocative solution to them, Ernest Renan's *Vie de Jésus* (1863).[44] Renan suggested that Jesus could be called divine simply because his unsurpassed, if hardly faultless, personality had done the most historically to help people toward the divine. In his conclusion, Renan confessed it was impossible to know whether future generations would follow in the paths laid down by Jesus or make their own; he was at least sure that tears would be shed over him long into the future.[45]

Renan's atheistic, lachrymose respect for Jesus savored too much of lapsed Catholicism to please British freethinkers. They preferred to see their pursuit of truth as an outgrowth of their Protestantism and were assisted in this by Carlyle, Froude, and other writers who presented the Reformation not as a return to an original gospel but as the initiation of a radical honesty that had made Protestant countries socially and economically progressive.[46] Freethinkers presented themselves as pilgrims who had advanced beyond the authority of the New Testament Christ. A startling expression of these convictions came in *The Earthward Pilgrimage* (1870) by Moncure Daniel Conway, a Hegelian, American ex-Unitarian, who had ended up ministering at Bloomsbury's South Place Chapel, where he preached a kind of radical theism. His book began with an allegory, telling how John Bunyan had led the narrator to his Celestial City. Yet he quickly came to find it too banal a place to be the end of his life's journey, particularly once it filled up with lukewarm liberal Christians. "The pilgrim could now travel in a first-class carriage" on a "Celestial Railway" that chugged by tunnel under the Hill of Difficulty, thanks to engineering works carried out by Professor Moonshine. Restless, the narrator undertakes a pilgrim's progress in reverse, fleeing the Celestial City for the lively, earthy Bothworldsburg—

Bunyanese for London.[47] In descriptions of visits to pilgrimage sites in Bothworldsburg, Conway voices his realization that the "utopian Pilgrim" requires no faith beyond confidence in humanity's capacity for self-reform. Clergymen and their dim chatter of "stiff-necked Jews and defunct Pilates" impeded that realization. One of the narrator's visits had been to that "Old Shrine," Canterbury Cathedral, to watch Stanley's friend Archibald Campbell Tait being enthroned as Archbishop. Canterbury struck him as an ornate "Conservatory" for an alien, Oriental plant, doomed to wither as thoroughly as Becket's shrine. The "great crowd of cultivated people" assembled "to witness the consecration of a plain old Scotch gentleman to the task of presiding over the work of maintaining in Britain the worship of a dead Jew" were remarkable for bad faith, not a pilgrim's zeal.[48]

Conway's quip was supposed to convince Christians that the future of religion, like the past, belonged to iconoclasts. Visiting Bunyan's restored tomb in Bunhill Fields, the narrator fancied that he heard the writer blessing troublemakers, for "not by the kissing of their bones, or the garnishing of their tombs, or the believing of their creeds, can the brave and free [of old] be honoured; but by an independence and fidelity like their own." Conway set out an alternative itinerary of pilgrimage sites sacred by association with the enemies of Christianity: Voltaire's house at Ferney; Shelley's at Christchurch; St. Peter's, Bournemouth, to which the remains of William Godwin, Mary Wollstonecraft, and Mary Shelley had been translated when old St. Pancras churchyard was carved up by the Midland railway (to be joined there by Shelley's heart). These heroes were dead but not buried, for they had been insurgents against "the Corpse-dynasty": the "Nazarene peasant" and his hordes of followers.[49] *The Earthward Pilgrimage* was ignored more than it was debated, but Conway was nonetheless making an impeccably Protestant argument that pushed the logic of Stanley's sermon at Jerusalem to its extremes: the truest pilgrim was least tethered by the historical origins or even forms of their progressive, universal faith.

: : :

Like John Betjeman's teddy-bear Archie, Joseph Ivimey was a Strict and Particular Baptist. A minister and an autodidact, Ivimey coped with (perhaps induced) crippling spells of depression by writing exhaustive histories of English Baptists. In 1815, he published a *Life of Mr. John Bunyan, Minister of the Gospel at Bedford: In Which Is Exemplified the Power of Evangelical Principles*, which insisted upon his Puritan and Baptist record as a protestor against the established church. He also produced an early example of fan fiction: *Pilgrims of the Nineteenth Century: A Continuation of the "Pilgrims' Progress" upon the Plan Projected by Mr. Bunyan* (1827). The narrator of Ivimey's sequel has a dream induced by late-night reading of Bunyan. In it, he learns from the grandson of the pilgrim's companion Great Heart the story of his "descendants, i.e., *Protestant Dissenters*," who had returned from the House Beautiful to found the town of Toleration. Toleration resembles Conway's Bothworldsburg in being a version of modern London, founded after the Glorious Revolution, accessed via the Street of Free Inquiry and guarded by the Gates of Liberty, Truth, and Conscience. It has at its center a thinly disguised Bunhill Fields, a "*Golgotha*" stuffed with dead Dissenters. Toleration was a noble but precarious polity. Rationalist ministers were active in its streets, painting over the windows of their chapels with designs that blocked out the "beams of the Sun of Righteousness," while Roman Catholics profited from Toleration's ethos to propagate idolatry.[50] When the dreamer awoke, he rushed to reaffirm his evangelical Dissent. Ivimey published his book at the very moment he was endeavoring to block civil equality for Roman Catholics in case they abused it in order to plot against the Protestant Constitution.[51] For Ivimey then, Bunyan was a Dissenting Protestant before he was an Englishman. The town of Toleration was not a staging post on the way to an inclusive national identity, but a sectarian stronghold to be defended against Catholics without and rationalists within.

Ivimey's extension of Bunyan's imagery to argue that Dissenters should cling to the partial toleration accorded to them by William III rather than marching onward to a civil equality for all Christians—lest Catholics exploit it—hints at the obstacles facing attempts to read the national past as a united march into a liberal future. With the political nation expanding in successive jumps—the Reform Acts of 1832, 1867 and 1885—the historians who interpreted the British to themselves sought to claim that they formed one people with a common ancestry. Thomas Babington Macaulay owed his prodigious popularity to convincing newly enfranchised constituencies that they too could feel a sense of descent from the feuding titans of the past.[52] Given that political conflict was generated or inflamed by religious differences, it was important to overcome them by emphasizing the common Christianity of the British or at least the English. Macaulay reimagined Westminster Abbey not as an Anglican redoubt but as a "Temple of silence and reconciliation." Even the great dead not buried there, claimed Macaulay's admirer Stanley, could be imagined as resting in "Chapels of Ease"

FIGURE 8.4. Stephen Ledger-Lomas (photographer), Bunhill Fields, London (2015).

linked to it by "invisible cloisters."[53] Stanley's belief that the graves of all persons of historical note formed one great whole went back to Godwin, whose *Essay* proposed a national committee independent of "party or cabal" to record and maintain the sites of the "illustrious dead."[54]

The nationalization of sectarian heroes and the erection or refurbishment of shrines to them duly became a feature of later nineteenth-century culture, with statues erected not only to Bunyan but Richard Baxter (Kidderminster), William Tyndale (Embankment Gardens), and Oliver Cromwell (Palace of Westminster). This was a British as well as an English enterprise: James Cameron Lees, the dean of St. Giles' Cathedral in Edinburgh, remodeled it as a capacious monument to "the story of our country's faith," complete with a memorial to his charitable friend Stanley.[55] Yet it is the fragility as well as the popularity of such projects that is remarkable, as well as the persistent lack of consensus on what qualified as a national shrine. In 1869, the City of London reopened Bunhill Fields—the heart of Ivimey's Toleration town and his own burial place—as a space for public recreation (fig. 8.4). Its representatives stressed that it contained "the dust of a greater number of notabilities whose memories have become the historic property of the nation than perhaps any other similar place of internment, if we except the venerable abbey of *Westminster*."[56] Yet it never took off as a popular national site. The Dissenter Alfred Light grumbled in his 1913 gazetteer to the Fields that to "the mere passer-by, this sacred plot in the City Road, London, is just an ordinary disused burying ground." Many graves were by then decayed or undocumented. His book was supposed to rectify this state of affairs, but as its excruciating poems reveal—"Oh, pilgrims of the past, we take fresh courage as we learn / Of *your* triumphant passage to the home for which we yearn"—it was aimed at provincial, nonconformist day-trippers, the only people interested in visiting the graves of such eclipsed worthies as John Gill, Samuel Chandler—or Joseph Ivimey.[57]

The Dissenters of Bunhill Fields thus depended for their memory on evangelical sectarians such as Light, who presented them as sectarian fighters against popery, while passing over the graves of those, notably William Blake, less easily presented as Christian crusaders. The division of Christianity into competing churches, sects, and denominations militated against a shared vision of England or Britain's religious past throughout the nineteenth century. Of course, all religious groups reproduce themselves through memory, but as political change—civil equality at home and the opening of ecclesiastically virgin territory in and beyond the British Empire—made membership in Victorian denominations a matter of choice rather than inheritance, it was more than ever important for their members to treasure their own saints and sacred topographies.[58] Unitarians offer a good example, not least because hostile observers considered that their rationalist rejection of Christ's divinity made them "cold" religionists, indifferent to historic associations. As a "sect everywhere spoken against," they kept up morale through memory and pilgrimage. The Unitarian journalist Robert Spears produced a *Record of Unitarian Worthies*, which picked a route through the Christian past, in which the primitive church, some Reformers, then Unitarian "martyrs and confessors" across Europe were oases in drifts of superstition and repression. Unitarians were forever hunting traces of such worthies precisely because they had been reviled or even killed by other Protestants. The search for a "Unitarian shrine" might lead them to a burial ground in Wrexham, where an "oblong recess . . . now choked with dirty fragments of pottery and brick" hid one "whose name will be remembered in our Churches with exceptional veneration so long as Unitarianism retains a place and a name"—the obscure Georgian divine Timothy Kenrick—or to the market square in Burton on Trent, where Edward Wightman had been the last person executed for heresy, joining the martyrs of every age who "shine as stars in the firmament."[59]

Other denominations present endless variations on the Unitarian pattern. On the Celtic fringe, the historiographical defense of the Covenanter martyrs and renewed acts of commemoration at the sites of their deaths, such as the column erected at Deerness, Orkney (1888), for Covenanter prisoners drowned while being transported to America, allowed Ulster and Scots Presbyterians alike to preserve their dissident reading of British history as the site of a national Reformation that had been frustrated, not fulfilled.[60] The Tractarian movement, whose scholars dominated ecclesiastical history in England in the later nineteenth century, was equally intent on resisting Stanley's liberal vision of religious history as fuzzy progress. When Conway visited St. Peter's, Bournemouth, its High Gothic reconstruction by G. E. Street was as yet incomplete. Street's patron, the Tractarian vicar Alexander Morden Bennett, had tried to stop the burial of the Shelley family's remains in his graveyard, viewing them as nasty flecks on his bastion of apostolic faith. His convictions were evident in the twelve windows in Street's north aisle, which depicted the apostles, each with a banner holding a sentence from the Apostles' Creed. They were paid for by donors in memory of deceased relatives, thereby associating their brief lives with undying pillars of the church. The Founder's Window, erected after Bennett's death in 1880, captures how Tractarians viewed themselves as a link in the chain of sacramental and priestly authority. Christ, bearing chalice and scepter, topped three pairs of priests and builders: Solomon and Aaron, two medieval bishops, and finally Bennett (in surplice, scarf, and hood, cradling a model of St. Peter's), and Bishop Flambard, who had built nearby Christ Church Priory. Such sermons in stone and glass were complemented by Tractarian historiography, which traced a continuous episcopal succession for their church back to the Apostles.[61] These notions of descent were, like Street's Gothic, compatible with "development" and adaptation to the English character, but also entailed indifference or hostility to avowedly Protestant sites of memory. One of the first public manifestations of Tractarianism had been to refuse to subscribe to the Martyrs' Memorial in Oxford, which commemorated Latimer, Cranmer, and Ridley's rejection of the "errors of the Church of Rome."[62]

: : :

We can still visit the sites created for Victorian pilgrims. Boehm's Bunyan is still on St Peter's Green, which is now a dog-eared patch of turf rather than the rural sward depicted in engravings of the Bunyan Festival. The

John Bunyan Museum still displays its contact relics, such as his tinker's anvil. Yet the Protestant and biblical culture which found these sites and objects of riveting interest was less durable. The Bunyan Meeting chapel, whose bronze doors with scenes from the Pilgrim's Progress were installed after Stanley's address, is still there, although its graveyard has been cleared. Elsewhere the decay of Dissent is obvious. Although the statue of the prison reformer John Howard (1890) still adorns the market square, the Congregational chapel whose facade is topped with his name closed in 1971 and is now a night club. That eclipse can obscure the fractious vitality of Protestantism, which made pilgrimage a central form of Victorian time-traveling. The practice and the idea of pilgrimage were always troubled by doubts about how and when it was proper to identify a spiritual and textual Christian faith with obeisance to bodies, places, monuments or even the historical Jesus. The archaeological or historical disciplines which promised to tidy up pilgrimage often exacerbated these disagreements or were captive to them. Yet if the desire to visit the place where *The Pilgrim's Progress* was written and the need to be a pilgrim were distinct, often contradictory urges, they were nonetheless both central to Victorian culture.

Notes

1. See Isabel Hofmeyr, *The Portable Bunyan: A Transnational History of "The Pilgrim's Progress"* (Princeton, NJ: Princeton University Press, 2003), chap. 10.
2. Arthur Penrhyn Stanley, "The Character of John Bunyan: Local, Ecclesiastical, Universal," in W. H. Wylie, *The Book of the Bunyan Festival. A Complete Record of the Proceedings at the Unveiling of the Statue Given by His Grace, the Duke of Bedford, June 10, 1874*, ed. W. H. Wylie (London: James Clarke, 1874), 50–51.
3. Stanley, "Character of John Bunyan," 51–52.
4. See the essays by Peter Mandler and Clare Pettitt in this volume (chaps. 2 and 10).
5. See Barry V. Qualls, *The Secular Pilgrims of Victorian Fiction: The Novel as Book of Life* (Cambridge: Cambridge University Press, 1982).
6. Simon Goldhill, "Ad Fontes," in this volume (chap. 4).
7. Clare Pettitt, "At Sea," in this volume (chap. 10).
8. See Robert Bartlett, *Why Can the Dead Do Such Great Things? Saints and Worshippers from the Martyrs to the Reformation* (Princeton, NJ: Princeton University Press, 2013), 1, and throughout, on sainthood; see also Thomas W. Laqueur, *The Work of the Dead: A Cultural History of Mortal Remains* (Princeton, NJ: Princeton University Press, 2015), for deep continuities in the importance accorded to the dead.

9. See Helen Brookman, "In the Beginning," in this volume (chap. 5).

10. William Godwin, *Essay on Sepulchres; or, A Proposal for Erecting Some Memorial of the Illustrious Dead in All Ages, on the Spot Where Their Remains Have Been Interred* (London: W. Miller, 1803), 71; Paul Westover, *Necroromanticism: Travelling to Meet the Dead, 1750–1860* (Basingstoke: Palgrave Macmillan, 2012); Laqueur, *Work of the Dead*, 49–54.

11. Nicola Watson, *The Literary Tourist: Readers and Places in Romantic and Victorian Britain* (Basingstoke: Palgrave Macmillan, 2006), 39–47.

12. See J. G. Davies, *Pilgrimage, Yesterday and Today: Why? Where? How?* (London: SCM Press, 1988), chap. 3 on critique of pilgrimage; Eitan Bar Yosef, *The Holy Land in English Culture, 1799–1917: Palestine and the Question of Orientalism* (Oxford: Oxford University Press, 2005) is the most provocative introduction to Victorian travel in the Holy Land.

13. Philip Edwards, *Pilgrimage and Literary Tradition* (Cambridge: Cambridge University Press, 2005), 5.

14. See Robert Pogue Harrison, *The Dominion of the Dead* (Chicago: University of Chicago Press, 2003).

15. See, for example, John Eade and Michael Sallnow, eds., *Contesting the Sacred: The Anthropology of Christian Pilgrimage* (London: Routledge, 1990); Simon Coleman, "Do You Believe in Pilgrimage?" *Anthropological Theory* 2 (2002): 355–68, and Anna Fedele, *Looking for Mary Magdalene: Alternative Pilgrimage and Ritual Creativity at Catholic Shrines in France* (Oxford: Oxford University Press, 2012).

16. Bartlett, *Great Things*, 425–33.

17. See Gareth Atkins, ed., *Making and Remaking Saints in Nineteenth-Century Britain* (Manchester: Manchester University Press, 2016).

18. On Stanley's reading of English history, see Michael Ledger-Lomas, "An Erastian Descent: History and Establishment in the Thought of Arthur Penrhyn Stanley and E. A. Freeman," in *Making History: Edward Augustus Freeman and Victorian Cultural Politics*, ed. G. A. Bremner and Jonathan Conlin (Oxford: Oxford University Press, 2015), 65–84.

19. Arthur Penrhyn Stanley, *Sermons Preached Mostly in Canterbury Cathedral* (London: John Murray, 1859), 352–53.

20. Stanley, *Sermons* 357.

21. Stanley, *Sermons*, 360.

22. Bartlett, *Great Things*, 411.

23. See Donald M. Lewis, *The Origins of Christian Zionism: Lord Shaftesbury and Evangelical Support for a Jewish Homeland* (Cambridge: Cambridge University Press, 2009), chap. 10.

24. See Edward Aiken, *Scriptural Geography: Portraying the Holy Land* (London: I. B. Tauris, 2009).

25. John Urry, *The Tourist Gaze: Leisure and Travel in Contemporary Societies* (London: SAGE, 1990).

26. Norman Macleod, *Eastward* (London: A. Strahan, 1866), 174.

27. William Henry Bartlett, *Footsteps of Our Lord and His Apostles in Syria, Greece and Italy* (London: Arthur Hall, Virtue, 1852), 168–69.

28. Godwin, *Essay on Sepulchres*, 84.

29. Anne Bush and Jane Garnett, "Rome," in *Cities of God: The Bible and Archaeology in Nineteenth-Century Britain*, ed. David Gange and Michael Ledger-Lomas (Cambridge: Cambridge University Press, 2013), 285–314.

30. Sarah Kochav, "The Search for a Protestant Holy Sepulchre," *Journal of Ecclesiastical History* 46 (1995): 278–301.

31. Kochav, "Protestant Holy Sepulchre, 294; Michael Ledger-Lomas, "Introduction," in *Dictionary of the Bible*, ed. William Smith (1860–63; repr., London, I. B. Tauris, 2016), xii–xiii.

32. Pierre Loti, *Jérusalem* (1895; repr., Paris: Payot, 2008), 28, 191.

33. On tourism, see James Buzard, *The Beaten Track: European Tourism, Literature, and the Ways to Culture, 1800–1918* (Oxford: Oxford University Press, 1993).

34. Glenn Bowman, "Christian Ideology and the Image of a Holy Land," in *Contesting the Sacred*, ed. Eade and Sallnow, 99–121; and Ruth and Thomas Hummel, *Patterns of the Sacred: English Protestant and Russian Orthodox Pilgrims of the Nineteenth Century* (London: Scorpion Cavendish, 1995).

35. Nicholas Tromans, "The Holy City," in *The Lure of the East: British Orientalist Painting*, ed. Tromans and Emily Weeks (New Haven, CT: Yale University Press, 2008), 164.

36. Loti, *Jérusalem*, 205, 207.

37. Arthur Penrhyn Stanley to Benjamin Jowett, September 29, 1852, Balliol College, Oxford MSS.

38. Colin Kidd, *The Forging of Races: Race and Scripture in the Atlantic World, 1600–2000* (Cambridge: Cambridge University Press, 2000).

39. Stanley, *Historical Memorials*, 190, 200, 238.

40. Stanley, *Historical Memorials*, 255, 259.

41. Simon Skinner, "'A Triumph of the Rich': Tractarians and the Reformation," in *Reinventing the Reformation in the Nineteenth Century: A Cultural History*, ed. Peter Nockles (Manchester: Manchester University Press, 2014), 69–92.

42. Stanley, *Historical Memorials*, 243.

43. For a penetrating exposition of this dilemma, see Johannes Zachhuber, *Theology as Science in Nineteenth-Century Germany: From F. C. Baur to Ernst Troeltsch* (Oxford: Oxford University Press, 2013).

44. On Renan's work, see Robert Priest, *The Gospel According to Renan* (Oxford: Oxford University Press, 2015).

45. Ernest Renan, *Vie de Jésus* (Paris : Michel Lévy, 1863), 558–59.

46. John Morrow, "The Real History of Protestantism: Thomas Carlyle and the Spirit of Reformation," in *Reinventing the Reformation*, ed. Nockles, 305–23.

47. Conway, *The Earthward Pilgrimage* (London: J. C. Hotten, 1870), 21–22, 29.

48. Conway, *Earthward Pilgrimage*, 66, 70.

49. Conway, *Earthward Pilgrimage*, 159, 241, 323.

50. Joseph Ivimey, *Pilgrims of the Nineteenth Century: A Continuation of the "Pilgrims' Progress" upon the Plan Projected by Mr. Bunyan* (London: Simpkin and Marshall, 1827), 48.

51. George Prichard, *Life and Writings of the Rev Joseph Ivimey, Late Pastor of the Church in Eagle St.* (London: George Wightman, 1835), 215–17.

52. See Theodore Koditschek, *Liberalism, Imperialism, and the Historical Imagination: Nineteenth-Century Visions of a Greater Britain* (Cambridge: Cambridge University Press, 2011); and Catherine Hall, *Macaulay and Son: Architects of Imperial Britain* (London: Yale University Press, 2011).

53. Arthur Penrhyn Stanley, *Historical Memorials of Westminster Abbey*, 2nd ed. (London: John Murray, 1868), 332.

54. Godwin, *Essay on Sepulchres*, 33–34.

55. See James Cameron Lees, "What Mean These Stones?," in *Service in St Giles University Upon the Reopening of the Church after Its Restoration by William Chambers* (Edinburgh: n.p., 1883), n.p.

56. *Proceedings in Reference to the Preservation of the Bunhill Fields Burial Ground* (London: Benjamin Pardon, 1867), 25–26. For this story, see Michael Ledger-Lomas, "'The Campo Santo of the Dissenters': Bunhill Fields and Sacred Space in Victorian London," in *Visualising a Sacred City: London, Art and Religion*, ed. Aaron Rosen, Ben Quash, and Chloe Reddaway (London: I. B. Tauris, 2016).

57. Alfred Light, *Bunhill Fields: Written in Honour and to the Memory of the Many Saints of God Whose Bodies Rest in This Old London Cemetery* (London: C. J. Farncombe and Sons, 1913), preface and 1.

58. See Danièle Hervieu-Léger, *La Religion pour Mémoire* (Paris: Cerf, 1993).

59. "A Unitarian Shrine," *Christian Life*, May 20, 1876, 8; Robert Spears, *Record of Unitarian Worthies* (London: E. T. Whitfield, 1876), 346.

60. James Coleman, *Remembering the Past in Nineteenth-Century Scotland: Commemoration, Nationality, and Memory* (Oxford: Oxford University Press, 2014), chap. 7.

61. James Kirby, "An Ecclesiastical Descent: Religion and History in the Work of William Stubbs," *Journal of Ecclesiastical History* 65 (2014): 84–110.

62. Andrew Atherstone, "The Martyrs Memorial at Oxford," *Journal of Ecclesiastical History* 54 (2003): 278–301.

9

Across the Divide

David Gange

> "There is nothing that distinguishes nations and ages more strongly than their attitude towards death. . . . The ancient Egyptian at any rate, being unacquainted with absinthe and cigarettes, was accustomed to look death very steadily in the face. . . . [He] would wonder at [modern Londoners] who for the most part trudge comfortably along their inevitable path with scarcely a glance at the dark door which ends it."[1]

It seems strange to see inhabitants of nineteenth-century London characterized as ignoring the "dark door" at the end of life, since death is usually portrayed as a characteristically Victorian preoccupation.[2] Material things pertaining to the dead, from grand mausoleums to treasured lockets of late relatives' hair, were so omnipresent that it can look to modern viewers like the Victorian door to the afterlife was always ajar. Twenty-first century novels set in the period often make a feature of these attitudes. Julian Barnes's *Arthur and George* treats the psychological peculiarities of Arthur Conan Doyle as the result of a childhood confrontation with the corpse of his grandmother.³ The stories historians tell about other famous Victorians place similar emphasis on encounters with death: Charles Darwin, it is often said, did not lose his religious faith because of the "dangerous idea" of evolution by natural selection, but because of the cruel loss of his daughter Annie to tuberculosis.⁴ Disease, and infant mortality in particular, meant that everyone had to be prepared to deal with loss. In a world of bad deaths, where disease put a rapid end to countless young lives, the "good death" was romanticized in art and poetry. This was not the "quick death" that has been seen as desirable in many cultures, but a "lingering death" that allowed time for earthly good-byes and spiritual preparation for the subsequent meeting with the creator.⁵

As the decades of the nineteenth century passed, several key principles of early Victorian attitudes to death were eroded and, by some, rejected. These included the idea of eternal punishment after death and even the existence of the site of that punishment, hell. By the 1870s, commentators were also beginning to question the idea of the soul. An increasing number of thinkers refused to accept definitions of the soul as "that part of the human being which survives death" without some clarity as to what (and even where in the human organism) the soul was before death.⁶ In tandem with these challenges to traditional concepts, new ideas and practices entered the Victorian lexicon of death. Among these, spiritualism and occultism are among the best known. Each, like Christianity, argued that death was not an ending but a bridge or door to something new; each, however, offered a practical mode for interacting with those beyond the threshold that traditional Christianity usually did not.

Both the established traditions of Christian death and the new phenomena of spiritualism and occultism have been explored in sophisticated historical studies.⁷ Yet there were other, less familiar, Victorian developments concerning death that tell us a great deal about the period. This chapter tackles one such theme, exploring how ancient attitudes to death, rediscovered during the nineteenth century, were interpreted and adapted to shape the age. As the traditions of Victorian Protestantism became less secure and authoritative, many commentators, from a diverse array of backgrounds, looked to other traditions that had given significance to ideas resembling the soul and afterlife. For a small minority, these were evidence that Christian ideas were derivative and therefore tainted; for many more, they provided evidence that a belief in the eternal soul was so widespread as to be undeniable: by exploring these traditions, they felt, deep truths that were lost or devalued in modern thought might be revealed.

The Victorian Cult of Death

Before turning to the impact of ancient ideas on attitudes to death, a little more should be said about conventions surrounding death with which Victorians (and modern students of the Victorians) have been familiar. The best-known themes, embodying all the Victorian tropes of sentimentality and earnestness, come from literature. Novels worked to establish conventions for approaching the "dark door." Among the most celebrated episodes was the death of Dickens's Little Nell. Nell is a character in *The Old Curiosity Shop* whose deathbed is described in detail, culminating in her passing: "She was dead. Dear, gentle, patient, noble Nell was dead. Her little bird—a poor slight thing the pressure of a finger would have crushed—was stirring nimbly in its cage; and the strong heart of its child-mistress was mute and motionless forever."[8] Many Victorians recorded responding in the "correct" manner to Dickens's orchestration of emotion. They expressed "anguish unspeakable" by "weeping copiously."[9] The seasoned Irish politician, Daniel O'Connell, was so upset that, bursting into tears, he threw the novel from the window of a train.[10] Yet Nell became such a cliché of typical Victorian sentimentality that when Oscar Wilde wished to disown Victorian attitudes, it was this scene he attacked: "One must have a heart of stone to read the death of Little Nell without laughing."[11]

It was not just realistic novelists who tried to teach Victorians how to handle death. In *Dracula*, Bram Stoker asked what it would mean for an unlucky few to escape the usual doorway to the afterlife and become eternal dealers of death rather than its subjects. His vision of death, and the language in which he described it, was inspired by Victorian social history, as is evident from a letter Stoker's mother once wrote to him: "In the days of my youth, the world was shaken with the dread of a new and terrible plague which was desolating all lands as it passed through them, and so regular was its march that men could tell where next it would appear and almost the day when it might be expected. It was the cholera, which for the first time appeared in Western Europe. Its bitter strange kiss, and man's want of experience or knowledge of its nature, or how best to resist its attacks, added, if anything, to its horrors."[12] Like many novels and short stories, *Dracula* gave Victorian fears of unnatural death a supernatural twist.

Technology, particularly photography, also promised immortality, allowing conventions surrounding death to spill into the images, as well as the literature, with which Victorians surrounded themselves. Death intruded, in ways modern viewers find uncanny, into Victorian family albums when dead children were posed as if alive for a final *mise en scène* before burial. Late in the period, ghosts, spirits, and souls, appear on photographs: every inconsistency on a plate was interpreted as the technological capture of an element of the afterlife imperceptible to human eyes.[13]

Books and photographs commemorated death and fostered new ways to engage with the beyond, but the cemetery was death's natural habitat. Early in the century, cemeteries had been associated more with scandal than with care for the public good. Infamous misdeeds had closely associated burial and criminality. This involved gruesome thefts of buried bodies for dissection. It also included speculative profiteering on the burial of corpses. In one instance, twelve thousand bodies were found in a small space beneath Enon Chapel in London.[14] The minister had made himself wealthy by charging fifteen shillings per "burial" before throwing the corpses into the open sewer beneath the building. Smells, seepage, body-snatching, and even the sight of rotting remains provided a macabre set of associations for the less-than-reputable cemeteries of the early Victorians. Increasingly strong associations between smells and the spread of disease made these shortcuts with corpses an ever more potent problem.

The result was that, by the mid-Victorian period, the cemetery had been reconceptualized. New cemeteries were no longer urban bone pits but cleanly suburban idylls. Commercial companies collaborated with politicians and religious authorities, including increasingly large bodies of nonconformists, to undertake vast ventures that combined Victorian religiosity with other stereotypically Victorian traits: entrepreneurship and market capitalism.[15] The suburban pleasure-garden cemetery became like the Egyptian tomb: a place in which those who had "crossed the divide" sustained their links with the living. Tombs and headstones could be intricate and spectacular, expressing the same civic pride that led well-to-do Victorians to build grand town halls, libraries, and swimming pools. These new conventions of the cemetery were closely connected to Victorian visions of continuity from life to death: heaven was envisaged as a community much like an earthly town or village, in which the dead continued to perform their functions as fathers, mothers, brothers, and sisters.

These cemeteries have a particular relevance to this book: they were time-travel theme parks. Combining the disciplines of engineering, architecture, and landscape gardening, designers of Victorian cemeteries recreated ancient Mesopotamia, Egypt, Greece, Rome, and Byzantium, as well as Gothic, medieval Europe, in the midst of British suburbs.

Architects chose specific ancient and medieval societies to emphasize the timeless and eternal nature of death. The result was that in the avenues of a large cemetery, a visitor could wander through millennia of history and witness a more diverse bricolage of historical styles than anywhere else in British culture. Death was historicized to weird effect.

Historical studies exploring the fascinations and fetishes of the Victorian "cult of death" establish a firm chronology. Books on the architecture of death, the poetry of mourning, and the artistic evocation of the afterlife tend to focus primarily on the mid-century decades.[16] They analyze the statistics, the institutions and the emotions that this cult encouraged during its heyday between the 1830s and the 1860s. The "hungry forties" witnessed major epidemics, not just of cholera but of typhus and tuberculosis, which saw life expectancies in urban areas plummet. The 1860s saw mass mourning after the death, in 1861, of Queen Victoria's husband, Albert. The whole royal household spent five years in an official state of mourning; Victoria herself never truly left that state.

Although these "High Victorian" attitudes toward death are very well known, less attention has been paid to what happened next, except through the lens of specific literary movements such as Decadence and Aestheticism. In fact, scholars have often seemed to assume that the Victorian cult of death was suddenly swept away when the mass killing of the First World War demanded new attitudes and new coping strategies.

The First World War was not, however, a watershed in this way: substantial changes had already taken place throughout the late Victorian period. Whether slain by the 1890s "absinthe and cigarettes" mentioned in the quotation that opens this chapter, or by dramatic changes in the theological and cultural landscape, the Victorian cult of death was itself long dead by 1914. This shift did not involve late-Victorians ignoring death or ceasing to discuss it; instead it was the end of a consensus. Rather than a shared Victorian vision for how to deal with death, these decades saw an increasing openness to other visions of death. Late Victorians were more willing to investigate the attitudes of other religious traditions; they asked what the sciences could tell of the fate of humans after death; and they asked how the established churches could continue to shape responses to death in a world where the number of shared beliefs about the afterlife was declining. Unsurprisingly, the late Victorian discovery of a range of traditions concerning death (such as the ancient Egyptian Book of the Dead) had a significant impact on already-mutating ideas about the nature of the afterlife.

Victorian Religion and the Afterlife

Even in societies where much of life is secular, the institutions and discourses of religion enfold social behavior concerning death. In Victorian Britain little of life was secular. Indeed, even the distinction between religious and secular is unhelpful when everything—morality, science, art, history—took place within an overarching religious framework.[17] Since religion dominated all of life, it is unsurprising that death was inseparable from faith.

Yet the Victorian period saw rapid religious change. In the 1830s, the influence of the biggest religious revival in modern British history could still be felt. The Evangelical Revival, beginning in the late eighteenth century, had swept through British religious denominations. It had intensified the religion of established churches, such as the Presbyterian Church in Scotland and the Anglican Church of England; it had also brought vigorously evangelical movements, such as Methodists and Baptists, from the margins to the center of British life.

Evangelicalism insisted on the pervasive role of religion in all aspects of life, but one of its most distinctive features was a fixation on the afterlife. Most early Victorians believed unswervingly in the physical reality of heaven and hell.[18] Across the divide, there lay either entire sanctification in the presence of God or eternal punishment amidst Satan's horrors. A good evangelical life was lived fixated on that divide: acting in *this* world so as to secure the soul's safety in the next. Evangelicals knew what happened after death, and their creed, their critics argued, was an instruction manual to ensure themselves the best deal in the beyond.

If the 1830s were a period of extravagant religiosity, then one way of narrating the Victorian period is to see a gradual decline of Evangelical intensity. This involved a backlash against evangelical fervor. It was a crisis of kinds, but it cannot easily be captured by well-worn phrases like "crisis of faith" or "secularization": most who lost their evangelical verve slipped either into a moderate Protestantism or a Catholic or pseudo-Catholic faith. This was therefore a case of "transformation" more than "loss."[19]

However, some things were lost. Among these was certainty concerning the afterlife. Eschatology (the study of the four last things: death, judgment, heaven, and hell) became increasingly controversial. One of the earliest casualties of the turn toward a less severe and authoritarian religion was the idea of real, eternal, and excruciating punishment in hell. In a series of mid-century scandals, several leading public figures and churchmen challenged the reality of eternal punishment.[20] And by the late Victorian period, large numbers of devout Christians did not believe in the reality of hell.

Another central Christian motif—Christ's triumph over death at the Resurrection—was challenged from a different angle. This was the tradition, originating in the German states, of higher criticism of the Bible, much of which sought to strengthen biblical religion by removing its "fanciful" or "superstitious" elements, creating a new rendering of Christianity for an age that found miracles hard to reconcile with its philosophical outlook. British believers tended to look with concern rather than respect at the claims of this higher criticism; they were now, however, aware of visions of Christianity that did not place the divine defeat of death at the center of their theology.[21]

Evangelical beliefs had been forged during the most intense period of urbanization, when death was most immediate and personal to dwellers in British cities. As urban conditions improved and life expectancy increased, the evangelical fixation on the afterlife began to look overblown and sanctimonious. This had been a religion of inky-skied shock cities such as Manchester or Glasgow, or of miners who risked death each day; it was less at home in an era of suburbanization, with urban outskirts peopled by clerks and actuaries. All these factors combined to create divergences in traditions surrounding death: those embraced by major British denominations were no longer persuasive to all British people.

Discovering Other Deaths

From the beginning of the century to its end, Britons experimented with traditions of death other than traditional Christian ones. Early in the period, these largely provided a stylistic vocabulary intended to express traditional ideas: it was rare for them to embody distinctive intellectual or spiritual content. Gradually, however, these other traditions gained some philosophical and even theological authority.

Ancient Egypt makes a fascinating case study on this theme. At the beginning of the century, following Napoleon's invasion of Egypt, the civilization gained new significance in European cultures. After the Battle of the Nile in 1798, when Nelson destroyed Napoleon's fleet in Aboukir Bay, Egypt gained patriotic associations for the British. Although the Egyptian hieroglyphs were deciphered in the early 1820s, it took several more decades for Egyptian literature to circulate among

European readers. Instead, Egyptian imagery became expressive of certain aspects of modern culture, among which death was predominant.

In 1829, with the post-Napoleonic craze for Egyptian fashion still strong and the crisis of London's overcrowded cemeteries looming large, an architect, Thomas Willson, proposed a new kind of cemetery. Willson's transformation of burial customs would begin with a stupendous pyramid on London's Primrose Hill. This would blend a timeless granite shell with modern brick construction and ultramodern steam-powered lifts. Rising more than ninety-four stories (with a base the size of Russell Square), Willson's pyramid would house five million dead. It would make profit from burial fees, but also from a kind of necrotourism: paying visitors would, Willson believed, inevitably flock to witness the Pyramid General Cemetery Company. Despite the fact that Willson soon became one of the most influential cemetery designers in Britain (this was no marginal eccentric), his pyramid was never built, and his revolution never enacted.[22]

It took some time for ancient Egyptian culture to gain greater resonance than the monolithic symbolism of the abortive Primrose Hill pyramid. This new significance arrived with the eventual publication of texts written by the ancient Egyptians themselves. Key among these was the Book of the Dead. This body of texts was recovered from millennia of obscurity in the late nineteenth century, and made its first real impression on the European public in the 1890s. Attempts to elucidate its meanings took diverse forms. Some readers saw it as barbaric and superficial. Others, however, assumed that it went to the heart of things, revealing human and metaphysical truths different in kind from those accessible through Greek philosophy, medieval mysticism, or modern science. The 1890s were particularly susceptible to the latter, positive, reading of Egyptian thought. The context for British engagement with Egypt had also changed dramatically since the 1830s. In 1882, British troops had occupied Egypt, and some British individuals had begun to claim a sense of custodianship over the nation's past, seeking to join French and German scholars, who had been digging for decades, in claiming the right to excavate and analyze ancient sites.[23] The first organized British excavations in Egypt therefore took place in the same year as the military occupation.

It was in that decade that "the Egyptian Book of the Dead" was established as a familiar phrase, even though very few individuals had any opportunity to read its texts. It was beginning to pick up advocates who saw it as a source of timeless wisdom. The great historical and literary surveys of the Victorian period written in the early twentieth century gave to the 1890s a series of epithets. In this "*fin de siècle*," this "decadence" and this "hinge" between two centuries, events such as the Oscar Wilde trial have drawn great attention, and a "revolt against Victorianism" has often been taken for granted. The macabre Book of the Dead might seem to fit easily into accepted portraits of this decade.

Yet the concepts associated with the *fin de siècle* are not actually very helpful in understanding how members of a new mass readership engaged with Egyptian funerary texts. The popularity of the Book of the Dead belongs to many different 1890s trends, but the most substantive were influenced by German, more than French, thought and, perhaps surprisingly, they were rooted in familiarity rather than exoticism. If the Book of the Dead belonged to either side of the stark divides the Victorians sometimes drew between Western and Eastern world systems, it was more often attributed to the West than the East. And the "new spirituality" of this moment was often deeply traditional; far from "revolting against Victorianism," readers of the Book of the Dead distilled distinctively nineteenth-century themes into particularly conservative forms. The Book of the Dead was one of several non-European texts that caused British readers to reflect deeply on their own religious traditions—not rejecting them but seeking to understand them in context. Buddhist, Zoroastrian, and ancient Mesopotamian literature were amongst the others.

The Death and Life of the Pilgrim Soul

The Book of the Dead, or the Book of Coming Forth by Day, was a central tradition in the funerary ritual of ancient Egypt. It consists of short "spells" for the protection and empowerment of the individual after death (echoing Bunyan's *Pilgrim's Progress*, this was rendered the "pilgrim-soul"). Coalescing from a long tradition dating back to the pyramid texts and coffin texts of the Old and Middle Kingdoms, the Book of the Dead in its conventionalized form emerged in the middle of the second millennium BCE, at the beginning of the New Kingdom. Even in their New Kingdom form, the spells of the Book of the Dead were not standardized. They exist in a bewildering array of combinations and a wide range of media, from short selections inscribed on the walls of tombs to much longer catalogues in New Kingdom papyri. As one 1890s commentator noted, "A belief that endured for forty-five centuries must needs undergo frequent changes." New Kingdom renderings, as in the case of the papyrus of Ani (acquired by the British Museum in 1888), can be lavishly and beautifully illustrated. These illustrations often depict some of the most familiar images from Egyptian art, such as the judgment of the dead, in which the human heart is weighed against a single feather, with "heaviness" figured as mortality. These late, extensive texts are a uniquely substantive source for reconstruction of the Egyptian conception of the afterlife and Egyptian belief in general.

In the short term the decipherment of hieroglyphs, usually dated to 1822, did little to facilitate European engagement with ancient Egyptian thought. The most important Egyptological works of the following decades relied more on art than text, more on Greek and Hebrew sources than on Egyptian ones. Only in the 1840s were substantial efforts made to force Egyptian inscriptions and papyri to give up their myths and metaphysics.[24] Even then, the first results—for instance, Karl Richard Lepsius's edition of the Ptolemaic papyrus of Iuefankh and Samuel Birch's attempt to produce an English version of Lepsius's text—drew little public attention (these efforts were later judged to have "failed to make reasonable sense" of a "misleading . . . original").[25] What these scholars revealed was not a literature to be enjoyed or even comprehended, but a mass of riddles that refused to yield its secrets. The histories of ancient Egypt that British readers did consume in the 1850s and 1860s owed little to the hieroglyphs and speculated little on this religious tradition.[26]

In the 1870s this began to change. This decade saw extravagant new claims for the power of archaeology to recover ancient worlds. In 1874 the Second Congress of Orientalists, which met in London, insisted that a proper, comparative translation of the Book of the Dead should be a new priority for Europe's scholars. This call to arms was marshalled by Lepsius, who instigated a breakfast-time meeting of international Egyptologists with the expressed aim of developing a rigorous translation of the Book of the Dead. This, it was intended, should be the life's work of a promising young linguist who would have all the resources of the world's orientalists placed at his disposal.[27]

The scholar chosen for this daunting task was Edouard Naville. Aged thirty, Naville had already devoted a decade to the language and culture of ancient Egypt: having published on the Horus myth, his new research on solar inscriptions was almost ready to go to press when he received this commission. Naville's devotion to ancient Egypt had been disrupted only by service in the Franco-Prussian war and his lifelong commitment to Bible translation.

The chief problem that confronted Naville in his new commission related to the huge diversity of different versions of the Book of the Dead: known editions were scattered through museums around the world and no one had much idea of what a comparative treatment of these might reveal. In December 1875 a circular was issued to the museums of Europe, signed by the leading Egyptologists of Germany (Lepsius), Britain (Birch), France (Chabas), and Switzerland (Naville himself), announcing the project and requesting information.[28] With Birch's help, Naville soon identified the British Museum's papyrus of Nebseni as a fruitful base text against which others could be compared. Originating from the city of Memphis, this had been bought by the Museum from Sotheby's in 1836. Following up the pioneering work of William Henry Fox Talbot in photography of hieroglyphic and cuneiform texts, Naville created a photographic record of this single papyrus. In 1877 he was ready to begin his tour of Europe's museums.

By the mid-1880s, Naville and his wife, who did much of the copying of texts ("one would have liked to see Mdme. Naville's name upon the title page," noted one British reviewer), had familiarized themselves with the collections and the reading rooms of most of Europe's major museums and had begun to construct their text.[29] The task had taken them through London, Paris, Berlin, Turin, Leyden, Florence, Bologna, Parma, Rome, Naples, Hanover, Marseilles, Liverpool, Dublin, and Cairo, and into at least six Egyptian tombs. Even in the 1870s—before two of the most frantic decades of collecting from the Nile—the task of compiling a comparative edition had proved impossible. The best that could be hoped for was a composite rendering of texts from just the eighteenth and nineteenth dynasties. The first fruits of the project, published by Asher in Berlin and printed entirely at the expense of the German government, were set before the 1886 Congress of Orientalists in Vienna.[30] "Breathless attention" and "prolonged applause" were noted in press reports of Naville's address.[31]

The publication appeared in two impressive and expensive volumes, complete with elegant hieroglyphic typesetting; these were still conceived as "introductory" to a larger project, although Naville presciently noted that "no one life" would be enough to complete it.[32] In a separate reminder of mortality, Lepsius never saw the product of his commission: one of the last letters he penned before his death congratulated Naville on completing his book's proofs.

The feat of producing this text was compared in the London periodical *The Academy* to Agamemnon's taking of Troy.[33] However, British requests that Naville's introduction be printed in English and French as well as German were not followed, and *The Academy*'s correspondent (Naville's closest ally in London) noted that the text would not reach "the learned world in general" but was destined for "a very limited circle of specialists. To all but Egyptologists *pur et simple*, M. Naville's new edition of the *Todtenbuch* is absolutely a dead letter."[34] Sir Henry Barkly (former Peelite minister and governor of Mauritius) did insist that a copy be made available to readers at the London Library, but *The Academy*'s commentator was correct: the work that made the Book of the Dead into a household phrase, and Egyptian ritual into a matter suitable for polite conservation, came later. Despite the lack of immediate public exposure, Naville's painstaking work provided the basis for all that happened during the following fifteen years.

Into the Light of Day

More than a decade had passed between the conception and consummation of Naville's project, and in that period much had changed. Britain had occupied Egypt. The new organization established to fund British excavations in Egypt had hired Naville to conduct them (his total lack of archaeological experience notwithstanding).[35] But changes had also taken place in British culture, and several of these concerned the presence and the meanings of death.

As the century wore on, engagements with death became increasingly diverse. From this Protestant country's strange fascination with Catholic Requiems in the concert hall, to the enormous impact of poems such as Tennyson's *In Memoriam*, the representations of suicide by artists such as Henry Wallis, or the ancient martyrs painted by figures such as Waterhouse, death was visibly present in far more than the framed photographs of the moldering corpses of relatives. By the late 1870s this fascination with death was not just increasingly diverse but increasingly global in inspiration. One small aspect of this was that the Book of the Dead, and Egyptian religious ritual more generally, had finally begun to enter European and American culture. This was not yet closely tied to the work of Egyptologists such as Naville: the Book of the Dead remained more an idea—a symbol—than a body of knowledge or a literary tradition. For instance, the Philadelphia poet George Henry Boker set out in 1880 to construct an American response to Tennyson's *In Memoriam*. He called this poem *The Book of the Dead* (1882) and stocked it with signature symbols of ancient civilizations. These were generic and classical, rarely resembling the contents of Egyptian ritual as it would soon emerge in public view.[36] Through the 1880s, literary engagement with Egyptian ritual increased. 1889 saw both H. Rider Haggard's most substantive engagement with the civilization and a popular treatment of Egyptian religion for children, G. A. Henty's *The*

Cat of Bubastes, which featured a well-meaning Egyptian priest as a central character and focused its plot around the nature of the mysteries in which Egyptian priests were initiates.[37] In texts like these, the central focus was always on Egyptians' knowledge of "that process of transmutation" that occurred to body, mind, and soul at the moment of death.

In a foreshadowing of the trends of the 1890s, ideas from the Book of the Dead were used increasingly as a resource for comparative religion. Between 1886 and 1890, Egyptian ideas of the afterlife were brought to bear in studies of Buddhist, Sufi, and Japanese belief. An article of 1888, for instance used the ritual of Osiris to demonstrate that the chroniclers of seventh-century Japan drew on a long-established tradition when they wrote of the moon as the abode of the Dead.

At the same time, museums and private collectors began acquiring papyri faster than ever before. The British Museum, for instance, received two celebrated recensions of the Book of the Dead in 1888 (papyrus of Ani) and 1890 (papyrus of Nu). Translating, publishing, and elucidating these texts was the work of the curator Peter le Page Renouf, who died with the task unfinished in 1896.[38] His successor, E. A. Wallis Budge, took up the work: his *Book of the Dead: The Egyptian Text According to the Theban Recension* (1897) has proved to be the most long-lived product of Victorian Egyptology.[39]

This activity made the 1890s into the moment when substantive British engagement with the Book of the Dead really begins. Between 1890 and Budge's 1897 text, dozens of book-length publications and scores of lengthy periodical articles gave Egyptian ritual its first period of public grace. Politicians, museum curators, librarians, clergymen, literary critics, anthropologists, social evolutionists, popular lecturers, schoolteachers, and art historians all contributed. Every major periodical, from across the whole political and theological spectrum, had soon featured at least some commentary on ancient Egyptian ritual. Some played up the presence of soul-destroying monsters in the Egyptian afterlife and generated salacious stories from the most innocuous of details, but for the vast majority, this was an exercise in spiritual discovery.[40]

There is a great deal we can learn from these texts beyond what they say of ancient Egypt. In one way or another, all of them contextualize ancient Egyptian ideas within 1890s debates concerning death. All of them put the ancients to use in thinking through the problems of the present. Among the most surprising features of these articles and books is the intense effort that was made to domesticate ideas of the Egyptian afterlife. This involved showing how Egyptian, Hebrew, and Christian traditions were part of one system rather than separate, contradictory, or mutually hostile cosmologies and theologies. Where Egyptian belief had been assumed to be polytheistic, it was now often rendered monotheistic, and sometimes used to demonstrate that the distinction between polytheism and monotheism was a false one. Where it had once been vulgar and literal, it was now mystical and symbolic; where it was once a mere curiosity to a Christian society, it could now aid interpretation of Old Testament texts whose meanings were increasingly disputed. British writers and readers, it seems, were actively looking for other spiritual traditions that might help them find ways out of their own metaphysical conundrums.

Higher Forms and Golden Thoughts

Two parts of the Old Testament were crucial in integrating the Book of the Dead into familiar traditions. The richest and most complex parallels were found with the Psalms, but the most immediate and straightforward was the resemblance between the Ten Commandments and the Book of the Dead's "Negative confessions." By 1889, some Egyptologists and biblical scholars were already expressing annoyance that the public were so convinced that "all of the ten commandments . . . may be found in the negative confession of the 125th chapter of the Book of the Dead."[41] Yet others were quite happy to perpetuate this idea.

In 1891 a text bearing uncanny resemblance to Psalm 104 was found on a wall of the tomb of a courtier of Egypt's "heretic" king, Akhenaten.[42] This text, and the "refined and really philosophic worship" illustrated in the images that accompanied it, contained "not a rag of superstition or of falsity" according to the most celebrated Egyptologist in Britain, William Matthew Flinders Petrie.[43] What was seen here was not a gloomy scene, but the radiant idea of the sun as image of God. The connection between this tomb text (known as the Hymn to the Aten) and Psalm 104 became a persistent site of dispute among Egyptologists, theologians, and litterateurs, including C. S. Lewis.[44] In the 1890s, however, the strength of this link was widely accepted. The result was that a few familiar metaphors linked Akhenaten into a whole tradition of wisdom literature and made of him a Christian saint. Once writers were attuned to these parallels in the tombs of Akhenaten's city, they found them elsewhere too. Psalm 22, for instance, was found to be present in inscriptions written for Thothmes II and Rameses II; "the inscription on

Cleopatra's Needle" now standing on London's Embankment was even "discovered" to be "a Psalm of David and written in Hebrew."[45]

This speculation, wild and apparently unorthodox as much of it was, often issued from positions of religious authority in 1890s Britain. Many volumes published on the Book of the Dead in this period were issued by the Religious Tract Society or the Society for Promoting Christian Knowledge. They speculate on Egyptian cyclical conceptions of life and death, yet they almost all emphasize the essentially Christian doctrines at the tradition's core. They emphasize Egyptian religion's "supposed inspiration, faith in the persistence of the human personality after death," trials of faith, apocalyptic beasts, the judgment seat of God, and the triumphant reception of the dead in Paradise, all of which are claimed to have made Egypt "eternally modern."[46] They found in the Book of the Dead a litany of "hymns and prayers and confessions," although Anglican readers were cautioned that there was no "Authorised Version."[47] Its most profound elements were an "esoteric monotheism," "almost Athanasian" in the mystical relationship of father and son (Ra and Osiris).[48] This biblical harmonizing wasn't just rhetorical: it was essential to drawing readers into Egyptian texts, and it defined how they read them.

Christology, typology, and other categories of Christian theology established patterns for identifying the "higher forms" behind the "surface appearances" of Egyptian cult. Praise for the contents of the newly popularized ritual ("a mine of golden thought and soul help") was effusive.[49] The best way to see this renegotiation at work is to look in detail at a single article, investigating how the juxtaposition of Hebrew and Egyptian ideas could give new perspectives on the Protestant present.

"The Book of the Dead and a Passage in the Psalms"

John Hunt Cooke was a religious butterfly. Having spent his early life as a devotee at London's Baptist Duke Street Chapel, he had, by the 1890s, converted (or, more likely, reconverted) to the Church of England. For reasons that are probably impossible to identify, this change of religious identity brought with it an apparently new interest in the multiplicity of the world's religions. This included fascination with modern Buddhism, but also with the religious ideas of the ancients. By the mid-1890s Hunt Cooke was a familiar face at the British Museum, assisting curators and delivering lectures in the Egyptian galleries. These talks were sometimes developed into articles for the periodical press, the longest being a study in the *Contemporary Review* entitled "The Book of the Dead and a Passage in the Psalms."

Hunt Cooke's article is typical of the new enthusiasm shown for Egypt. His purposes were threefold. The first was to show that the failure of modern Europeans to find philosophical insight in the Book of the Dead was due to a fault of their own, not a fault of the ancient Egyptians: they had assumed that the ancient Egyptians were incapable of abstract thought and so had read literally texts that are rich with symbolic meaning.[50]

His second, related, purpose was to review translations published between 1890 and 1894, but also to champion a new kind of rendering: he insisted that as long as translators looked for literal meanings, unhelpful versions would continue to pour forth. He championed poetic translation like that through which Edward Fitzgerald produced a work of modern literature from the *Rubaiyyat of Omar Khayyam*.[51] To this end, Hunt Cooke adapted everything he quoted from the Book of the Dead into rhyming couplets.

Third, and most significantly, Hunt Cooke set out what might be learned when the tradition was recognized as "not simply imagination but philosophy, not simply poetry, but ontology."[52] He worked through each aspect of the multiform Egyptian self, demonstrating its parallels in Christian and Buddhist tradition. The Khaib, he speculated, was the glory about the heads of saints in elder days; the Ka, the Devas of Buddhism; the Sah, the spiritual body referred to by St. Paul when writing to the Corinthians.[53] In this way, familiar texts became keys for unlocking passages in the Book of the Dead. But the passage in the Psalms referred to in the title demonstrates, Hunt Cooke insisted, that the Book of the Dead itself has an elucidatory potential: once the scriptures have helped make sense of Egyptian ritual, that ritual can cast light on scripture. Passages in Psalm 16 had been read as poetical, but comparison with the Book of the Dead would reveal ontological meaning.[54] Hunt Cooke's brief exposition (he offers little elaboration on this final point) treats these traditions as mutually intertwined.

Philosophy and Ontology

Hunt Cooke's article focuses primarily on the nature of the self. In particular it addresses the question of how the human self is changed by death. This is the common theme of most serious treatments of the Book of the Dead in the 1890s. To explain the new power of this theme, we need to broaden our scope by considering the period's ideas about "time" as well as "death" and "the self." One of the key developments in the literature of death between 1890 and 1914, elaborated by literary scholars such as John Holmes, is a new fascination with the primeval.[55] This involves geological and evolutionary time as well as the earliest civilizations. When Thomas Hardy wrote an elegy for his wife, for instance, he filled it with the language of deep time: "primeval rocks" that have witnessed "the transitory in earth's long order" now witness Hardy's dying love.

This intertwining of ideas concerning death and concepts of unending time was not just British. Its most famous instance is Nietzsche's principle of "eternal recurrence," which by the late 1890s had found its way into British philosophical disquisition, literature, journalism (via George Bernard Shaw), and even London's concert halls (in the "endless melody" of Richard Strauss's *Also Sprach Zarathustra*). Where Hardy tied his many speculations on death and time to the primeval history of the British Isles, Nietzsche, like many others, found inspiration in the "ancient East": it is no coincidence that the idea of "eternal recurrence" is voiced by Nietzsche's Persian alter ego, Zarathustra. These ideas also framed much debate over Buddhist thought in the 1890s. Stretching deeper into the ancient past, it was also during this period that a tradition previously received as a biblically inspired romance, the "Tale of Ishtar and Izdubar," was transformed into a deep reflection on *Todesfurcht*, the fear of death. This was the *Epic of Gilgamesh*. When the first fragments were discovered in the 1870s, scholars had no inkling that connections could be drawn between this text and profound questions concerning the nature of death. In the version familiar by the 1910s, when Rilke celebrated the epic as a "truly colossal happening and being and fearing," Gilgamesh was a culture hero and timeless model for struggles with mortality.

Many leaders of late Victorian and Edwardian culture embraced the comparative potential of these traditions. Collecting became a fascination not just for the antiquarian but for the most avant-garde of thinkers. The most famous example of this comes from Vienna—Sigmund Freud's 1890s collecting of Egyptian, Greek, and Mesopotamian symbols of death and the afterlife—but this was shaped by British comparative religion, including James Fraser's *Golden Bough*.

What such examples represent is a new syncretism associated with the decay of ideas that had seemed like permanent Christian certainties but might better be read as themes specific to evangelical revival. The mid-Victorian period had seen the collapse of certainty regarding evangelical notions of the afterlife, leading to the idea that modern

European Protestantism had taken a wrong turn. The quest for new philosophical input that could provide new clarity about what death meant for the human organism and what, if anything, survived it was a key reason why the period saw such enthusiasm for ancient Egyptian traditions.

Notes

1. "The Book of the Dead," *Saturday Review* 88 (1899): 415–16.
2. James Stevens Curl, *The Victorian Celebration of Death* (London: Sutton, 2000).
3. Julian Barnes, *Arthur and George* (London: Vintage, 2005).
4. Randal Keynes, *Annie's Box: Charles Darwin, His Daughter and Human Evolution* (London: Fourth Estate, 2001).
5. Michael Wheeler, *Death and the Future Life in Victorian Literature and Theology* (Cambridge: Cambridge University Press, 1990).
6. See, for instance, Frederic Harrison et al., *A Modern Symposium: The Soul and Future Life* (Detroit, MI: Rose Belford, 1877).
7. Alex Owen, *The Place of Enchantment: British Occultism and the Culture of the Modern* (Chicago: University of Chicago Press, 2004); Janet Oppenheim, *The Other World: Spiritualism and Psychological Research in England, 1850–1914* (Cambridge: Cambridge University Press, 1988).
8. Charles Dickens, *The Old Curiosity Shop* (London: Chapman & Hall, 1841), 209.
9. Thomas Dixon, *Weeping Britannia: Portrait of a Nation in Tears* (Oxford: Oxford University Press, 2015).
10. Patrick Geoghegan, *Liberator Daniel O'Connell* (London: Gill & Macmillan, 2010), 36.
11. Quoted in Hesketh Pearson, *Oscar Wilde: His Life and Wit* (New York: Harper, 1946). For further coverage of Little Nell and Victorian sentimentality, see Nicola Bown, "Crying Over Little Nell," *Interdisciplinary Studies in the Long Nineteenth Century* 4 (2007): 1–13; DOI: http://doi.org/10.16995/ntn.453.
12. Abby Bardi, "'Labours of Their Own': Property, Blood and the Szgany in *Dracula*," in *Bram Stoker and the Gothic: Formations to Transformations*, ed. Catherine Wynne (Basingstoke: Palgrave Macmillan, 2016), 78–91.
13. Clement Cheroux et al., *The Perfect Medium: Photography and the Occult* (New Haven, CT: Yale University Press, 2005).
14. Peter Jupp, "Enon Chapel: No Way for the Dead," in *The Changing Face of Death: Historical Accounts of Death and Disposal*, ed. Glenys Howard and Peter Jupp (Basingstoke: Palgrave Macmillan, 1997), 90–104; Walter Thornbury, *Old and New London: A Narrative of Its History, Its People, and Its Places* (London: Cassell, 1872).
15. J. S. Curl, "The Architecture and Planning in the Nineteenth Century Cemetery," *Garden History* 3, no. 3 (1975): 13–41.
16. Michael Wheeler, *Heaven, Hell and the Victorians* (Cambridge: Cam-

bridge University Press, 1994); Patricia Jalland, *Death and the Victorian Family* (Oxford: Oxford University Press, 1996).

17. Frank Turner, *Contesting Cultural Authority* (Cambridge: Cambridge University Press, 2003); Timothy Larsen, *A People of One Book* (Oxford: Oxford University Press, 2011).

18. Wheeler, *Heaven, Hell and the Victorians*, esp. 175–218.

19. Timothy Larsen, *The Crisis of Doubt: Honest Faith in Nineteenth-Century England* (Oxford: Oxford University Press, 2006).

20. Jeremy Morris, *F. D. Maurice and the Crisis of Christian Authority* (Oxford: Oxford University Press, 2005).

21. John Rogerson, *Old Testament Criticism in the Nineteenth Century: England and Germany* (Philadelphia, PA: Fortress, 1985); Thomas A. Howard, *Religion and the Rise of Historicism: W. M. L. De Wette, Jacob Burckhardt, and the Theological Origins of Nineteenth-Century Historical Consciousness* (Cambridge: Cambridge University Press, 2000); David Katz, *God's Last Words: Reading the English Bible from the Reformation to Fundamentalism* (New Haven, CT: Yale University Press, 2004); Michael Legaspi, *The Death of Scripture and the Rise of Biblical Studies* (Oxford: Oxford University Press, 2010); and John Rogerson, *The Bible and Criticism in Victorian Britain* (Sheffield: Sheffield Academic Press, 1995).

22. Bodo Neuss, "The Funerary of Thomas Willson," *Architectural Association Files* 68 (2014): 88–92.

23. Gange, *Dialogues with the Dead*, 151–237; Elliot Colla, *Conflicted Antiquities: Egyptology, Egyptomania, Egyptian Modernity* (Durham, NC: Duke University Press, 2008); Donald Malcolm Reid, *Whose Pharaohs? Archaeology, Museums and Egyptian National Identity* (Oakland: University of California Press, 2002).

24. Gange, *Dialogues with the Dead*, 53–120.

25. J. Hunt Cooke, "The Book of the Dead and a Passage in the Psalms," *Contemporary Review* 70 (August 1896): 277.

26. These included texts such as Samuel Sharpe, *The History of Egypt from the Earliest Times till the Conquest by the Arabs*, 2 vols (London: George Bell, 1846); and John Kenrick, *Ancient Egypt under the Pharaohs* (London: Redfield, 1852); but also John Gardner Wilkinson, *Manners and Customs of the Ancient Egyptians* (London: John Murray, 1837), which remained among the most popular texts on ancient Egypt for several decades.

27. "The Book of the Dead," *Athenaeum* 2521 (1876): 265.

28. "Book of the Dead," 265.

29. "Naville's Edition of the Book of the Dead," *The Academy* no. 32 (1887): 171.

30. "Naville's Edition," 170.

31. "Naville's Edition," 170.

32. "Naville's Edition," 170.

33. "Naville's Edition," 170.

34. "Naville's Edition," 170.

35. Gange, *Dialogues with the Dead*, 151–236.

36. George Henry Boker, *The Book of the Dead* (Philadelphia: J. B. Lippincott, 1882).

37. G. A. Henty, *The Cat of Bubastes* (London: Blackie & Son, 1889).

38. Peter le Page Renouf, *The Book of the Dead: Translation and Commentary* (London: Society of Biblical Archaeology, 1904).

39. E. A. Wallis Budge, *The Book of the Dead: The Egyptian Text According to the Theban Recension*, 3 vols. (London: Kegan, Paul, 1897); this is the only text of late nineteenth-century Egyptology that has been in print ever since, now as a Penguin Classic.

40. See, for example, "The Dawn of Civilisation," *Graphic*, 50 (1894): 654.

41. F. C. H. Wendel, "The Value of Egyptological Study," *Old and New Testament Student* (1889): 278–84, 281.

42. See, for example, Arthur Weigall, *The Life and Times of Akhnaton, Pharaoh of Egypt* (London: Thornton Butterworth, 1910), 134–36.

43. [W. M. F. Petrie], "The Excavations at Tel el Amarna," *The Academy*, 41 (1892): 356–57.

44. Dominic Montserrat, *Akhenaten: History, Fantasy and Ancient Egypt* (Abingdon: Routledge, 2000).

45. Margaret Benson, "The Plain of Thebes," *Edinburgh Review* 186 (1897): 454–82; [Petrie], " Excavations at Tel el Amarna," 356.

46. Benson, "Plain of Thebes."

47. Benson, "Plain of Thebes."

48. Grant Allen, "The Gods of Egypt," *Universal Review* 8 (1890): 51–65.

49. Cooke, "Book of the Dead," 285.

50. Cooke, "Book of the Dead," 277.

51. Cooke, "Book of the Dead," 277.

52. 278.

53. Cooke, "Book of the Dead," 283.

54. Cooke, "Book of the Dead," 283.

55. John Holmes, *Darwin's Bards: British and American Poetry in the Age of Evolution* (Edinburgh: Edinburgh University Press, 2009), 121–29.

10 At Sea

Clare Pettitt

Introduction: Against Chronology

From the 1830s, the emerging sciences of geology, electrical engineering, telecommunications, marine zoology, and oceanography were all plunging down to the depths of the sea. Unexpectedly, what was washed up from the oozy seafloor by these very modern scientific inquiries was a new perspective on the past. This essay suggests that the sea in the nineteenth century became a multidisciplinary, multitemporal space of the imagination, vast and unknown enough to accommodate a slew of syncretic theories of mythography, religion, history, and evolution (fig. 10.1). By using the Victorian lens of syncretism and putting aside our own anachronistic assumptions about "disciplinization," this essay suggests that we can attend instead to the Victorian desire to identify a global unity of form. The pull toward philosophical monism and the resultant reliance on syncretized history were as useful to science as they were to politics, to religion, and to more esoteric forms of crypto-religion. In the face of the increasing complexity, disciplinization, and nationalization of knowledge, the sea could still do useful symbolic work in maintaining the possibility of a universalized global history.

FIGURE 10.1. Illustration of Poseidon, from "The Kings of Atlantis Become the Gods of the Greeks," in *Atlantis: The Antediluvian World*, by Ignatius Donnelly (London: Samson Low, Marston, 1889), 305. Rare Books Room MF.23.7. By kind permission of the Syndics of Cambridge University Library.

Byron "Upon the wide, wide sea"[1]

> ... as my bark did skim
> The bright blue waters with a fanning wind,
> Came Megara before me, and behind
> Aegina lay, Piraeus on the right,
> And Corinth on the left; I lay reclined
> Along the prow, and saw all these unite
> In ruin ...
> (Bryon, *Childe Harold's Pilgrimage*,
> canto 4, st. 44, 138–39)

Byron's *Childe Harold's Pilgrimage* is a poem that is crucially at sea. The Childe reclines on the prow of his boat surveying from this moving vantage point on the Mediterranean the ruins of past civilizations as his "stately vessel glided slow / Beneath the shadow of that ancient mount" (2.41). History for Byron seems to operate along the distant shoreline as the recumbent Childe descries "[t]he colouring of the scenes which fleet along"(3.112).

For much of this extended series of poems, published across six years, the Childe floats somewhat aimlessly past the remains of history on "his weary pilgrimage" (1.10). When he does disembark, it is to tour the battlefields of Europe, both ancient and modern. Byron hired a guide at Waterloo in Belgium in the Spring of 1816 and recalled that "I went on horseback twice over the field, comparing it with my recollection of similar scenes. As a plain, Waterloo seems marked out for the scene of some great action, though this may be mere imagination: I have viewed with attention those of Platea, Troy, Mantinea, Leuctra, Chaeronea, and Marathon."[2] The battle of Waterloo had happened only the summer before, whereas the battle of Marathon had taken place in 490 BCE, and the Trojan wars, if they happened at all, happened in the twelfth or eleventh century BCE. Byron is experiencing these events out of any historical sequence but nevertheless is able to contemplate them with equal "attention" as "similar scenes."

The lines "I . . . saw all these unite / In ruin" express the central paradox of a poem that constantly attempts to create coherence and historical sequence across time and space only to break it up again. William Hazlitt recognized this pattern of abandoned sequence in *Childe Harold's Pilgrimage* and did not like it: "There is here and in every line an effort at brilliancy, and a successful effort; and yet, in the next, as if nothing had been done, the same thing is attempted to be expressed again with the same effort of labour as before, the same success, and with as little appearance of repose or satisfaction of mind."[3] Byron's poem refuses historical and metrical continuity. Not just in its famous concluding encomium to the ocean, but throughout all its four cantos, this is a work at sea that crashes as rhythmically and relentlessly as the waves, using its patterns of swell and collapse to break up a teleological or serial view of history. For "Art, Glory, [and] Freedom fail" (2.87) again and again in this history. Byron's Spenserian stanza allows a "broken and more impassioned movement," as William Wordsworth noticed with disapproval.[4] And by its constant return from the land to the sea ("Once more upon the waters! yet once more!" (3.2), *Childe Harold's Pilgrimage* rejects settling in favor of moving and repeatedly reprises a maritime perspective from which nationhood swiftly comes to look like landlocked introspection. As old regimes such as the Ottoman Empire fragment and collapse, the Childe apprehends the emergent global order as "the shattered links of the world's broken chain" (3.18), and Byron seems to celebrate these "shattered links" and to revel in the uncertainty of being cast adrift from teleology.

In *Childe Harold's Pilgrimage* literary form becomes, then, a category of historical knowledge. The poem dramatizes the problematics of historical sequence and collapses times together so that Marathon and Waterloo are superimposed one upon the other. The poem was famously influential at a time when the aftermath of the French Revolution, an event that had "seemed to outstrip all previous experience," was still ricocheting around Europe in the confusion of the Napoleonic Wars. Byron frequently sought out and reported evidence of these wars on his European travels when he was writing *Childe Harold*.[5] The poem does not merely bear witness to the beginning of a break from the teleological and serial history of the Enlightenment, but partly creates this break, plunging its readers into a disorienting seascape in which it is difficult to make out landmarks: even as the Childe leaves Dover "the white rocks faded from his view, / And soon were lost in circumambient foam" (1.12). Flung out into the middle of the ocean, the topographical mode of eighteenth-century poetry is undone and gives way instead to a Romantic *mobilité*, or, as Francis Jeffrey put it more sourly in *Blackwood's*, the poem "floats and fluctuates in cheerless uncertainty."[6] History imagined from the sea seems to be composed of overlapping and imperfectly aligned layers and spaces that are as much disconnected as connected by the salt water between them.[7] While the Childe's experience of being at sea reinforces the sense of distance between one place and another, the poem at the same time dramatizes his restless search for the emergence of an alternative connective order. The constantly collapsing seriality of *Childe Harold's Pilgrimage* mimics the struggle for a "joined-up" political revolution across Europe. Byron is already dramatizing the difficulty of aligning diverse national histories with a "European" revolutionary identity.

Byron understood the power of what he called the "midland ocean" (4.175) to undo nationhood and to unbind sequence. Yet the question of historical agency remains fraught throughout the poem, and the Childe frequently disavows any control, figuring himself as a piece of seaweed tossed around by the tide:

> "Still must I on; for I am as a weed,
> Flung from the rock, on Ocean's foam, to sail
> Where'er the surge may sweep, or tempest's breath prevail."
> (3.2)

In January 1821, predicting the probable failure of a Risorgimento insurrection in Ravenna, Byron returned to the image of the determinative force of the ocean, although this time in an explicitly political simile: "It is not one man, nor a million, but the *spirit* of liberty which must be spread. The waves which dash upon the shore are, one by one, broken, but yet the *ocean* conquers, nevertheless. It overwhelms the Armada, it wears the rock, and, if the *Neptunians* are to be believed, it has not only destroyed, but made a world. In like manner, whatever the sacrifice of individuals, the great cause will gather strength, sweep down what is rugged, and fertilise (for *sea-weed* is *manure*) what is cultivable."[8] Here, Byron is partially referring to "Neptunism," a stratigraphic theory of geology that proposed that the core of the earth was originally formed by a process of sedimentation from the waters of the primeval ocean. Also known as the Wernerian theory, this idea was current in the 1820s when Byron was writing, and its leading British proponent was Robert Jameson, the Regius Professor of Natural History at Edinburgh University.[9] The theory had been robustly rejected by James Hutton, another eminent geologist in Edinburgh, who had argued instead for uniformitarianism in his *Theory of the Earth* (1788), in which he claimed that the earth's crust was created and altered by natural processes over an enormous span of geologic time, creating "a succession of worlds." And in fact, later in his career, Jameson came round to Hutton's beliefs and abandoned Neptunism. When Byron writes, "if the *Neptunians* are to be believed," he makes it clear that the deep history of the earth and the sea is represented in the 1820s by a series of hotly contested and competing narratives. In deploying a complex geological image here, Byron is able to represent the potential of revolutionary collective action both to destroy and to create. The image is particularly useful to his purpose, as it makes no claim to immediacy, usefully marrying the uncertainties attendant upon imagining the deep past with those attendant upon imagining the future. The erosive work of the sea is long and slow. Throughout the nineteenth century, the sea was to remain the imagined site of powerful alternative histories. In 1830, in *Principles of Geology*, Charles Lyell suggested that "[i]t is not too much to say that every spot which is now dry land has been sea at some former period, and every part of the space now covered by the deepest ocean has been land."[10] Lyell was deeply indebted to Byron for his imaginative vision of time and change, and in his *Principles of Geology*, he quotes both Byron's image of himself as "flung seaweed" and the famous encomium to the sea in canto 4 of *Childe Harold*. By the 1830s and 1840s a "vast new field of research" was revealing itself: "the bottom of the sea."[11]

It has been commonly claimed that a major shift occurred in the kinds of historical knowledge that might be offered up by the sea between the years of the publication of Byron's *Childe Harold's Pilgrimage* and the "scientific" mid-nineteenth century, which saw the rapid institutionalization of geology and oceanography. In fact, as Adelene Buckland has argued, Byron's "Romantic" sensitivity to broken sequence and imperfectly aligned histories, arbitrarily formed boundaries, and the collapse of entire systems of social organization underpin and inform the construction of the "science" of geology in essential and still under-recognized ways.[12] Byron's iconoclastic unbinding of historical sequence and his emphasis on mobility and changefulness not only reflected the revolutionary upheavals around him, but also actively formulated and delivered a challenge to linear chronology that would resonate through European intellectual life throughout the nineteenth century.

Ooze, Slime, and Mud

No man born and truthful to himself could declare that he ever saw the sea looking young. But some of us . . . have seen it looking old as if the immemorial ages had been stirred up from the undisturbed bottom of ooze.
Conrad, *The Mirror of the Sea* (1904–1906)[13]

Ooze was, indeed, being stirred up in the formerly undisturbed bottom of the ocean in the nineteenth century as electrical cables were paid out along the sea floor to carry telegraphic messages between continents. Modern commercial and imperial imperatives drove the exploration of the seabed: good data on the geography and geology of the sea floor was urgently needed to enable submarine telegraph cabling. It was therefore almost as a side effect of the most cutting-edge electrical technologies that the sea became newly understood as "old" and as harboring the secrets of the earth's deep past. Electrical engineering became the midwife of oceanography. Technologically state-of-the–art cables found themselves nestling in ooze on the ocean bed that sustained life forms previously believed to be extinct.

The deep ocean had long been imagined as "a waste of utter darkness, subjected to such stupendous pressure as to make life of any kind impossible" but now it was suddenly endowed with a history of its own.[14] Until the late 1850s, a theory of an azoic zone below three hundred fathoms, an abyss in which no life could exist, was lodged firmly in scientific thinking.[15] T. H. Huxley in 1857 had asked, "How can animal life be conceived to exist under such conditions of light, temperature, pressure, and aeration as must obtain at these vast depths?"[16] But it was in these depths that "the numberless transitorial links which must formerly have connected the closely allied or representative species found in the several stages of the great formations" now began to appear and to fill in some of the gaps in the fossil record that Darwin had previously found it difficult to explain.[17]

The professor of natural history at Edinburgh University, Charles Wyville Thomson, befriended the newly appointed professor of engineering, Fleeming Jenkin, when he arrived in Edinburgh in 1868. Jenkin told his new colleague about his experience as a young man, when he had been "employed about the ocean cables" and had trawled the Mediterranean and Red Sea in 1860 to recover lost cable from depths of fifteen hundred fathoms and more, and he recalled that "[n]o portion of the dirty black wire [of the cable they dredged up] was visible; instead we had a garland of soft pink, with little scarlet sprays and white enamel intermixed" and that "worms ate through the insulation; and it has been laid taut on an uneven sea bed, so that when it became encrusted with barnacles the extra weight caused it to break."[18]

Wyville Thomson was intrigued by this circumstantial evidence that life might exist on the sea floor, and in 1872 he was named the captain of the celebrated *Challenger* expedition of 1872–76, an expedition the "principle object of [which was] . . . to investigate the physical and biological condition of the great ocean basin."[19] This was an immensely well-publicized expedition on the specially fitted-up science ship *HMS Challenger*. *Challenger* circumnavigated the entire globe and systematically sounded and dredged the depths of the sea for the first time, measuring undersea temperature, currents, and pressure, and dredging the sea floor for life and geological data.

The depths of the sea turned out to be much deeper and sometimes much warmer than had previously been thought, and, surprisingly, were also discovered to be teeming with previously unknown forms of life. Dredging the ocean bed of the Atlantic, Wyville Thomson, declared that "[e]very haul of the dredge brings to light new and unfamiliar forms—forms which link themselves strangely to the inhabitants of the past periods in the earth's history."[20] One neurologist and invertebrate zoologist, William Carpenter, was also enthusiastic about the survival in the sea of prehistoric life forms: "We have great reason to believe that we shall here find at considerable depths a large number of Tertiary species which have been supposed to be extinct."[21] Publications and images from the *Challenger* expedition flowed steadily into the public domain from 1877 to 1895 and were widely reported in the popular as well as the scientific press (see fig. 10.2). The scientific reports catalogued more than four thousand previously unknown species. Thomson remembered that on his *Challenger* voyages, he was always "on the watch . . . for missing links which might connect the present with the past."[22] Kipling's 1893 poem "The Deep-Sea Cables" dives right down to the bottom of the sea and the submarine "plains of ooze":

> There is no sound, no echo of sound, in the deserts of the deep,
> Or the great gray level plains of ooze where the shell-burred
>   cables creep.
> Here in the womb of the world—here on the tie-ribs of earth,
> Words, and the words of men, flicker and flutter and beat.[23]

By the 1890s, then, the deep sea was no longer perceived as sterile and empty, but as amniotic, "the womb of the world," pulsating to a new human beat. The imagery of the poem means that Kipling's cables seem not only to be transmitting coded signals, but also to be actively gestating them.

And it was the *Challenger* expedition that was eventually to discredit one of the most seductive of the alternative histories that washed up from the ooze and slime on the sea floor.

FIGURE 10.2. "At the Bottom of the Sea," in *Under the Waves; or, Diving in Deep Waters*, by R. M. Ballantyne (London, 1876), 199. Rare Books Room 1877.7.78. By kind permission of the Syndics of Cambridge University Library.

Primeval Protoplasm

In 1857 the British ship *Cyclops* had gone out sounding and dredging in preparation for the laying of the Atlantic cable. The scientists on board discovered that the bottom of the ocean "to our amazement [was not] bound round with thick ribs of granite" but was found instead "to be as soft as a silken vest."[24] This "silken vest" was composed of "an ooze" that reappears in Jules Verne's 1869–1870 novel, *20,000 Leagues Under the Sea*, when Dr. Pierre Aronnax and his fellow prisoners aboard the *Nautilus* are taken on an undersea hunting expedition by Captain Nemo, and find themselves walking on "a layer of sticky mud" or "ooze" (fig. 10.3): "These lawns of fine-woven material were soft under our feet and could be compared to the silkiest rugs made by man's hand."[25] The scientists on board the *Cyclops* discovered that samples of this ooze, "when carefully preserved and subjected to a powerful microscope" appeared to be composed of vast numbers of tiny shell-like structures. This was a most exciting find, and the specimens of mud were sent to the scientist Thomas Huxley upon the ship's return.[26]

It was only when Huxley procured a new and powerful microscope ten years later that he pulled out some of these old specimens of "particularly viscid mud" to take a better-magnified look at them.[27] What he saw, he declared with excitement to be a new "protoplasmic" organism that he named *Bathybius haeckelii* after his friend and fellow supporter of Darwin, Ernst Haeckel.[28] Huxley ran a series of tests on different samples, and when *Bathybius* was apparently found in both the Indian Ocean and the South Atlantic it began to appear that it might be a uniform and globally distributed protoplasmic life-form.[29] W. B. Carpenter understood it as offering a continuous link back to the Cretaceous period: "As Prof. Huxley has proved the existence of Bathybius through a great range not merely of depth but of temperature, I cannot but think it probable that it has existed continuously in the deep seas of all Geological Epochs."[30]

Haeckel himself was quickly convinced that this simple organism had come into being through spontaneous generation, or heterogeny, and therefore might represent the origin and unit of life that was identical in plants and animals.[31] A global protoplasm that "covers the sea-bottom with a living layer of slime" seemed to throw "a new light . . . upon one of the most difficult and most obscure problems of the history of creation—namely the question of the origin of life upon the earth."[32]

Huxley expounded on this idea in his Edinburgh lecture of 1869, "On the Physical Basis of Life," when he asked "What hidden bond

FIGURE 10.3. "A Walk Under the Waters," Jules Verne, *20,000 Leagues Under the Seas* (London: Ward, Lock, and Tyler 1876), 86. Rare Books Room 1876.7.568. By kind permission of the Syndics of Cambridge University Library.

can connect the flower a girl wears in her hair and the blood which courses through her youthful veins?" He concluded that, "a three-fold unity—namely, a unity of power, or faculty, a unity of form, and a unity of substantial composition—does pervade the whole living world."[33] The religious or esoteric language is offered in cautious mitigation of the extreme materialism of this proposition of a uniform material that constitutes the essential "building block" of life on earth.

Throughout the 1860s the status of this deep-sea protoplasmic substance was hotly debated in the periodical press. The *Westminster Review* announced of protozoa that "these creatures. . . . have performed, and continue to perform, one of the chief parts in the world's history"; although the *Athenaeum* remained unconvinced of the "mechanistic" implications of this theory, protesting that "[u]nder what modification or combination of the general polarizing force, the slime of mud or ooze is first condensed into a protoplasmic centre of such low vital force is still undetermined."[34] Darwin entered the debate too, warning that, "the nature of life will not be seized on by assuming that Foraminifera are periodically generated from slime or ooze." He concluded by rejecting the idea of spontaneous generation and asserting that "our ignorance is as profound on the origin of life as it is on the origin of force or matter."[35]

The *Bathybius* story started to unravel when Wyville Thomson wrote cautiously to Huxley on June 9, 1875, while on board the *Challenger*, suggesting that a reaction of alcohol with seawater might have produced "the flocculent precipitate" that Huxley had interpreted as *Bathybius*: "It seems to me possible that a trace of organic matter may combine with the lime sulphate to give it that very peculiar form," wrote Thomson.[36] Huxley was quick to acknowledge publicly that he had made a mistake, admitting that "I thought my young friend Bathybius would turn out a credit to me. But I am sorry to say, as time has gone on, he has not altogether verified the promise of his youth."[37] Professor Mobius was soon astonishing audiences in Germany by simply mixing alcohol with a glass of sea water and causing "Bathybius" to appear before their eyes.[38]

Philip Rehbok has explained that "[i]ndependent but concurrent developments in cytology, protozoology, submarine geology, Precambrian paleontology, microscopy, and evolutionary biology converged to produce an intellectual environment congenial to Bathybius in the 1860s. That environment had not existed a decade before; it ceased to exist a decade later."[39] But this interpretation is partial in that it reads events back through a later model of highly organized disciplinization that can-

FIGURE 10.4. J. W. Dawson, "Life in the Primordial Sea," *The Story of the Earth and Man* (London: Hodder & Stoughton, 1873), 40. Rare Books Room MF.22.15. By kind permission of the Syndics of Cambridge University Library.

not fully explain the culture of the mid-nineteenth century. The desire to find a global unity of form, the pull toward philosophical monism, and the resultant reliance on syncretized history together created a centripetal force in European intellectual life and culture at this time, as useful to science as it was to politics and religion, as well as to more esoteric forms of crypto-religion. *Bathybius* emerged at a moment when structural monism was highly fashionable, especially in continental Europe.

Monism is pre-disciplinary and even antidisciplinary in its model of convergence over competition, seeking "the one ancestor common to all."[40] A principle of unity was to become increasingly unattractive to scientists, as competing and complex systems of disciplinization emerged, each of which extrapolated from its own field of inquiry its own specialist theories (fig. 10.4). Nevertheless, in Germany Ernst Haeckel refused for some time to give up on *Bathybius*, the "ill-famed primordial slime of the sea-depths," and in 1892 he lectured on "Monism as Connecting Religion and Science," avowing "the indissoluble connection between energy and matter, between mind and embodiment—or, as we may also say, between God and the world."[41] The implied slippage in the definitions of energy, mind, and God is again partly designed to counter accusations of radical atheistic materialism. The interest in system-building, which is clear from the 1840s in the philosophy of Auguste Comte and the theology of David Friedrich Strauss, powered syncretic thinking throughout the century. It might be more productive to view the organization of knowledge in the mid-nineteenth century not at a distance through the lens of "disciplinization," but using the viewfinder that was being used at the time: that of syncretism.[42]

Primordial ooze was not therefore merely a geological curiosity but part of a larger cultural imagination of shared origins in deep time. The nineteenth-century historical and mythographic imagination collapsed geological time with the recuperative purpose of reconnecting the modern to the origins of life on earth. As one of the only "blank space[s] of delightful mystery" left on the map, the sea offered the final frontier and a useful site of unexplored possibilities.[43] As Captain Nemo puts it in *20,000 Leagues Under the Sea*, "The sea is everything. It covers seven tenths of the terrestrial globe.... The sea is the environment for a prodigious, supernatural existence; ... it is a living infinity.... It was through the sea that the globe as it were began, and who knows if it will not end in the sea?"[44] In the face of the increasing complexity, disciplinization, and nationalization of the physical and biological sciences, the sea could still do useful symbolic work in allowing the possibility of a universalized global space.

Cities Undersea

Increasingly and perhaps paradoxically, it came to be "science" that licensed speculation in this period. So many strange and previously unseen new forms had appeared on the *Challenger*'s dredging platform that, as *All the Year Round* put it, there was no knowing what might be discovered next in the deep-sea mud: "We are discovering new forces every day. What if some enterprising Briton, or keen Yankee, should yet dredge from out its unfathomed slime the lost Atlantis?"[45] In Jules Verne's exhaustively topical novel of the 1860s, the lost city of Atlantis does indeed appear to Dr. Aronnax: "Right there in front of my eyes—ruined, broken, collapsed, ... a whole Pompeii sunk beneath the waters; ... my hands were touching ruins hundreds of thousands of years old, contemporary with the early geological periods! I was walking on the same spot where the coevals of the first man had walked! I was crushing under my heavy soles skeletons of animals from those fabulous times, which the now mineralized trees used to cover with their shade!"[46] Plato's story of the lost continent beyond the Pillars of Hercules enjoyed a new currency in the mid-nineteenth century with

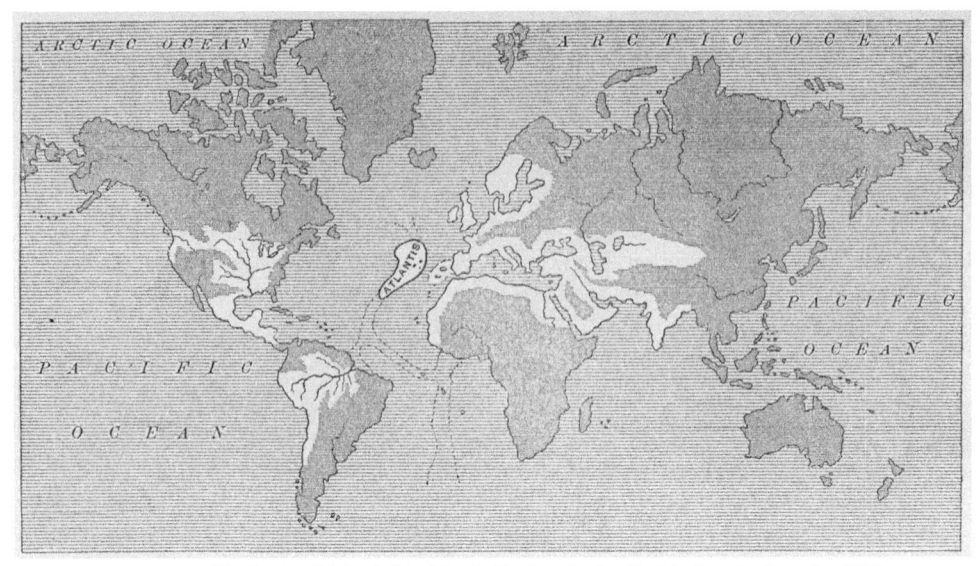

FIGURE 10.5. "The Empire of Atlantis," in *Atlantis: The Antediluvian World* by Ignatius Donnelly (1889 edition), 295. Rare Books Room MF.23.7. By kind permission of the Syndics of Cambridge University Library.

the emergence of a popular syncretic mythography. Geologists were as vulnerable to the *zeitgeist* as anyone else, as the *Popular Science Review* reminded its readers in 1878: "We are all familiar with that favourite speculation among geologists as to the existence of the former continent of Atlantis."[47]

Ignatius Donnelly's sensational and influential 1882 book, *Atlantis: The Antediluvian World* argued that all ancient civilizations were descended from the high-Neolithic culture of the lost city of Atlantis (fig. 10.5). Donnelly proposed a syncretic view of "traditions of the deluge" that allowed him to place the "face of the waters" in the creation story of Genesis and Noah's flood alongside more up-to-date geological theories: "[W]hether we turn to the Hebrews, the Arysans, the Phoenicians, the Greeks, the Cushites, or the inhabitants of America, we find everywhere the traditions of the Deluge; and we shall see that all these traditions point unmistakeably to the destruction of Atlantis."[48] In the chapter "The Testimony of the Sea," Donnelly suggests that the *Challenger* soundings have located "the backbone of the ancient continent which once occupied the whole of the Atlantic Ocean," adding that "[t]he officers of the *Challenger* found the entire ridge of Atlantis covered with Volcanic deposits; these are the subsided mud which, as Plato tells us, rendered the sea impassable after the destruction of the island."

Donnelly's syncretic assumption was that the essential facts of the flood had been gradually lost through transmission but could still be recuperated. Syncretism is a "positivistic assumption regarding the accessibility and nature of the past [that] informed a view of religions as more or less coherent systems of belief in which a primordial or original 'hierophany' became progressively diluted or contaminated through historical transmission."[49] Syncretism is structurally analogous to imperialism in its impulse to idealize "complete" systems, while dissembling the hierarchy of their architecture. Both issue from a misapprehension of space as well as time as essentially and perpetually materially available.

Spiritualism is often represented as an oppositional reaction to materialism in the late nineteenth century, but it could also be described as an emanation of materialist thinking that fantasized a trans-temporal embodiment of experience. The curiously materialist imagination of spiritualism early fastened upon communications technologies and was very quick to use the telegraph, for example, as one of its "structuring analogies."[50] The medium Kate Fox is recorded to have explained, "From the first working of the spiritual telegraph by which invisible beings were enabled to spell out consecutive messages, they [the spirits] pointed to the ultimation of a science whereby spirits, operating upon and through matter, could connect in the most intimate relations of the worlds of material and spiritual existence."[51] Celebrity occultist and founder in 1875 of the Theosophical Society, Madame Blavatsky, had read Donnelly on Atlantis and subsequently wrote in *The Secret Doctrine* (1888) that the lost lands of Lemuria and Atlantis were linked as the homelands of the third and fourth "Root-races." Her editor and convert, Annie Besant, agreed that these regions would one day return from the sea.[52] The cosmology of theosophy involves a complicated system of progressive embodiment, where disembodied magical beings gradually take on bodily form and become gendered. Atlantis operates in esoteric and in more mainstream thinking in this period as the tantalizing possibility of the rematerialization of a hidden and submerged past in the present.

The little chimney sweep, Tom, in Charles Kingsley's 1862–63 *The Water-Babies*, also visits the sunken city that "Old Plato called . . . Atlantis," here called "St. Brandan's fairy isle."[53] In a book much concerned with transformation and a recuperation of purity, Kingsley's description of the ruins draws out the way in which geology is itself a syncretic science in this period: "Now when Tom got there, he found that the isle stood all on pillars, and that its roots were full of caves. There were pillars of black basalt, like Staffa; and pillars of green and crimson serpentine, like Kynance; and pillars ribboned with red and white and yellow sandstone, like Livermead; and there were blue grottoes, like Capri; and white grottoes, like Adelsberg; all curtained and draped with seaweeds, purple and crimson, green and brown."[54] The jumbled accumulation of all these highly localized geological identities in one hybridized, techno-natural space (the images of pillars and caves confuse the boundaries between man-made and naturally formed) creates a kind of materialization of the topographic confusion of the fossil record and the illegibility of geological evidence as a straightforwardly chronological narrative. Sumathi Ramaswamy has suggested that lost lands are "place[s] made to cope with modernity's preoccupation with loss," but here Atlantis seems to appear as a displaced trope of the plenitude that constantly threatens to overwhelm syncretic thinking.[55] The syncretic model produces a problematics of surplus. What happens to the bits (of history or culture) that are not successfully enfolded into a dominant system? What happens to the finitude of each element once it comes into contact with others? Kingsley and Verne, both writing in the 1860s, fill their fiction with extraordinary lists of names and objects. Roland Barthes observed that, for Verne, "The world is full of numerable and contiguous objects. The artist can have no other task than to make catalogues, inventories; ... his work proclaims that nothing can escape man, that the world, even in its most distant part, is like an object in his hand."[56] Indeed, Kingsley's description of Atlantis itself takes the form of a list, attempting to perform the bounded fixity of categories that geology and, by implication, any kind of history can itself never achieve.

Henri Bergson understood and wrote about the dangers of attempting to objectify or embody the past: "The difficulty that we have in conceiving [the past] comes simply from the fact that we extend to the series of memories, in time, that obligation of *containing* and *being contained* which applies only to the collection of bodies instantaneously perceived in space. The fundamental illusion consists in transferring to duration itself, in its continuous flow, the form of the instantaneous sections which we make in it."[57] What the sea could offer as a counter to the containment of chronology was the suggestion of dissolution of form and a return to formless duration. When Byron's Childe contemplated history from the deck of his boat, he did not address the sea as outside time:

> "Time writes no wrinkle on thine azure brow—
> Such as creation's dawn beheld, thou rollest now."
> (4.182)

Byron's sea is not timeless; here, instead, it represents pure duration, standing, or rather *rolling*, as Bergson's "continuous flow" that is "unwrinkled" and thus unmarked or unbound, even through the passing of time.

Forms and Formlessness

"Of all confusions, the ocean is the most indivisible and the most profound. . . . It is the universal recipient, a reservoir for the germs of life, and mould of transformations."
Victor Hugo, *Toilers of the Sea* **(***Les Travailleurs de la Mer***, 1866)**[58]

Darwin was characteristically severe about the potential of "ooze," pointing out that "[a] mass of mud with matter decaying and undergoing complex chemical changes is a fine hiding-place for obscurity of ideas."[59] But the obscurity of ideas was just what was so compelling about this history and why it spawned so many speculative responses. The wonder over the magic of the electrical cable, the "eighth wonder of the world," and the wonder over the new forms of life at the bottom of the ocean washed together in the public imagination from the 1860s onward (fig. 10.6).[60] Hans Christian Andersen reflects this generalized deep-sea wonder in his 1871 story about the Atlantic submarine cable, "The Great Sea Serpent," in which the deep-sea creatures try to identify "this immense, unknown sea-eel" that suddenly appears among them on the ocean floor: "Sponges, polyps and gorgonia swayed above it and leaned against it, sometimes hiding it from view. Sea-urchins and snails climbed over it; and great crabs, like giant spiders, walked tight-rope along it. Deep blue sea cucumbers—or whatever those creatures are called who eat with their whole body—lay next to it; one would think that they were trying to smell it."[61]

In Victor Hugo's *The Toilers of the Sea*, the fisherman Gilliatt spends a sleepless night churning about the "incomprehensible promiscuity" of the reef that, ""caus[es] the mineral to become vegetable; the vegetable to rise to higher life."[62] The sea fauna comprises "monstrous creatures," and "[v]ague forms of antennae, tentacles, fins, open jaws, scales and claws, float about there."[63] Such ambivalences were clearly unsettling. Kingsley plays with the grotesque possibilities of marine bodies in *The Water-Babies*, describing a fantastic creature: "for every wing above it had a leg below, with a claw like a comb at the tip, and a nostril at the root."[64] Of the large numbers of previously unknown species that were discovered in the nineteenth century most were lower invertebrates, such as single-celled foraminifera; echinoderms, such as starfish; and

FIGURE 10.6. "The Diver in Search of the Atlantic Cable Gets into Hot Water." *Punch*, January 6, 1866. By kind permission of the Syndics of Cambridge University Library.

the sponges, whose place in the scheme of life, then as now, was far from clear. The debate over whether a sponge should be classified as a plant or an animal was never resolved in the nineteenth century. In Verne's novel, too, category errors are all too easy to make in the sea: "For a while, . . . I involuntarily mixed up the kingdoms, taking zoophytes for hydrophytes, animals for plants. And who wouldn't have made such a mistake? Flora and fauna were so close in this submarine world!"[65]

"It has no scales," said the polypus of the Atlantic cable, in Andersen's story, "it has no skin." But in fact the armoring, insulation, and core of the telegraphic cable were often described in terms of organic form at the time. Telegraph cables were popularly described as "the nerves of the world," and vice versa.[66] Defining nervous energy in 1880, for example, neurologist J. R. Gasquet wrote that "[i]t seems to resemble most nearly electricity; and the varying electrical conditions of nerves seem at first sight to justify the popular illustration which likens the brain to a galvanic battery and the nerves to telegraph wires."[67] Neurologists were soon using newly discovered species of deep-sea jellyfish to experiment on the nervous system, as their transparent bodies displayed the neural networks with convenient clarity. The cable, like the sponge and the mermaid, was uncomfortably suspended in the public imagination somewhere between matter and spirit (fig. 10.7). Kipling's cable "flicker[s] and flutter[s] and beat[s]" with the words of men, and Andersen's cable "lay perfectly still, as if it were lifeless; but inside, it was filled with life: with thoughts, human thoughts."[68] The crossover between the disembodiment of electrical messaging and the materiality of its apparatus challenged ideas of containment and of the synchronicity of content and form. It suggested possibilities of hybridity, fluidity, and disbursement of form that were at once attractive and frightening.

Marine zoologist Edward Forbes believed that it would be by studying "the world of waters" that we would be able to "unveil the mystery which yet hangs over the early history of our globe."[69] Taking the view of the past from the sea over the long nineteenth century, albeit in the briefest of ways, can usefully change the focus of inquiry and can show how the empiricist materialism of nineteenth-century "science" and indeed "historicism" was not contrapuntal or oppositional to Romantic modes of thinking, but developed out of those very ideas of fragmentation and retrieval, of form as variable and plastic, of sequence as broken and jumbled that are already present in Byron's poetry. What the sea offers up most particularly in this period is the possibility of new forms, through the vast oceanographic surveys undertaken by *Challenger* and other marine scientific expeditions. Undersea telegraphy extended this challenge to form by demanding a new way of thinking about dematerialized and rematerialized experience, spawning new morphologies in art and literature, as well as "science." The notion of the sea as an ahistorical space turns out to have a complex history of its own.

FIGURE 10.7. Edward Burne Jones, *The Depths of the Sea* (1887). Wikimedia commons.

Notes

1. Bryon, *Childe Harold's Pilgrimage*, canto 1, Song, verse 9, in *Lord Byron: The Complete Poetical Works*, ed. Jerome McGann, vol. 2 (Oxford: Clarendon Press, 1980), 15. All references are to vol. 2 of this edition and are given hereafter by canto and stanza in parentheses in the text [e.g., (2.41).

2. Byron, note to *Childe Harold's Pilgrimage*, canto 3, in McGann, ed., *Lord Byron*, 303.

3. [William Hazlitt], "Literature: Childe Harold's Pilgrimage. Canto the Fourth. By Lord Byron. Murray," *Yellow Dwarf*, May 2, 1818, 144. Reprinted in Donald H. Reiman, ed., *The Romantics Reviewed* (9 vols.), Part B, vol. 5 (New York: Garland, 1972), 2338.

4. Letter from Wordsworth to Catherine Grace Godwin, [Spring 1829], letter no. 423. *The Letters of William and Dorothy Wordsworth: V, The Later Years, Part Two: 1829–1834*, ed., Ernest de Selincourt and Alan G. Hill, 2nd ed. (Oxford: Clarendon Press, 1979), 58.

5. Reinhart Koselleck, *Futures Past: On the Semantics of Historical Time*, trans. Keith Tribe (Cambridge, MA: MIT Press, 1985), 33.

6. Presbyter Anglicanus [Francis Jeffrey], "Letter to the Author of *Beppo*," *Blackwood's Edinburgh Magazine* 3 (1818): 326.

7. "*CHP* is a highly moralized travelogue very much in the tradition of eighteenth-century topographical poetry" (*Lord Byron: The Complete Poetical Works*, ed. Jerome McGann [Oxford: Clarendon Press, 1980]), 270. Anne K. Mellor describes Byron's "exuberant mobilité" in *English Romantic Irony* (Cambridge, MA: Harvard University Press, 1980).

8. Byron, journal entry for January 9, 1821; reprinted in Thomas Moore, *Letters and Journals of Lord Byron: With Notices of His Life* (London: John Murray, 1860), 476. Emphases original.

9. The theory had first been propounded by Abraham Gottlob Werner, a German geologist.

10. Charles Lyell, *Principles of Geology; or, The Modern Changes of the Earth and Its Inhabitants Considered as Illustrative of Geology*, 2 vols., 10th ed. (London: John Murray 1867), 1:257. For more on Lyell's use of Byron, see Adelene Buckland, *Novel Science: Fiction and the Invention of Nineteenth-Century Geology* (Chicago: University of Chicago Press, 2013), chap. 3.

11. Charles Wyville Thomson, *The Depths of the Sea* (London: Macmillan, 1873), xi.

12. Buckland, *Novel Science*.

13. Joseph Conrad, "The Mirror of the Sea, No. III: Gales of Wind," *Pall Mall Magazine* (March 1905): 363–68. This and other essays were later collected as *The Mirror of the Sea (London: Methuen, 1906)*.

14. Thomson, *Depths of the Sea*, 1. Edward Forbes (another Edinburgh natural historian) had put forward the idea that "below 300 fathoms (1,800 feet) no life could exist in the ocean." Richard Corfield, *The Silent Landscape: The Scientific Voyage of "HMS Challenger"* (Washington, DC: Joseph Henry Press, 2003), 2.

15. See Corfield, *Silent Landscape*, 3.

16. T. H. Huxley, *Deep Sea Soundings in the North Atlantic Ocean in June and July 1857* (London: Lords Commissioners of the Admiralty, 1858), appendix A.

17. David Brewster, "The Facts and Fancies of Mr. Darwin," *Good Words*, 3 (December 1862): 7.

18. Gillian Cookson and Colin A. Hempstead, *A Victorian Scientist and Engineer: Fleeming Jenkin and the Birth of Electrical Engineering* (Aldershot, UK: Ashgate, 2000), 48.

19. John C. Galton, "In the Wake of the *Challenger*," *Popular Science Review* 15 (January 1876): 3.

20. Thomson, *Depths of the Sea*, 280.

21. "Carpenter's Letter to the President of the Royal Society: 24 March 1870," in Thomson, *Depths of the Sea*, 199.

22. Thomson, *Depths of the Sea*, 434.

23. Rudyard Kipling, "The Deep-Sea Cables," first published in the *English Illustrated Magazine* (May 1893) as one of the six sub-sectional poems to "A Song of the English" and then collected in *The Seven Seas* (London: Methuen, 1896), 9–10.

24. Henry Martyn Field, *Story History of the Atlantic Telegraph: To the Return of the Expedition of 1865*, ([1866]; New York: C. Scribner's Sons, 1892), 66.

25. Jules Verne, *Twenty Thousand Leagues under the Seas*, trans. William Butcher (Oxford: Oxford University Press World's Classics, 1998), 110.

26. See Donald J. McGraw, "Bye-Bye Bathybius: The Rise and Fall of a Marine Myth," *Bios* 45 (December 1974): 164–71.

27. See W. B. Carpenter, "Preliminary Report of Dredging Operations in the Seas to the North of the British Islands, Carried on in Her Majesty's Steam-vessel *Lightning*," *Proceedings of the Royal Society* 17 (1868): 168–200.

28. Philip F. Rehbock, "Huxley, Haeckel, and the Oceanographers: The Case of Bathybius haeckelii," *Isis* 66 (1975): 504–33. See also T. H. Huxley, "On Some Organisms Living at Great Depths in the North Atlantic Ocean," *Quarterly Journal of Microscopical Science*, n.s. 8 (1868): 205; and T. H. Huxley, "On Some Organisms Which Live at the Bottom of the North Atlantic, in Depths of 6000 to 15,000 Feet," *Report of the Thirty-Eighth Meeting of the British Association for the Advancement of Science Held at Norwich in August 1868* (London: John Murray, 1869), 102.

29. See Gerald L. Geison, "The Protoplasmic Theory of Life and the Vitalist-Mechanist Debate," *Isis* 60 (1969): 272–92.

30. Carpenter, "Preliminary Report of Dredging Operations," 190–91.

31. Haeckel had published "*Monographie der Moneren*" in *Jenaische Zeitschriftfur Medicin und Naturwissenschaft* 4 (1868): 64–137. Rehbock explains that "[f]rom the seventeenth through the mid-nineteenth centuries 'spontaneous generation' usually referred to the apparent development or 'heterogenesis' of intestinal worms, infusorians, yeasts, bacteria, and certain insects within an organic environment where no such organisms were thought to have been previously present" (Rehbock, "Case of Bathybius haeckelii," *Isis* 66 (1975): 509.

32. Ernst Haeckel, "Bathybius and the Moners" [sic], [Translated from the German] *Popular Science Monthly* (October 1877): 645.

33. T. H. Huxley, "On the Physical Basis of Life," *Fortnightly Review*, n.s. 5 (February 1, 1869): 130.

34. "The Dawn of Animal Life," *Westminster Review* 22 (July 1862): 173; "Introduction to the Study of the Foraminifera," *Athenaeum*, March 28, 1863, 417.

35. Charles Darwin, "The Doctrine of Heterogeny and Modification of Species," *Athenaeum*, April 25, 1863, 555, 554. Heterogeny is also known by its older name, "spontaneous generation."

36. Thomson's letter is held with the Huxley Papers (Imperial College), vol. 27, MSS 315–17. This letter is quoted in Rhebock, "Case of Bathybius haeckelii," 528–29.

37. "Professor Huxley on Bathybius," *Popular Science Monthly* (October 1879): 862.

38. Haeckel, "Bathybius and the Moners," 649–50.

39. Rehbock, "Case of Bathybius haeckelii," 533.

40. McGraw, "Bye-Bye Bathybius," 170."

41. Haeckel, "Bathybius and the Moners," 641; Ernst Haeckel, *Monism as Connecting Religion and Science: The Confession of Faith of a Man of Science*, trans. J. Gilchrist (London: Adam and Charles Black, 1894), 4–5.

42. Luther H. Martin has argued that syncretism is linked to nineteenth-century historicism; see "Historicism, Syncretism, Comparativism," in *Religions in Contact*, ed. Iva Doležalová, Břetislav Horyna, and Dalibor Papoušek (Brno: Czech Society for the Study of Religions, 1996), 39–50.

43. The phrase is Marlow's in Joseph Conrad, *Heart of Darkness*, ed. Owen Knowles (Harmondsworth: Penguin, 2011), 9, although Marlow is thinking of Africa here as no longer as blank a space on the map as he had pictured it to himself in childhood. The Arctic was the other unexplored region that offered itself up to speculation in his period.

44. Verne, *Twenty Thousand Leagues*, 68.

45. "The Lost Atlantis," *All the Year Round*, June 3, 1882, 373.

46. Verne, *Twenty Thousand Leagues*, 260–61. Verne based the route of the *Nautilus* on Matthew Fontaine Maury's *The Physical Geography of the Seas*, 2nd ed. (New York: Harper, 1855).

47. J. Starkie Gardner, "How Were the Eocenes of England Deposited?," *Popular Science Review* 2 (January 1878): 290.

48. Ignatius Donnelly, "The Destruction of Atlantis Described in the Deluge Legends," in *Atlantis: The AnteDiluvian World* (New York: Harper & Brothers, 1882), 65–66.

49. Luther H. Martin, "Syncretism, Historicism and Cognition: A Response to Michael Pye," in *Syncretism in Religion: A Reader*, ed. Anita Maria Leopold and Jeppe Sinding Jensen (London: Routledge, 2004), 286.

50. Pamela Thurschwell, *Literature, Technology and Magical Thinking, 1880–1920* (Cambridge: Cambridge University Press, 2004), 87.

51. Emma Hardinge on the medium Kate Fox, quoted in Jeffrey Sconce,

Haunted Media: Electronic Presence from Telegraphy to Television (Durham, NC: Duke University Press, 2000), 36.

52. Sumathi Ramaswamy, *The Lost Land of Lemuria: Fabulous Geographies, Catastrophic Histories* (Berkeley: University of California Press, 2004), 63.

53. Charles Kingsley, *The Water-Babies*, ed. Brian Alderson (Oxford: Oxford University Press World's Classics, 2014), 102 and 101, respectively.

54. Kingsley, *Water-Babies*, 102.

55. Ramaswamy, *Lost Land of Lemuria*, 7.

56. Roland Barthes, "The Nautilus and the Drunken Boat," in *Mythologies*, trans. Annette Lavers (London: Jonathan Cape, 1972), 65.

57. Henri Bergson, *Matter and Memory*, trans. Nancy Margaret Paul and W. Scott Palmer (London: George Allen Unwin Ltd, 1919), 193.

58. Victor Hugo, *The Toilers of the Sea*, trans. W. Moy Thomas (London: J. M. Dent & Sons, 1910[?]), 193.

59. Darwin, "Doctrine of Heterogeny," 554.

60. "The Eighth Wonder of the World: The Atlantic Cable": a print issued in 1866 by Kimmel & Forster, 254–256 Canal St., New York.

61. Hans Christian Andersen, "The Great Sea Serpent," in *The Penguin Complete Fairy Tales and Stories of Hans Christian Andersen*, trans. Erik Christian Haugaard (Harmondsworth: Penguin, 1985), 1011.

62. Hugo, *Toilers of the Sea*, 239. See also Margaret Cohen, *The Novel and the Sea* (Princeton, NJ: Princeton University Press, 2010), 197, for a discussion of this novel.

63. Hugo, *Toilers of the Sea*, 131.

64. Kingsley, *Water-Babies*, 154.

65. Verne, *Twenty Thousand Leagues*, 112.

66. J. Munro, "The Nerves of the World," *Leisure Hour* (January 1895): 183–87.

67. J. R. Gasquet, "Recent Research on the Nerves and Brain," *Dublin Review* 3 (1880): 373.

68. Andersen, "Great Sea Serpent," 1011.

69. Edward Forbes and Robert Godwin-Austen, *The Natural History of the European Seas* (London: John Van Voorst, 1859), 288.

Part Four: Unfinished Business

11 Looking Forward

Jocelyn Paul Betts

It remains convenient shorthand for historians to refer to a Victorian faith in progress.[1] Yet such statements invite much qualification. The Victorians looked to the future in myriad ways. Millenarians anticipated the imminent second coming of Christ, supporting their beliefs with scripture and prophetic interpretation of recent history; novels imaginatively transported readers to worlds in which humanity was extinguished by disease or the inexorable march of evolution; while political economists traced the inevitable limits of economic growth in the petering out of capital accumulation and the finitude of land and fuel.[2] Ideas of progress also coexisted throughout the nineteenth century with a range of fears about moral and political decay, decline, and degeneration, while the possibility of stasis or retrogression was frequently a natural corollary of hopes for improvement.[3] Theories of progress were not simply matters of faith, but often forms of political argument designed to collapse facts and values and distinguish positive from negative developments, while giving certain outcomes an air of probability. In this last guise

My thanks go to Lawrence Klein and the other contributors to this volume for discussion of drafts. Any errors remain mine.

their theoretical underpinnings and political valences came in such a variety of forms that lumping together the writings of John Stuart Mill, Henry Maine, and Macaulay in a single box marked "progress" can be to miss their significance as arguments.[4]

Yet the idea of mid-Victorian optimism also continues to make sense in qualified ways. Developmental ideas proliferated in Britain from the second quarter of the century onward, as Boyd Hilton has recently synthesized. This involved not just the insurgence of new ideas of history and nature, but the retreat of an evangelical idea of society and the natural world as repetitive, mechanistic expressions of divine order and justice, and of a relatively timeless and universalizing conception of reason that had animated utilitarian jurisprudence, political economy, and many Enlightenment-era histories. Political and social debate was moving toward interpretation of the laws of history and evolution; apparently irrational past beliefs were increasingly viewed as reflecting necessary, interconnected stages in the development of the human mind; while space was opening for a sense of cosmic optimism coming not from the static regularity of things or the possibility of the world's rational reconstruction, but from ever-increasing perfection, or what Herbert Spencer in his 1851 *Social Statics* called the "evanescence of evil."[5]

An emphasis on the development of society, morality, and ideas was not entirely new. As Peter Mandler points out in chapter 2 of this volume, ideas of civilizational growth had been bequeathed from the Scottish Enlightenment, while certain meanings of "progress" and "improvement" had been common currency long before the Victorians.[6] Yet in the widening field of discussion of progress at mid-century, civilizational advancement both gained in significance and shifted in emphasis. A combination of liberal theology, evolutionary ideas, and British political stability after 1848 combined to create sanguine hopes for the growth of ordered freedom.[7] Under these conditions, as mid-Victorians struggled with questions about the coming of democracy, the progress of civilization was oriented more closely toward the moral development of the populace, particularly among liberals. Ideals of domesticity, economic self-reliance, and popular education were woven into a vision that found a linchpin in ideas of strengthening individual male character and helped to define the mission of the period's intellectual classes as "public moralists" working for a virtuous polity by keeping contemporaries to their professed standards of behavior.[8] The period to the 1880s was characterized by a "measured confidence" in the possibility of this moralizing agenda.[9] Alongside ideas of technological and scientific progress, and a strident sense of global superiority that both

Sadiah Qureshi and Peter Mandler treat in their chapters in this volume, this moral confidence helped to form a particular atmosphere for the mainstream of educated opinion.

This chapter is primarily about a little-known figure who in one sense belonged squarely within such an atmosphere: the dramatist and philosopher William Henry Smith.[10] During the 1830s and 1840s Smith had worked as a barrister in London while trying his hand as a poet and novelist, as well as a playwright, eventually finding his *métier* as a prolific reviewer for *Blackwood's* magazine. During this time, despite an essentially reclusive nature, he moved in circles at the center of the growing interest in the progress of ideas and morality. He was friends with John Stuart Mill, whose journey from Benthamism toward the search for a science of history is indicative of the trend and did much to foster it. He was close to Frederick Denison Maurice, whose theology emphasized the connection of humanity and God in Christ and the possibility of social meliorism rather than man's intractably fallen nature, the increasing popularity of which was a bellwether of change.[11] Smith also had regular discussions with George Henry Lewes, an early conduit for the ideas of Auguste Comte that were to profoundly shape arguments over the nature of progress, and an important figure in the radical milieu around John Chapman's publishing house and the *Leader*, who propagated ideas of the movement of human beliefs toward a new Reformation going beyond Protestantism.[12] In 1857 Smith published a work entitled *Thorndale; or, the Conflict of Opinions* that was in keeping with these intellectual bearings. *Thorndale* could be read as a case study in the way that, at mid-century, ideas of the progress of civilization from its origins to the nineteenth century could combine with hopes for popular moral improvement to give history a spiritual feeling of purpose deliberately pitched against static ideas of reason and an evangelical focus on the next life. Yet the greater interest of Smith's text lies in the struggle that it reveals with the idea of how moral and intellectual progress might be fostered.

Among Smith's acquaintances, the problem of progress was not simply ethical but also a search for the intellectual basis on which a moral community might rest. The vision of being adrift in a period of intellectual anarchy between two epochs of consensus was a powerful one, shaping intellectual personas as forcefully as the mission to keep up existing moral standards. It was a vision lent force by the shifting relations between religious and secular authority, as well as the aim of ensuring the virtuous qualities of the masses. The problem of aiding a movement toward restored intellectual community animated Coleridge's ideas, fueled by images of ancient Israel, about the need to reunite philosophy and theology, and for the people to be cultivated under the aegis of a national clerisy or intellectual class, a proposal that Mill seriously considered, thanks to his friendship with Maurice and John Sterling.[13] It was a problem shared by Comte, whose systematic works presented the age as moving from an era of outworn metaphysical speculation toward the Positivist triumph of scientific principles and the religion of humanity, overseen by a new spiritual power of savants deliberately echoing the power of the medieval church.[14] The question was also central to the work of Thomas Carlyle, whose self-consciously prophetic and anti-systematic voice sought to neuter contemporaries' analytical modes of argument and reinstate an age in which fractious philosophizing would be subordinate to a shared spirituality that could support social unity and common virtue.[15] Mill's own sense of "intellectual anarchy" formed the basis for his 1831 essay on the "Spirit of the Age" as one of transition, and at the same time, he conceived of what would become his *System of Logic* as an attempt to unite Britain's "advanced intellects" by clearing up the realm of philosophical method.[16] Mill never doubted that freedom of thought and discussion was part of a route toward future unanimity, and mused on the possibility of returning to an age of belief through a "philosophy of life," his moralizing linked to his empiricism through a conception of rational virtue.[17]

These authors differed drastically on vital questions about the connections between philosophy, religion, and science. Yet from their highly influential writings flowed a common sensibility—that the conflicting opinions of the age were in a temporary flux. This was a sensibility strongly linked to the idea of an intellectual vanguard mapping a route toward new sources of consensus that contemporaries were invited to follow. It was linked too with competing histories of ideas, as authors sought to legitimize their visions of the trajectory of the public mind. As we shall see, both Lewes and Maurice wrote histories of philosophy, while Mill has been rightly regarded as an important fountainhead for the writing of intellectual histories that grappled with the idea of rationality expressed by authors such as Mark Pattison and Leslie Stephen.[18] Such histories had a sometimes oblique but real relationship with the effort to define and shape the further progress of thought.

Thorndale provides a counterpoint to Smith's more famous acquaintances, because of its questioning of the possibility of guiding such progress. Smith was read by contemporaries as having created an unusual work that broke with the normal course of Victorian argument, and the ways in which he did so are revealing of the ambivalences that intellectuals could face toward their own role in the history of opinion. Smith was a free-thinking theist who firmly believed that the ideas of the age were out of joint and in need of reinvigoration. Yet his didacticism was tempered by a series of qualms. Smith viewed the history of thought as shaped by a combination of social institutions and imaginative resources alongside philosophical or theological doctrines in a way that made systematic attempts to alter ideas redundant. At the same time, he believed that if progress were to be achieved, it could only come from stirring the reflective powers of all members of society to find their own livable truths, rather than any form of direct preaching. Yet Smith also worried that the philosophy he regarded as a component of the progress of human ideas was a self-limiting affair, incapable of certainty, plagued for each individual by the temptations of total skepticism or retreat to dogma.

These concerns were far from unique, being echoed, for instance, in Carlyle's skepticism toward philosophical systems and Mill's sense of the rising importance of educating all members of society. Yet Smith explored their combined ramifications to produce an intricate, self-contradictory work about the difficulties of didactically reshaping popular beliefs. Smith broke his authorial voice into several characters, so that *Thorndale* appears at first sight as a novel of ideas. Inspired in part by Carlyle's *Sartor Resartus*, *Thorndale* is presented by an editor as a discovered manuscript of several parts, left in a Neapolitan villa in 1850. The deathbed diary of the young, tubercular philosopher Charles Thorndale gives an account of his life and thoughts, his confusion over the nature of progress, and his failure to find philosophical certainty. Discussions follow with friends of Thorndale, who disagree about the plausibility and means of bringing the conflict of opinions to an end. The text ends with a cautious but optimistic philosophical tract on the progress of the individual mind and civilization by one of these characters—Clarence. Clarence attempts to make sense of the other characters' disagreements as aspects of a progressive whole, yet it is made clear in closing that he has not fully succeeded.

Before proceeding to examine this polyphony, it is useful to understand the vision of progress that informed Clarence's tract. Smith's theistic religion was linked to an idea of the harmonious development of human history. As the Scottish divine John Tulloch summarized after Smith's death, he had put forward a highly abstract variant on the argument from design, that the universe displayed the observable marks of divine intelligence. Smith emphasized the ordered relation of parts to whole that emerged when one conceptualized the world as evidence that reality was an emanation of a divine idea.[19] This order was above all temporal: the relationship of past, present, and future. Like others, Smith saw a new theodicy, or explanation of the apparent flaws of creation, in ideas that nature moved gradually and did nothing in vain.[20] Smith pushed this to its limit in his treatment of history's tendencies as constituting evidence of a beneficent mind behind creation, and his sense of the universe's wholeness as fostering a relationship between creature and creator. This view was inevitably haunted by the question of how exactly to grasp as a whole a world that was still coming into existence (further explaining why Smith was prone to self-probing and doubt).

Smith's views informed his treatment of the links between psychology, moral philosophy, and social relations as he tried to understand the historical malleability of morals without reverting to relativism or materialism. An 1839 pamphlet was couched as refuting ethical intuitionism, humans being molded through their sympathies for pleasures and pains, and their desire for esteem, toward upholding the values of the society in which they were raised, but also crucially having the power to transcend them.[21] Smith's mature psychological viewpoint was characterized by an emphasis on judgment, or the perception of relations, as the only innate faculty separate from sensation and as connected to a necessarily immaterial element of consciousness, saving the soul from physiology. Such judgment turned sensations into awareness of space and time, into memories, and ultimately led to the self to ethical life. As individuals developed toward selfhood and moral possibility, so over time this process could vary in its results, leading to moral change. Yet there was a continuous imperative underlying such change: that humans should act in accordance with the good of the whole of society, to the best of their knowledge, and according to society's stage of development—an idea that, in its emphasis on conscious appreciation of the needs of society, had much in common with Mill's marriage of utilitarian consequentialism with virtue, and with the period's growing culture of altruism.[22] Throughout Smith pursued the idea that mental phenomena required histories rather than definitions, and that later developments were fuller expressions of divine will than points of origin.[23]

When it came to the history of civilization, Smith saw an intertwined development of social organization and religious conceptions, both affecting what constituted the good of the whole. Society had gradually moved from slavery, via feudalism, toward wage labor, and was progressing toward an era of partnership, a view drawn from the period's debates over profit-sharing and cooperative production, in which the Christian Socialism of Maurice and the political economy of Mill were both important. In religion, the ancient idea of a despotic deity had been replaced with ideas of regular divine justice, and the continued movement was toward a God of love and forgiveness rather than strict judgment after death. In this unfolding history, the elements of past life were necessary moments in the movement to higher stages of society, with the time now ripe for changes that were dependent on the extension of conscious reflection on society's needs by a larger portion of the public.[24] Yet if this constituted Smith's idea of progress, it was not the only truth that *Thorndale* contained, and we must turn from the ideas that Smith put in Clarence's tract toward the doubts that he gave the other characters in the work about the prospects of such progress.

Thorndale himself epitomizes the insufficiency of purely abstract thought to provide stable foundations for consensus. In love with philosophy from an early age, he has led the life of a solitary thinker, writing and rewriting without ever reaching a finished manuscript: "The great problems of life lie around me unsolved—in hopeless confusion. I must leave them thus! . . . I hear my contemporaries boast often of the enlightened age they live in. I do not find this light. To me it seems that we state our problems somewhat more distinctly than heretofore; I do not find that we solve them. We are very luminous in our doubts."[25] Such a train of thought was a favorite of Smith's. In a short piece denigrating German idealism, he wrote,

> Philosophise men will—men must. Even the darkest paths, and the most labyrinthine of metaphysics, must be perpetually trodden. In vain is it proclaimed that they lead back only to the point of ignorance from which they started; in vain is it demonstrated that certain problems are indemonstrable; . . . each new generation finds them as fresh and attractive as if they had never been touched, never probed and tortured by fruitless examination; . . . to each generation must their solution at least be shown to be unattainable. . . . It is futile, therefore, to think of discarding metaphysics; if a good system is not adopted, its contrary will speedily prevail. "A good physician," says Paul Richter "saves us—from a bad one—if from nothing else." And a rational method of philosophising has, at all events, the same negative merit.[26]

Smith saw rationality in the aspirations of Scottish philosophers like William Hamilton to avoid both materialism and the excesses of idealism, but this did not amount to a full endorsement.[27] Metaphysical curiosity remained an incurable disease—expressed in Thorndale's congenital tuberculosis and early death.

Thorndale's confusion thus represents a point about the inconclusiveness of speculation, but also extends to problems of social progress. "All moral progress finally resolves itself into a public opinion wise and unanimous—which unanimity implies a certain degree of similarity in tastes, desires, passions, and a certain general level of intelligence; and lo! This inveterate diversity of development! Inseparable from our very industry, our productive acts, and social organisation."[28] This fear that unanimity might be impossible in a world of social divisions is elaborated by a specific example, for Thorndale is particularly concerned that belief in progress might destroy beliefs in the soul and the afterlife, as superstitions fostered by a flawed social system. Added to the fear that philosophy is incapable of settling questions is the fear that a philosophy of progress might be actively damaging to a society that is unprepared for it.

These themes are developed in discussions between the optimistic landscape painter Clarence and the skeptic Seckendorf, a German baron who has fled to England in order to become a doctor after a duel with a fellow soldier while a member of the Prussian forces in the Napoleonic wars. Clarence unites ideas of religious and social progress in hopes for the spread of a rational appreciation of the needs of society. Yet Clarence faces Seckendorf, who combines Catholicism, aristocratic *hauteur*, and materialism to argue against the chance of society's conscious improvement. His medical profession connects with a tendency to reduce mental phenomena to their physiological conditions and to describe society as both a developing organism and a beneficial struggle for survival. For Seckendorf, there is little chance of improving on social relations that have stemmed from unconscious developments, particularly economic exchange. Deliberate attempts to tinker with such a system are doomed to failure—society's growth means an increasing division of labor that renders any individual less capable of understanding society as a whole. In making a similar argument to combat the utopianism of Proudhon during the 1840s, Smith had drawn from Jean-Baptiste Say the idea that the division of labor necessarily reduced the mental capacities of workers who were restricted to shrinking spheres of activity.[29] Associated with this corrosion of universal experience is Seckendorf's argument that institutional religion is indispensable to social order. His religion is born out of fear and the desire to see amends made in the afterlife for terrestrial injustice. To weaken the bonds of a religion based on the doctrine of Hell, he felt, would be to destroy it entirely, and leave the masses with no reason to tolerate their position. Thus society is a fortunate congeries of humanity's errors and instinctive desires, which risks falling apart if one trusts to the reasoning powers of its members.

Seckendorf shows a detached tolerance of the hubbub and folly of life. He accepts men as they are rather than humoring Clarence's utopian leanings. His aristocratic background fuels his aloofness and chimes with his desire to propagate a dual doctrine whereby the masses are to believe what is useful rather than following the freedom of thought of those higher up the social scale. He is fond of quoting Voltaire's maxim on the usefulness of error, a reminder of the *philosophe*'s desire not to discuss atheism in front of servants, later targeted by John Morley in order to uphold popular rationalism.[30] In a dry run of Seckendorf that Smith had published in 1851, Voltaire's ghost visits the Great Exhibition to argue with a range of English archetypes, consistently ridiculing their evidences of progress (with the exception of a Lucifer match), and leaving quite content to have lived in eighteenth-century Paris. Here Smith was playing on familiar representations of Voltaire as a destructive influence with little positive doctrine, in order to fathom the worth of nineteenth-century claims to advancement.

Through Seckendorf, Smith voiced his keenly felt sense of the potential closeness of skepticism to orthodoxy. If one could not trust in intellectual progress to produce moral order, one was potentially doomed to sacrifice sincerity for the sake of social cohesion. In a notebook from the early 1860s, Smith wrote the following entry:

> HYPOCRISY. On the greatest subjects on which human beings can think, there ought not to be an habitual systematic hypocrisy. Nor is there amongst the multitude. But in our educated classes there is. But this is not to be wondered at. A free-thinker who does not see an absolute gain to society by the substitution of his own faith or opinion for the popular faith can have no motive for sincerity. There will be this hypocrisy till the moment comes when a new and simpler faith brings in some new enthusiasm on the side of virtue.[31]

There is an echo here of Clarence's account of the history of religious ideas: the development of philosophy had often diverged from institutionalized religion, the two being reunited only through the work of religious reformers prepared to face down public opposition.[32]

It would be tempting in all this to view Seckendorf as the imaginary opponent of Smith's efforts to propagate this simpler and more sincere faith, yet the truth was more complicated. Seckendorf's quiescence expressed the difficulty of founding either certainty or didacticism on philosophical reflection. In reviewing Ernest Renan, the controversial biographer of Christ, Smith wrote positively of Renan's appreciation for the social influences of Roman Catholicism and his awareness that his own opinions were less important than the "manifest wants, tendencies, and aspirations of mankind." Renan saw that the "attitude of mind of the incessant inquirer after truth—by which the philosopher is supposed to be distinguished—can belong only to a few." Such inquirers needed freedom, but should not be made into teachers of the masses, since they might communicate doubt but never the spirit of inquiry itself.[33] Smith did not condemn those who turned to dogmatic authority as a way out of their own, unsettled intellectual world. One of *Thorndale*'s characters is Cyril, a Cistercian monk driven toward the Roman church because its doctrine of purgatory offered a solution to his qualms about eternal punishment, a journey treated with frank sympathy. Moreover, one finds an important aspect of *Thorndale*'s treatment of intellectual conflict in Seckendorf's pleasure at contemplating the "vast variety of opinions," and his claim that "I have not taken upon myself to remodel the world upon my own convictions," as well as his statement that "it is *life* that is the end of life, not *truth*."[34] Caught between existing doctrines that appeared to just about function, and philosophy's inability to offer a definite, livable certitude, there were risks in intellectual didacticism.

Smith was not simply advocating hypocritical silence, but thematizing the need and the difficulty of equipping those led to probe their beliefs with their own means of moving through the philosophical labyrinth rather than accepting fragile new sources of authority. One of the ways that Smith thought about this office was through his reflections on poetry. Poetry introduced people to mental activity by inviting them to continue to think about what had been presented to them, and to do so not only in terms of "that truth of form and outline which the dry light of reason reveals, but also in that charm and allurement of *colour* which it is the office of the imagination and the passions to supply."[35] The poet thus taught in the same manner as the painter or sculptor, or (as Smith put it) in the manner of the world. For poets to appeal directly to reason, as Milton had at points, opened them to censure for didacticism in the manner of a "school divine." Poets and writers of dialogue had the privilege of self-contradiction—the truth they sought was in part faithfulness to the inconsistencies of the human mind. Poetry's aim was to "accompany, not to direct, our progress," by providing a variety of views that helped to awaken the intelligence.[36] Smith defined the difference between the novel and poetry in terms of the novel's inevitable arousal of curiosity; thus, novels were read with too much of a propelling interest to also excite reflection.[37] It is thus not surprising that Smith made plain to his future wife, the translator Lucy Caroline Cumming, that *Thorndale* was not a novel but a "diary" and hence could combine the depiction of people and events with the invitation to thought.[38]

Among the various relationships of individuals to truth depicted in *Thorndale* is one of a poet, Luxmore. While Thorndale searches unsuccessfully for certainty, Seckendorf denies its possibility, and Cyril ends his search within the walls of a monastery, Luxmore suggests that poetry has a special part to play, saying,

> Thorndale, I could not write prose, if by prose you mean a didactic expression of settled systematic opinions. I have no systematic opinions. It is not that, in general, I am indisposed to believe; but one belief destroys another. *There are too many truths* [emphasis in original]. And there are truths of negation as well as of affirmation. I cannot help it. There are some subjects on which the more I read, and the more I think, the more bewildered I become. To me it seems that our world is veritable poetry—suits admirably, and suits only, the poet's verse; for there are things most beautiful and grand in it, and these individually may be faithfully reflected in the poem. But philosophy attempts to embrace the world as a whole scheme, and philosophy fails—only impresses on it an irremediable confusion, the result of her own limited intelligence.

Luxmore thus asks, "What if one *use* of the poet be to give some notes and fragments of truths which he himself, as little as any other, can yet harmonise into a complete system? [emphasis in original]."[39] This chimed with the editor's statement as to the purpose of *Thorndale*, a work whose "indecisions and contrarieties" might give the reader some "hints and helps to the formation of that settled and consistent scheme of thought which he is doubtless building up for himself."[40] It also echoed Clarence's closing hopes that readers might be able to reconcile the work's conflicting messages. To the extent that *Thorndale* taught reflection rather than truth, it was a poetic prelude to readers pursuing for themselves the difficult exercise of philosophy, which Smith defined as the harmonization of convictions.[41]

Thorndale's commitment to reexamining questions, Seckendorf's aloof skepticism, and Luxmore's poetic multiplicity all reflect different ways of reading *Thorndale*'s open-ended self-contradiction; yet, from another perspective, they were united in an idea of the relationship of character to thought. At the close of *Thorndale*, it is not only from Clarence's philosophical statements that readers are to learn, but from the grounded optimism, aided by bonds of duty and family, that allows him to maintain beliefs in progress and the afterlife. Toward the end of *Thorndale*, Clarence becomes an adoptive father, and only then writes his final tract, echoing the way that Thorndale, when he tries to explain his views to his beloved cousin, Winifred, is briefly moved to his own species of optimism. Notably, Smith had written two prior stories that had dealt with young philosophers trapped by abstraction until the discovery of a more sociable way of life.[42] While Thorndale's questing solitude finds its apotheosis in the fugitive Seckendorf's view of human beliefs as excusable untruths, Clarence crafts his views from within society's developing fabric. Paradoxically, only via this abandonment of detachment can any convincing attempt at philosophy be made. This conclusion appears related to Smith's early engagement with the third Earl of Shaftesbury, whose *Characteristics* Smith had defended in the *Athenaeum* in one of his earliest articles. Smith praised Shaftesbury's view that the truth was better served by philosophizing with gaiety rather than gloominess.[43] He also echoed both Shaftesbury's deism, based on the universe's observable harmony, and the idea that all rational thought contained an implicit level of skepticism, while *Thorndale*'s dialogic construction and concern with the relation of poetry to philosophy was likewise reminiscent of *Characteristics*.[44]

The parallel is most suggestive if we regard Smith as following Shaftesbury in attempting to avoid an overly magisterial authorial persona. While *Thorndale* was riddled with concern over the future of public doctrine and social cohesion, this was held in a tense balance with questions of how to both stimulate inquiry and show its limits in ways that could aid readers' personal self-fashioning. As Smith wrote in a self-disciplining notebook entry, "PROGRESS. I no more wish you to be eternally occupied with progress than forever occupied with your immortal state. Live your best—do your best—progress and immortality will take care of themselves."[45] Despite the period's rage for virtue and character, the idea that these might take precedence over the search for a foundational creed was not one Smith's contemporaries found easy to handle. The clearest contrast lies with Lewes. The idea of the circular nature of metaphysical speculation, of "the paths of philosophy . . . leading back to my own footmarks in the sand," was one that Smith and Lewes shared.[46] Lewes's *Biographical History of Philosophy* (1845) sought to show the history of thought culminating in Comte's rejection of metaphysics, and the establishment of positive science as the intellectual bedrock of a new era: "Philosophy only moves in the same endless circle. It has made no progress, although in constant movement. Precisely the same questions are being agitated in Germany at this moment as were being discussed in ancient Greece. . . . The history of science on the other hand is the history of progress. . . . In this constant circular movement of Philosophy and the constant linear progress of Positive Science, we see the condemnation of the former. . . . No progress is made, because no certainty is possible."[47] Where Smith had sought to embed an inevitable and interminable philosophy in questions of how to manage perplexity, Lewes sought to abandon the non-empirical altogether and to usher society out of intellectual anarchy toward positive authority. It is not entirely surprising that Lewes, who helped Herbert Spencer to decide not to write for the *Leader* in dialogue,[48] remarked on *Thorndale* to John Blackwood that "the age of doubt and vacillation has passed, and I want something solid or seeming solid whereon to stand and look about me," and thus that the "structure" of *Thorndale* was a "mistake."[49] Mill was more judicious, writing to Smith that he was pleased at *Thorndale*'s success, and that it was "a valuable contribution to the floating elements out of which the future moral and intellectual synthesis will have to shape itself."[50] Yet one might detect in this remark a barbed reference to the need for such a synthesis to be more actively pursued.

Competitors to Lewes's history are useful in fleshing out the picture. John Daniel Morell's highly successful *Historical and Critical View of the Speculative Philosophy of Europe* (1846) followed Victor Cousin's eclecticism to show philosophy's history as a series of competing, incomplete systems and impulses that it was possible to reconcile in a way that preserved their separate truths and expunged their errors. Morell specifically addressed Lewes's question of whether philosophy progressed, claiming that its disagreements were reflections of a path toward truth and that, as with the sciences, its course was akin to the tacking of a vessel against the wind: "a series of movements, each one of which seems to bear it away from the true direction, yet brings it in fact so much further on its destined course."[51] Yet, if Smith shared Morell's sense that metaphysics was not dead, and termed Clarence an "eclectic," he had less faith in philosophy as the essence of intellectual progress.

More similar in one sense was Maurice, who published his own multivolume history, entitled *Moral and Metaphysical Philosophy*, from 1847 to 1862. Maurice ended his mammoth work with Hamilton's assertion of man's inability to reason on the infinite, observing that if the thought of the nineteenth century had achieved anything, it was to show that discussions of philosophy could not be insulated from the masses. Thus, if leading thinkers were to treat philosophy as silent on the topic of divinity or even morality, it was all the more important that Christianity should maintain a strong existence independent of philosophical argument.[52] Smith in certain ways echoes Maurice's distaste for system, his desire for a "progress" that "must somehow lead us beyond 'isms,'"[53] and Tulloch's similar statement that "dogma splits rather than unites from its very nature. It is the creature of intellect, and intellect can never rest."[54] Yet this Broad Church treatment of system's limits was indistinguishable from the need to encourage a living, personal relationship with a revealed Christian God, which remained wedded to the project of theology. Smith, then, was a distinctive figure in this picture, maintaining metaphysics (unlike Lewes), unsure of its definite advance (unlike Morell), yet unable to fall back on Christianity, in ways that shaped his difficulty with tracing a path toward consensus.

Smith's fractured voice and his calling into question of his own authority unsettled reviewers. A perceptive puff piece in *Blackwood's* described it as a modern Book of Ecclesiastes, puncturing the intellectual vanity of the nineteenth century.[55] Yet for the most part reviews were tellingly torn between suggesting that Smith had either failed in his duty as an author to present a clear viewpoint, or was being underhanded in his preaching of heterodoxy.[56] The combination was put most succinctly in the *London Review*'s claims that Smith had been "unfaithful to his vocation as one of the priests of literature, and guilty of a cowardly attempt to shirk the responsibility involved in the power to form and publish an opinion," but that he was also a tactician in a wider conflict, turning "open, manly" debate into a "war of ambuscades and subtle stratagems."[57] *Thorndale* was thus critiqued as a failure of coherence or sincerity, in ways that tried to contain it within a framework of established and understandable intellectual conflict over the direction of intellectual development.

Reviewers did, however, appreciate *Thorndale*'s style and depth, and the work was far from being ignored: it comforted Laurence Oliphant in the wake of a *rōnin*'s violent attack on the British legation in Japan in 1861; Smith was commemorated in the poetry of Edward Carpenter; and the town of Gravenhurst, Ontario, takes its name from the work that followed *Thorndale*, which was being read by a Canadian postal official.[58] Tulloch thought him "almost certain to hold a higher niche in the Temple of Philosophy than he yet enjoys."[59] Yet this was not to be. Smith's vision of philosophy as uncertain of independent advance and intertwined with literature and moral bearings would not translate easily into the lecture theater or the specialist journal. The suggestion of philosophy as a highly personal quest with a complex relationship to the movement of popular ideas was in keeping with the Wordsworth-inspired tendencies to seclusion that led Smith to give up on London and his life as a barrister in the late 1840s to seek places of natural beauty, and shortly afterward to reject duties in the chair of moral philosophy at Edinburgh suggested due to the failing health of the incumbent, fellow *Blackwood's* luminary John Wilson.[60] Yet it was also part of a considered intellectual attitude that indicates something of what was foreclosed in philosophy's halting late Victorian journey toward being thought of as one or several specialized disciplinary fields.[61]

In the Summer of 1842 the Temple Church in London had been in the midst of a restoration project intended to reverse the work of iconoclastic Puritans, and was supposed to be closed to the public. Yet Smith prevailed upon a "weak brother" to gain admittance for a small party, including Sterling, Mill, and Caroline and Robert Barclay Fox. Sterling and Mill praised Correggio, Sterling quoted Coleridge on dead knights, and the conversation turned to Templar effigies: was it true that all Templars had their legs crossed, or was it that none did? As the conflicting interpretations were exchanged, Smith remarked that he was gradually coming to disbelieve everything that had ever been asserted.[62] Smith's witticism played on the fear that conflicting authorities were signs of an age at risk from an all-consuming skepticism. His conflicted didacticism shows a more troubled attitude toward the possibility of an intelligentsia taking control of more intellectual anarchy than we have been used to. It suggests further comparisons—for example, with George Eliot's presentation of clashes of opinion and the search for unified knowledge in *Middlemarch*, with its ultimate dissolving of historical progress into the varied thoughts and moral actions of individuals. Lurking behind broadly consensual moral bearings is a diversity to Victorian intellectuals' attempts to situate themselves in the moving stream of public opinion, as history, philosophy, and literary form combined to shape authorial voices and strategies that are far from having been exhaustively studied.

Notes

1. Mark Bevir, "Historicism and the Human Sciences in Victorian Britain," in *Historicism and the Human Sciences in Victorian Britain*, ed. Mark Bevir (Cambridge: Cambridge University Press, 2017), 3.

2. Robert H. Ellison and Carol Marie Engelhardt, "Prophecy and Anti-Popery in Victorian London: John Cumming Reconsidered," *Victorian Literature and Culture* 31 (2003): 373–89; Mary Shelley, *The Last Man*, ed. Morton D. Paley (Oxford: Oxford University Press, 2008); H. G. Wells, *The Time Machine*, ed. Roger Luckhurst (Oxford: Oxford University Press, 2017); Fredrik Albritton Jonsson, "Political Economy," in Bevir, ed., *Historicism and the Human Sciences*, 186–210.

3. On fears of retrogression and their connection with ideas of progress, see, for example, the discussion in K. Theodore Hoppen, *The Mid-Victorian Generation, 1846–1886* (Oxford: Oxford University Press, 1998), chap. 13; Daniel Pick, *Faces of Degeneration: A European Disorder, c. 1848–1918* (Cambridge: Cambridge University Press, 1989); and on empire, see Duncan Bell, *Reordering the World: Essays on Liberalism and Empire* (Princeton, NJ: Princeton University Press, 2016), chap. 5.

4. For a warning on the diversity of theories of progress, see John W. Burrow, "Henry Maine and Mid-Victorian Ideas of Progress," in *The Victorian Achievement of Sir Henry Maine: A Centennial Reappraisal*, ed. Alan Diamond (Cambridge: Cambridge University Press), 55–69, esp. 58–60.

5. J. W. Burrow, *Evolution and Society: A Study in Victorian Social Theory* (Cambridge: Cambridge University Press, 1966); Duncan Forbes, *The Liberal Anglican Idea of History* (Cambridge: Cambridge University Press, 1952); Boyd Hilton, *The Age of Atonement: The Influence of Evangelicalism on Social and Economic Thought, 1785–1865* (Oxford: Oxford University Press, 1986); Boyd Hilton, *A Mad, Bad, and Dangerous People? England, 1783–1846* (Oxford: Oxford University Press, 2006), chaps. 5, 7, and 10; Herbert Spencer, *Social Statics; or, the Conditions Essential to Human Happiness Specified, and the First of Them Developed* (London: John Chapman, 1851).

6. Roy Porter, *Enlightenment: Britain and the Creation of the Modern World* (London: Allen Lane, 2000), chap. 19; David Spadafora, *The Idea of Progress in Eighteenth-Century Britain* (New Haven, CT: Yale University Press, 1990).

7. Jonathan Parry, *The Politics of Patriotism: English Liberalism, National Identity and Europe, 1830–1886* (Cambridge: Cambridge University Press, 2006), 71–72.

8. Peter Mandler and Susan Pedersen, "Introduction: The British Intelligentsia After the Victorians," in *After the Victorians: Private Conscience and Public Duty in Modern Britain*, ed. Peter Mandler and Susan Pedersen (London: Routledge, 2004), 2–5; Stefan Collini, *Public Moralists: Political Thought and Intellectual Life in Britain, 1850–1930* (Oxford: Oxford University Press, 1991).

9. J. P. Parry, "Liberalism and Liberty," in *Liberty and Authority in Victorian Britain*, ed. Peter Mandler (Oxford: Oxford University Press, 2006), 73.

10. Smith is mentioned in Charles D. Cashdollar, *The Transformation of Theology, 1830–1890: Positivism and Protestant Thought in Britain and America* (Princeton, NJ: Princeton University Press, 1989), 31–33; David E. Latané, "William Henry Smith and the Poetics of the 1830s," *Wordsworth Circle* 20, no. 3 (1989): 159–65; Richard D. McGhee, "William Henry Smith,"

in *British Short-Fiction Writers, 1800–1880*, ed. John R. Greenfield, vol. 159 of *Dictionary of Literary Biography* (Detroit, MI: Gale Research, 1996), 294–98; Milton Millhauser, *Just Before Darwin: Robert Chambers and Vestiges* (Middletown, CT: Wesleyan University Press, 1959), 153–56; J. M. Robertson, *History of Freethought in the Nineteenth Century* (London: Watts, 1929), 204–5; Rick Rylance, *Victorian Psychology and British Culture, 1850–1880* (Oxford: Oxford University Press, 2000), 128, n. 51, 263, 289–90; J. B. Schneewind, *Sidgwick's Ethics and Victorian Moral Philosophy* (Oxford: Clarendon Press, 1977), 150–51, 163, n. 1, 177; James A. Secord, *Victorian Sensation: The Extraordinary Publication, Reception, and Secret Authorship of Vestiges of the Natural History of Creation* (Chicago: University of Chicago Press, 2003), 470; Roger Smith, "The Physiology of the Will: Mind, Body, and Psychology in the Periodical Literature, 1855–1875," in *Science Serialized: Representations of the Sciences in Nineteenth-Century Periodicals*, ed. Geoffrey Cantor and Sally Shuttleworth (Cambridge, MA: MIT Press, 2004), 81–110.

11. Jeremy Morris, *F. D. Maurice and the Crisis of Christian Authority* (Oxford: Oxford University Press, 2005).

12. Hock Guan Tjoa, *George Henry Lewes: A Victorian Mind* (Cambridge, MA: Harvard University Press, 1977); Rosemary Ashton, *142 Strand: A Radical Address in Victorian London* (London: Vintage Books, 2008); Mark Francis, *Herbert Spencer and the Invention of Modern Life* (Stocksfield: Acumen, 2007), part 2.

13. See Samuel Taylor Coleridge, *The Collected Works of Samuel Taylor Coleridge*, vol. 10, *On the Constitution of Church and State*, ed. John Colmer (Princeton, NJ: Princeton University Press, 1976); Ben Knights, *The Idea of the Clerisy in the Nineteenth Century* (Cambridge: Cambridge University Press, 1978); John Morrow, *Coleridge's Political Thought: Property, Morality, and the Limits of Traditional Discourse* (London: Macmillan, 1990).

14. See Auguste Comte, *Early Political Writings*, ed. and trans. H. S. Jones (Cambridge: Cambridge University Press, 1998); Mary Pickering, *Auguste Comte*, 3 vols. (Cambridge: Cambridge University Press, 1993–2009).

15. See especially "Characteristics," in *The Works of Thomas Carlyle*, ed. Henry Duff Traill, 30 vols. (London: Chapman and Hall, 1896–99), 3:28. Also, for a general view, see John Morrow, *Thomas Carlyle* (London: Hambledon Continuum, 2007).

16. John Stuart Mill, *The Collected Works of John Stuart Mill*, ed. John M. Robson, 33 vols. (Toronto: University of Toronto Press; London: Routledge and Kegan Paul, 1963–1991), 22:233; 12:79.

17. On unanimity and freedom of discussion, see Mill's *On Liberty*, in Mill, *Collected Works*, 18:250–52. On an age of belief, see Mill, *Collected Works*, 27:645–46. On reason and virtue, see H. S. Jones, "John Stuart Mill as Moralist," *Journal of the History of Ideas* 53, no. 2 (1992): 287–308.

18. Eldon Eisenach, "Mill's Reform Liberalism as Tradition and Culture," *Political Science Reviewer* 24 (1995): 71–146; H. S. Jones, *Intellect and Character in Victorian England: Mark Pattison and the Invention of the Don* (Cambridge: Cambridge University Press, 2007), chap. 6; Marco de Waard, "History's (Un)Reason: Victorian Intellectualism from J. S. Mill to Leslie

Stephen," *Victorian Studies* 53 (2011): 457–67; Jeffrey Paul von Arx, *Progress and Pessimism: Religion, Politics, and History in Late Nineteenth-Century Britain* (Cambridge, MA: Harvard University Press, 1985).

19. See John Tulloch, "The Author of *Thorndale*," *Contemporary Review* 25 (1874): 377–96.

20. James R. Moore, "Theodicy and Society: The Crisis of the Intelligentsia," in *Victorian Faith in Crisis: Essays in Continuity and Change in Nineteenth-Century Religious Belief*, ed. Richard J. Helmstadter and Bernard Lightman (Stanford, CA: Stanford University Press, 1990), 153–86.

21. William Smith, *A Discourse on Ethics of the School of Paley* (London: William Pickering, 1839).

22. Collini, *Public Moralists*, chap. 2; Thomas Dixon, *The Invention of Altruism: Making Moral Meanings in Victorian Britain* (Oxford: Oxford University Press, 2008).

23. William Smith, *Thorndale: or, the Conflict of Opinions* (Edinburgh: W. Blackwood, 1857), 406–501; William Smith, *Knowing and Feeling: A Contribution to Psychology* (n.p., 1874).

24. Smith, *Thorndale*, 502–607.

25. Smith, *Thorndale*, 18.

26. William Henry Smith, "The Visible and Tangible," *Blackwood's* 61 (1847): 580.

27. Smith, "Visible and Tangible"; see also William Henry Smith, "Sir William Hamilton," *Blackwood's* 86 (1859): 499, 501–2; William Henry Smith, "Psychological Inquiries," *Blackwood's* 77 (1855): 420.

28. Smith, *Thorndale*, 25.

29. William Henry Smith, "M. Prudhon" [sic], *Blackwood's* 65 (1849): 309.

30. John Morley, *On Compromise* (London: Macmillan, 1886), 45–46.

31. George S. Merriam, *The Story of William and Lucy Smith* (Boston, MA: Houghton, Mifflin, 1889), 423.

32. Smith, *Thorndale*, 587–88.

33. William Henry Smith, "M. Ernest Renan," *Blackwood's* 90 (1861): 627.

34. Smith, *Thorndale*, 285–86.

35. William Henry Smith, "A Prosing Upon Poetry," *Blackwood's* 46 (1839): 194.

36. Smith, "Prosing Upon Poetry," 195–200; William Henry Smith, "Works of Mr. W. S. Landor," *Quarterly Review* 58 (1837): 125–26.

37. Smith, "Prosing Upon Poetry," 197.

38. Merriam, *William and Lucy Smith*, 224.

39. Smith, *Thorndale*, 184–85.

40. Smith, *Thorndale*, 14.

41. Smith, "William Hamilton," 496. Smith could be as pessimistic about such a harmonization in science as in philosophy: see William Henry Smith, "Physical Geography," *Blackwood's* 66 (1849): 457–58.

42. William Smith, *Ernesto: A Philosophical Romance* (London: Smith, Elder, 1835); William Henry Smith, "Wild Oats—a New Species," *Blackwood's* 47 (1840): 753–62.

43. William Henry Smith, "The Woolgatherer VI" *Athenaeum* (1828): 746.

44. Anthony Ashley Cooper, Earl of Shaftesbury, *Characteristics of Men, Manners, Opinions, Times*, ed. Lawrence E. Klein (Cambridge; Cambridge University Press, 1999), 370, 87–91.

45. Merriam, *William and Lucy Smith*, 423.

46. Smith, *Thorndale*, 26.

47. George Henry Lewes, *Biographical History of Philosophy*, 2 vols. (London: Charles Knight, 1845), 1:20–21.

48. Herbert Spencer, *Autobiography*, 2 vols. (New York: D. Appleton, 1904), 1:447.

49. George Eliot and George Henry Lewes, *The George Eliot letters*, ed. Gordon S. Haight, 9 vols. (London: Oxford University Press; New Haven, CT: Yale University Press, 1954–1978), 8:182.

50. Mill, *Collected Works*, 15:542.

51. J. D. Morell, *An Historical and Critical View of the Speculative Philosophy of Europe in the Nineteenth Century*, (London: William Pickering, 1846), 18.

52. Frederick Denison Maurice, *Modern Philosophy; or, a Treatise of Moral and Metaphysical Philosophy from the Fourteenth Century to the French Revolution, with a Glimpse into the Nineteenth Century* (London: Griffin, Bohn, 1862), 673–76.

53. Frederick Denison Maurice, *The Life of Frederick Denison Maurice: Chiefly Told in His Own Letters*, ed. Frederick Maurice, 3rd ed., 2 vols. (London: Macmillan, 1884), 2:339.

54. John Tulloch, *Movements of Religious Thought in Britain during the Nineteenth Century: Being the Fifth Series of Saint Giles' Lectures* (London: Longmans, Green, 1885), 335.

55. R. H. Patterson, "*Thorndale*," *Blackwood's* 83 (1858): 238.

56. "*Thorndale*," *Athenaeum* 1567 (1857): 1389–90; "*Thorndale*," *British Quarterly Review* 27 (1858): 36–62; S. B. A. Harper, "*Thorndale*," *Dublin Review* 44 (1858): 299–335; "Strictures on *Thorndale*," *Titan* 28 (1859): 397–405; "*Thorndale*," *Saturday Review* May 21, 1859, 626–27.

57. "*Thorndale*," *London Review* 11 (1859): 446.

58. Margaret Oliphant, *Memoir of the Life of Laurence Oliphant and of Alice Oliphant, His Wife*, 2 vols. (New York: Harper & Brothers, 1891), 1:262; Edward Carpenter, "William Smith," in *Narcissus and Other Poems* (London: H. S. King, 1873), 234.

59. Tulloch, "Author of *Thorndale*," 377–78.

60. Merriam, *William and Lucy Smith*, 140–42.

61. See Collini, *Public Moralists*, 211–12.

62. Caroline Fox, *Memories of Old Friends*, ed. Horace N. Pym, (London: Smith, Elder, 1882), 165–66.

12

How We Got Here

Daniel C. S. Wilson

By the close of the Victorian era Britain was the world's dominant industrial power, and as the first nation to have mechanized its economy, it had the first opportunity to reflect *historically* on its social and economic journey. On surveying the scene in the years around 1900, many Britons were prompted to wonder how it had come to this. The modern world, illuminated by bright electric light and moving at ever greater speeds, bore so little resemblance to that of 1800 as to demand explanation. The possibilities of motor travel, of transatlantic crossings completed in days rather than weeks and, before long, of flight itself were the symbols—although real enough—of the fundamentally new economic base that underpinned a society not yet ripe for evaluation. The optimism arising from these famous material advances was tempered not only by the *fin-de-siècle* gloom enveloping the whole continent in worries about its degenerating culture, but also by the harsh realities of Europe's first megacity. The existence of slums in London's East End had to be made sense of: was the "stagnant pool" described by Friedrich Engels—this "immense haunt of misery"—a necessary corollary of the new economy? Was it the hallmark of a productive system transformed by machines and thus a template for the future of the world? Was it a blip on the road to modernity, which would soon be smoothed over as people adjusted to the new exigencies of work? Or was it the entirely predictable result of an unfair political economy, surely discredited by a depression lasting for almost the whole of the final quarter of the nineteenth century? Whichever perspective one adopted, one could not fail to be struck by the paradox of a system at once offering the utopian promises of progress while producing what, with understatement, was known as "the social problem."

The perceptible changes to the lives of workers, to consumption, and to communication were attributed, above all, to the role of machinery and produced a widespread disorientation, hardly lessened by the approaching milestone of the year 1900. There was much to explain, but who could explain it? Social prophecy could be found in works of fiction, above all that of H. G. Wells, to whom turn-of-the-century readers flocked in huge numbers, making the author of "scientific romances" into the world's most famous writer. Wells's speculations about the shape of things to come allowed an imaginative engagement with radical change, transposed into other times and places. Ostensibly projected into the future, Wellsian fantasy nonetheless engaged the disorientation of the present, rendering science in particular as a drama of human transcendence and achievement. The opposite sentiment could be seen at work, for example, in E. M. Forster's pungent reaction to news that the first manned flight was imminent: "It's coming quickly, and if I live to be old I shall see the sky as pestilential as the roads . . . and can't expect to feel anything but despair. Science, instead of freeing man . . . is enslaving him to machines." After the turn of the century, one enterprising publisher collected a series of writings by leading social critics and philosophers, such as Henri Bergson, G. K. Chesterton and, of course, H. G. Wells. These twelve "prophets of today" were cast in the role of a modern-day Hosea, Joel, or Zephaniah, who would respond to the public desire for guidance—echoing biblical Jerusalem—through iniquitous and tumultuous times.[1] While this turn-of-the-century testament included a healthy blend of scientists, literary authors, and philosophers—often ruminating outside their areas of expertise—it did not include any thinker perhaps better equipped to attempt a material explanation of change; that is, an economist or, better, a historian.

The reason for such a choice may be rooted more in style than in substance, yet such decisions have long implications. A focus on prophecy and science fiction has shaped a view of the *fin-de-siècle* as a period in which technology was considered solely in terms of its possible future. However, both imaginative projections of possible futures and historical inquiry have often stemmed from a singular impulse to understand the present, and it is with the latter form of writing that this chapter is concerned. At issue in understanding the world of 1900 were the ongoing effects of the transition set in train somewhat earlier in the century, which came to fall under that loose term used to account for the shift from premodern to modern life; namely, the Industrial Revolution. This term, as unsatisfactory as it is pervasive (and thus unavoidable), itself only germinated in the 1880s and so carries its own accumulated meanings for us to consider. For *fin-de-siècle* thinkers facing either forward or backward, the Industrial Revolution was an important touchstone: a starting point whose importance stared them in the face. What had caused this change? Was it a one-off, or would there be more of the same? What was its particular character? And why Britain?

While certain writers produced prophecies—and criticisms—of the new social order in the bright public gaze, similar questions of its provenance and its direction were springing up elsewhere in the intellectual ecology of late Victorian Britain. In a set of debates taking place at one remove from the popular press, the field of Victorian social critique was being carved up into its more familiar twentieth-century contours, as represented by the new academic disciplines of history and economics. These two subjects—cornerstones of social investigation—were still newly minted, with only a handful of university-based practitioners even as late as 1890. Even so, disputes over method were rife, nowhere more visibly than in political economy where (following the German example) two camps faced off, favoring respectively a more mathematical versus a more historical approach to the subject. Political economists, for example, following Stanley Jevons, imagined themselves as applied mathematicians whose subject was money, and whose aim was to study the dynamics of its movement between human subjects—its accretions and dissipations—in order to deduce general laws governing its nature and to use this understanding to control its effects. The group that opposed them believed that each economy was unique and—like all human activity—had to be seen in its proper context and as a temporal sequence: in order to be understood, it should be read like a narrative from beginning to end. Regardless of their stated position in this so-called *Methodenstreit*, economic thinkers were remarkably consistent in their focus on the *historical* question regarding the changes in industry that occurred around 1800, since it provided the richest source material for both historical and theoretical inquiry.

The question of machinery and of the Industrial Revolution had not at all been addressed by historians as they became professionalized during the 1870s. Historians had been concerned with the distant past, the medieval period, and, to an extent, England's revolutionary early modern history. This changed from around 1880, when certain historians began to approach political economy both as a tool and as an object of study in itself, and so their interests converged with those of the historically inclined economists. The tectonic plates of the disciplines shifted, partly because that more general concern for the "social problem" outlined above prompted certain individuals—such as Arnold Toynbee (1852–1883)—to study more closely those changes in industry that, in hindsight, began to appear revolutionary and prompted a palpable sense of historical process.

Although not the first to do so, and certainly not the last, Arnold Toynbee was perhaps the most influential writer to adopt a historical approach to the economy. Toynbee believed that to make sense of the social malaise required a history of British industry that would go beyond the hyperbole of Thomas Carlyle's famous attack on hard-hearted capitalists, as well as the soft-headed tradition of Victorian social critique that followed in its wake. The task was to investigate properly the nature of the changes wrought by machinery.[2] How did they affect employment patterns, the concentration of capital, land ownership, and the relations between bosses and their workers? If it seemed to contemporaries that machines had arrived from nowhere, it became imperative to tell their story. The idea of writing a machine history was epitomized by Toynbee's lecture series, published posthumously as the "Industrial Revolution" in 1884, but this was an endeavor already set in train by other contemporaries, such as William Cunningham (1849–1919), as well as Toynbee's student, William Ashley (1860–1927). These two historian-economists were of a generation now forgotten. Whereas the first cohort of historians—famous Regius Professors at Oxbridge as well as popularizers of historical lore, from Lord Acton to J. R. Green—held public and political sway, much like the generation at work around and after the First World War (such as Lawrence and Barbara Hammond), the position of figures such as Ashley and Cunningham was, by contrast, awkward: their work was difficult to read and involved the hard graft of statistical data, but it repays our attention nonetheless.[3] Ashley and Cunningham were academic outsiders who at first struggled to find university positions, yet their scrupulous research has outlasted them: for example, Ashley's landmark history, *The Economic Organisation of England* (1915) was still to be found on economics syllabi late into the twentieth century.[4]

The approach taken by Ashley and Cunningham in addressing the question of how we got here was different from those adopted by most contemporaries. There was no speculation, whether fictive or theoretical, and the emphasis was on detail and historical accuracy. Unlike Alfred Marshall—at that time Cambridge professor of Political Economy and figurehead for the mathematical tendency—history was not to be used merely as the backdrop for establishing a predetermined theoretical belief.[5] History was itself the thing at stake, and so, for Cunningham and, *a fortiori*, for Ashley, their endeavor was seen as a science, with the respect for evidence, method, and definition that they felt was implied.[6] The failure of both Ashley and Cunningham to win university appointments at the hands of neoclassical economists helped to eclipse the historical approach in the discipline by 1890. This came despite the historical turn taking place more generally at that time in a range of intellectual fields, such as sociology and anthropology. Even as economic histories continued to generate healthy sales for publishers, political economists attempting a historical approach—in the cases of Ashley and Cunningham at least—were forced abroad in search of employment. Both men found a more sympathetic environment in North America: Ashley taught first at Toronto before a chair was created specifically for him at Harvard in 1892, where he became the first professor of Economic History in the English-speaking world. Such experiences greatly influenced both men's thought and encouraged their increasingly comparative approach, which became the hallmark of their style. The single feature, above all, that distinguishes Cunningham and Ashley when thinking about Victorian temporality is their preoccupation with transitions: transitions in general but also that one in particular—the Industrial Revolution, as it tentatively came to be known—which had brought machines into the daily lives of millions of their countrymen and women.

The almost universal triumph of the historical method in today's intellectual world makes it difficult to imagine the ways in which the first professional historians set about constructing their problematics. This generation was concerned to establish chronology, to raise questions of causality, but also to convey an impression of change to their readers in ways they could imagine or at least conceive by the force of analogy. To a society that prided itself on having arrived as modern— such as the Britain that approached 1900 under electric power—it was hard to imagine the sort of future shock that must have accompanied the first coming of steam a hundred years earlier. The experience of new machinery has been the focus for many latter-day theorists and historians of technology. For the most part, such writers have sought variously to characterize this reaction as a type of cognitive dissonance, strangeness, sublimity, or "fear and wonder"; in short, modern historians have found—or at least assumed—that, across vast swaths of time and place, there has existed an analogous and discomfiting "shock of the new."[7] Whether or not this was actually the case is another question, but in the late nineteenth-century attempts, on the part of Ashley and Cunningham, to develop a narrative of machinery, this effect was crucial in conveying a dramatic sense of change. These writers sought to narrativize something—the rise of the machine economy—that it was not obviously possible prima facie to place into narrative order because of the multidimensional nature of the changes involved, their opaque connections, and the lags between cause and effect, themselves often displaced across time and space in illegible ways.

William Cunningham's gargantuan work, *The Growth of English Industry*, was a detailed empirical study of industry and the state. Packed with data, interpretation, and narration, its thousand pages were the standard work on the subject for many decades, and it provides an excellent example of contemporary thinking, with the simultaneous crudity and complexity that are entailed, both historical and literary. In the course of the text, a set of different timeframes and transitions are run alongside each other to create sophisticated temporal effects. For all of the drama of the British Industrial Revolution, it had involved long, drawn-out changes, which, at the time of writing in the 1880s, were occurring around the world in a much accelerated manner, telescoping decades of material changes into the space of years. The comparison in Cunningham's text thus drew together the British Industrial Revolution of c.1800 with the more widespread and visible industrial revolutions now taking place elsewhere. The changes of the late nineteenth century thus provided a lens through which to glimpse certain historical features of the original Industrial Revolution. In particular, the shock and disempowerment felt by Britons around 1800 are illustrated through a comparison with those around the world at the time Cunningham was writing: "The introduction of machinery continues slowly, but surely, to revolutionise the habits and organisation of industrial life in all parts of the globe. Half-civilised and barbarous peoples are compelled to have recourse, as far as may be, to modern weapons and modern means of communication; they cannot hold aloof, or deny themselves the use of such appliances. . . . England was the pioneer of the application of mechanism to industry, and thus became the workshop of the world, so that other countries have been inspired by her example."[8] This passage suggests, among other things, a general view of how technical change has been experienced. By describing the incredulity of "barbarous peoples" when faced with modern machinery around 1900, Cunningham gave his readers an insight into the same incredulity experienced by their own grandparents, and suggested a parallel between the Britons of 1800 and the "half-civilised" peoples of 1900. With regard to new technology at least, this passage posits those around the world as phylogenetically equivalent to the Britons of 1800. Cunningham made no normative judgment on either side of the comparison: far from condescending, he commented merely that, as people become subject to technical change beyond their control, they could not "hold aloof."[9]

Modern theorists of technology have been preoccupied with questions of agency; such a passage may therefore strike today's readers as crude in its suggestion that machines act straightforwardly upon society and determine its course, thus eliding the problem of causation entirely. However, Cunningham should not be underestimated, since the task he was attempting was one that few contemporaries had even considered, and he was humbly aware of the difficulties involved, beginning his account of industrial change by noting that even to frame the terms of reference would be prohibitively complex. The difficulty was such that Cunningham decided to fix on a single aspect of the changes in question—the main actor—and began by examining the nature of a machine. "A machine, as commonly understood, does not assist a man to do his work, it does the work itself, under human guidance; its characteristic feature is that it is an application of power, and not of human exertion."[10] Unlike definitions that focused on the source of motive power, Cunningham focused more on the effect; namely, the relief of exertion on the part of the worker.[11] As exceptional examples of machines, Cunningham included bicycles and spinning jennies, because they directed human power in new ways. Neither historical nor philosophical, Cunningham merely stipulated his definition, which fell back on a commonsense understanding rather than ascertaining its characteristic features. As far as a definition goes, Cunningham said—in effect—that if it seems like a machine, then it probably is one.

In his own seminal text on the history of industrialization, *The Economic Organisation of England*, William Ashley paid similarly close attention to this term, which would be so central to the story. Seeking greater clarity still, Ashley reflected that "[t]here has been a great deal of discussion" as to the distinction to be drawn between the "tool" and the "machine": the one, it has been said, can be owned by the workman, the latter is too expensive."[12] It is unclear which particular discussion Ashley had in mind, but he began his own with Karl Marx, who had set the framework for those that followed.[13] The question of expense was central to the way Ashley sought to define *machine* and related to, while remaining analytically distinct from, the question of capitalist organization. Marx had named the congregation of many workers under one roof "manufacture" and claimed that this practice, which saved money for the employers and was thought to increase productivity, was already prevalent between 1550 and 1750. Ashley's historical thesis disputed that this had been the case and questioned Marx's reading of the historical sources, arguing instead that the shift to manufacturing was not inherently more productive and, in fact, only took place because of government inducements in the period that he described as "Parliamentary Colbertism."[14] The state, rather than that mythical brace of invention and free-moving capital, had been the driver of industrialization. It was the use of water power that had caused initial accumulations of capital, and this process was merely intensified, Ashley claimed, by the use of steam underwritten by governmental support.

Both Ashley and Cunningham were driven by a thoroughgoing historicism that was as ready to question the precepts of Karl Marx as to question those of Alfred Marshall, notwithstanding the fact that Ashley, at least, regarded himself as a socialist. The key to understanding the role of the machine for both of these authors was, therefore, of a piece with their intellectual approach more generally: to situate the subject within an accurate narrative, while seeking neither a total explanation nor to delimit the extent of its wider influence on events. Nonetheless, taking up machinery as the subject of historical inquiry did produce some special effects in their historiographical texts, in which the idea of the machine came to be marked by a complex rhetorical pattern, appearing not only as the object of historical study but also breaking out as an explanatory figure. At different times in these texts, the working of a machine is posited as a symbol of history itself and, as such, it is hard to imagine another subject inveigling a new metaphorics into the articulation of its own history in this manner.

Ashley's account of the Industrial Revolution depicts the onset of mechanization as dramatic and difficult, but not a nefarious process. Mechanization is explained as the inevitable outcome of a process of capital accumulation, which Ashley described in terms of Protestant parsimony.[15] As for its results, Ashley delineated two important spheres in which important "facts" can be observed. First was the impact of machine-made goods on the market for hand-made goods. The problem arising from this "fact" was one "which no country so far has had the wisdom to solve satisfactorily: in England it was the long-drawn agony of the handloom weavers."[16] This seemingly universal tendency of technology to cause unemployment was first seen in Britain and demonstrated the power of the machine to cause specific internal changes in a given industry. The most important aspect of this change was a stark new set of industrial relations. The cost of the machinery being so high, its introduction "necessarily created a wide social cleavage between employers and employed. Although what we may call 'patriarchal' conditions of intimacy . . . tended to be replaced wherever large bodies of workpeople were brought together by a purely 'cash nexus.' This was not, as Carlyle might lead us to suppose, due to any peculiar hardness of the heart on the part of the employers: it was due to the necessities of the situation."[17] Ashley was preoccupied with this aspect—the human consequences—of mechanization and concluded that the introduction of technology must necessarily be traumatic for many of those involved, even at the best of times. By contrast, the Industrial Revolution was the worst of times, being marked by preexisting hardship and distress among the working class, which mechanization merely compounded. This substantive conclusion is, however, buttressed by an emerging metahistorical idea of the machine as a model for change: a nexus through which existing tendencies are amplified and exaggerated but allow little contingency. This convergence between the substantive subject of research (machinery) and an overarching intuition about its historical inevitability appears as a special feature of techno-*logy*—that is, in the story of machinery—which has marked its consideration until the present day. As with the handloom weavers mentioned by Ashley, the Luddites have archetypically been presented—negatively—as the embodiment of this historical law of progress. Cunningham gives them, perhaps, their first historical treatment in English, but it may seem ironic that his earlier discussion of seventeenth-century England outlines the legislative interventions that disallowed the use of certain machines precisely on the grounds of the economic and social harm that might be caused.[18] The mere possibility of these interventions—in the form of proclamations made by Stuart monarchs—reveals the political contingency of technical progress, even though such restrictions would have been anathema in Victorian Britain, as indeed they seem today.

The very first section of the first edition of *The Growth of English Industry* had the title, "The Industrial Machine." In what was Cunningham's first work of history, it was the metaphor of the machine that opened a general examination of the industrial system as a whole. As we reflect on its profound complexity, wrote Cunningham, "we naturally come to regard the industrial system of our time as a great machine, wonderfully delicate, and wonderfully powerful, which provides the necessaries and comforts and luxuries of our people." However, Cunningham wanted to discourage this comparison and suggests that "the analogy between our industrial system and any machine, however wonderful, is apt to mislead. For there is one very important difference between the two that should never be overlooked:—our industrial system is not pieced together out of inert matter and set agoing by mechanical forces, but its various parts are intelligent human beings with infinite capabilities of co-operation and self-development. . . . We shall understand our industrial system better if we think of it not as a machine, but as an organism,—a body economic."[19]

Cunningham opened his work with the idea of the machine, but it was a ruse: the image was offered to his readers but then rejected and inverted, according to the long-standing dyad of machine and organism.[20] The analogy of a machine was not an apt one for the economy: "The conception of a growing body is more instructive than that of a machine. . . . A great nation is not a mere machine for producing the greatest amount of wealth with the greatest amount of speed, it is an organism which cannot be healthy unless the conditions of distribution are satisfactory as well as those of production." Indeed, any growth under such defective relations "can only be an unhealthy swelling which proves the existence of some deadly disease."[21] With this model of the industrial system as an organism, Cunningham cast the political economist into the more demanding role of a physician rather than a mechanic. Any assumption that the economy could be modeled as a machine would have obscured the crucial role of humans in the industrial system and encouraged an excessive focus on artifacts. Cunningham concluded that while machines certainly gave the economy its character, a different sort of analysis was needed to understand its complex workings and movements more deeply.

The same first edition of *The Growth of English Industry* ended with a section entitled "Past, Present and Future," in which Cunningham posed the topical question of whether, by aligning present conditions with those of the past, it was possible to produce an informative trajectory into the future. His answer was firmly in the negative because, he argued, history is unlike a mechanism and "there is no uniformity like that of physical nature in the course of human history."[22] For Cunningham, any suggestion that history was like physics would have approached the unpalatable conclusions of determinists, fatalists, or worse; whereas the more appropriate natural-scientific role model was in fact biology.[23] Instead of Newtonian predictability, history involved discovering something closer to animal and vegetable patterns of interconnection. The machine metaphor was therefore inappropriate as a heuristic model for human action: historical knowledge was better advanced through the subtler use of analogy, as Cunningham demonstrated through his counterposition of different times and places.

Addressing the Economics Section of the British Association for the Advancement of Science as its president in 1891, Cunningham made a special point of welcoming the Indian economists in attendance.[24] This reflected his internationalism but amounted to a criticism of the theoretical school: by ostentatiously commending the foreign outlooks, which, he said, could enrich and correct their own, Cunningham drew attention to the limitations of classical economic theory, which was marked by its implausible pretensions to universality.[25] Cunningham recognized the cosmopolitan nature of capital as a historical force when operating in conjunction with machinery: "Capital introduces new methods of production, for it brings machinery and thus induces the labourer to spend his whole time in working for wages. There are still many lands where the old system which was general in this country at the close of last century holds good; and where the artisan supports himself partly by labour on his land and partly by his trade. But industrial capital, and expensive machinery, and factory towns are not compatible with such conditions of life."[26] Cunningham's innovation was this comparative perspective, which placed the historical changes that had occurred in Britain around 1800 alongside the contemporary conditions around the world. In this passage, capital is endowed with the agency to unleash machinery; but, whereas the agency of capital is clearly metaphorical, the way that the machine operates in this discourse remains oblique. The machine is also a metaphor for change, but the idea is complicated by the nature of machines as literally motive, and so the meaning is achieved less through metaphor and more through the intuition of a directing force or engine of history. Capital is now seen as the prime mover and machinery the determining force of change, which finally renders ways of life no longer "compatible." Cunningham's paper is an example of his attempt to illuminate general trends through a sensitive historical comparison. This was not something he felt that neoclassical economists were interested in doing, and, worse, they used the data of history dishonestly to support their preformed theories.[27]

Cunningham's 1905 Presidential Address, "Unconscious Assumptions in Economics," developed this perspective further and was praised for its lucidity by an editorial in the *Times* (London) and drew on his wide experience of travel and study.[28] Once again, Cunningham used the idea of the Industrial Revolution as a device for thinking simultaneously about contemporary India and British history. With a deliberate turn of phrase, Cunningham described how, during the 1880s,

> many of the changes which had revolutionised English industry and internal traffic were beginning to make themselves felt throughout India.... The results of the age of mechanical invention had begun to invade the changeless civilisation of the East. Still the persistence of the old order was also noticeable.... The highly developed gild system at Ahmedabad was the very image of much that I had read of regulated industry in mediaeval towns. On every side it seemed as if the survivals of the past had been preserved in the East, so as to make the story of bygone ages in the West alive before my eyes.[29]

The power of machinery was such that it could reshape even a "changeless" society, and yet, for Cunningham, the persistence of pre-industrial forms made India into a living archaeological specimen. It allowed the historian to enter a sort of time machine in order to witness at first hand the medieval Britain that preoccupied so many of the historical economists.[30] Cunningham refined the Toynbeean question of the "revolutionary" nature of industrialization through the use of this comparative rhetoric. While the industrial transformation of Britain was dramatic, it was at least gradual. For those countries undergoing the same process around 1900, the experience had been intensified, and so the contrast was extreme: "The transition from the old to the new, which had gone on steadily in England for centuries, seemed to be ready to sweep over Hindustan like a flood, that would disintegrate existing institutions, while it showed little constructive power."[31] Cunningham used this rhetoric of revolution to introduce a new distinction with the US and so to advance a more general conclusion.

Although similar to Britain in many ways, the US had a very different mechanism for negotiating change. In contrast to India, the "native races" of North America, Cunningham claimed, had exerted very little influence on its history or economy and so, "there had been room for the development of a new country pure and simple, unhampered by the traditions and customs of bygone days, except in so far as their wisdom was confirmed in present experience."[32] Unlike the Britain of the Industrial Revolution, the US of 1900 had no defense whatever against the growth of mechanized industry. It is there, Cunningham continued, "that we can note most clearly the lines on which modern industry and commerce develop with the full employment of modern appliances and the minimum of control from traditional habits and institutions."[33] In other words, the US provided a pristine laboratory for the creation of an industrial society unfettered by retarding cultural or mercantile traditions, as had been the case in Britain (and to an extent in India).[34] This historical method consisted in constructing mutually informative comparisons between the past and present; thus, for anybody thinking about the impact of machinery, the Industrial Revolution became the cardinal store of wisdom for tackling the political and economic challenges facing not only the Britain of 1900—now in its second industrial revolution—but also the rest of the world.[35]

How might we assess Cunningham and Ashley—their aims, their motives, and their disciplinary style? The two men were, in many ways, highly orthodox figures, certainly if we consider their institutional roles as, variously, Cambridge Don, clergyman, professors at times of Economics, History and Commerce; however, they remained for various reasons outsiders—to an extent radicals, at least in their ambition—who engaged in a battle over the proper method for a major new discipline (i.e., economics) that would perhaps have ended up looking very different in the twentieth century if their approach had won the day. The reality was that to address economic questions from the standpoint of polities and their histories was a task being diverted into the new field of economic history, leaving economics proper free of such messy considerations.[36]

Much like the better-known Arnold Toynbee, both men were motivated—it is clear—by aspects of their experiences and personal dispositions. Ashley came from a working-class family and became a historian who was concerned to explain commerce and industry simultaneously from both above and below. As the vicar of Great St. Mary's in Cambridge, Cunningham was in close contact with a more diverse range of people than many Dons and was reputed to have known every child in the parish by name.[37] Both Cunningham and Ashley were interested in the Industrial Revolution in order to explain contemporary poverty. However, their histories cannot be seen as part of what has become known as the "catastrophist" tendency; that is, a uniquely pessimistic reading of industrialization. Rather, they were produced in the service of a broader interest in social conditions and the nature of work.[38] To this end, they interwove different types of source material in their texts, combining quantitative analysis with the history of ideas, often drawing on foreign-language authorities (such as the unknown Max Weber), which resulted in a cosmopolitan and erudite style. This was Cunningham's approach in 1882, and it remained so in 1905 and beyond. As against the approach being practiced by political economists, Cunningham stressed repeatedly the "interconnectedness" of industry with politics. One could indeed study one without the other, but this would be pointless, because the two were mutually determined to such a large extent. This was an approach we may wish to call "interdisciplinary"—not because Cunningham located himself at the boundaries of multiple fields, but because the boundaries of the fields shifted while he was standing still, carving up his unitary questions along new disciplinary lines. The direction of travel is palpable in Cunningham's plaintive request for political economists to recognize the uses of history and to be holistic in their attitude: he felt that "more need to insist on this interconnexion . . . because the fact seems to be so imperfectly recognised in ordinary histories. The manner of treatment conveys the impression that facts about industry and commerce can be easily distinguished from the rest, and dealt with in separate chapters; this can never be a thorough way of working."[39] This thoroughness that Cunningham called for—the recognition that a broad-based historicism was a prerequisite for economic understanding—was beyond the scope of the emergent Marshallian form of political economy, renamed merely *economics* for the twentieth century, in a move that Cunningham deplored.

What becomes clear from this examination of Cunningham and Ashley's treatment of industrial history is, first, a sense of the trajectory that took a mid-Victorian tradition of general social inquiry from Friedrich Engels to Arnold Toynbee, who, between them, coined the term "Industrial Revolution," to the treatment of that inquiry as a staple subject of historical discourse so familiar to students after the First World War—in the works of the Hammonds, R. H. Tawney, and others, through to the present flood of interest in the British Industrial Revolution (from all sides of the political spectrum). Just as the field of political economy had emerged from debates about the "machinery question" in the 1820s and 1830s, so this key strand of the historical discipline was forged as a form of social inquiry through the late Victorian interest in machinery.[40] In Ashley's case, in particular, it was *only* historical research that could link Toynbeean concerns about work and poverty (which had developed in the 1880s) with new fears about the deployment of machinery, which, by the 1910s, came to have real urgency in the face of increasing mechanization, engineering strikes, and the emergence of scientific management. It was Ashley's long-standing interest in the history of machinery—not previously recognized—that qualified him to write a preface to Edward Cadbury's seminal book on industrial organization in 1912 and to intervene in the famous transatlantic debates about Taylorism, Ford, and the engineering trades.[41]

Issues concerning the Industrial Revolution were, in the 1880s and 1890s, a cutting-edge field of historical research, which was constantly being updated in light of new information, as shown by the flurry of edition after edition of historical works. The publishing history of Cunningham's major text shows the ascent of that concept to prominence, notwithstanding the fact that its very coherence as a category was always being questioned. In addition to the many full editions, Cambridge University Press published in 1908 an extract from Cunningham's *Growth of English Industry* as a volume on its own. This volume was printed cheaply for a mass audience and comprised the two chapters, "Parliamentary Colbertism" and "Laissez Faire"—that is, the chapters that did most to undermine the historical narrative that straightforwardly coupled British liberty to its inventive genius. The title given to this short book by the press was *The Industrial Revolution*—a phrase that did not appear in the original table of contents but would appeal to the emerging audience with an interest in that issue. In a prefatory note, Cunningham outlined the particular reasons for the new edition. "The importance of studying the causes, through which the existing state of affairs has come into being, is being recognised by many of those who take an interest in the social and industrial questions of the present day."[42] The publication of this edition is notable not only for demonstrating the growing interest in the idea of the Industrial Revolution, but also for the implied suggestion that if readers only had time for one section of Cunningham's work, then this should be it.[43] If the idea of the "Industrial Revolution" did not exist at the time of the first edition of 1882, it had begun its ascent by the second edition of 1890, in which the expression appeared three times in the course of the book. The 1908 extract marked its growing currency by its title, *The Industrial Revolution*, and by the time of the sixth and final edition in 1917, the phrase appeared nineteen times in the index of the three-volume edition—at that time the most comprehensive English-language account of industrialization available.

Because Cunningham and Ashley were attempting to produce more holistic accounts of economic change than had been given previously, and because their focus was primarily on the period culminating in industrialization, they required some account of the machine that included its economic and historical function. To this end, we can see different conceptions of the machine developing in their work, not only as a historical actor and an object of study, but also as a figurative device and heuristic model for the project of historical explanation itself. The role of the machine sometimes appears as self-evident in their narratives, while at others they attempted a closer analysis; nonetheless, neither economics nor history alone were enough to account for its position, which remained ever-present and yet only partly examined. The focus on machinery was not in the service of prophecy or speculation; rather, it resulted from their thorough and holistic method. In this respect—unlike H. G. Wells and those lauded in public for their present and future insights, however wild—Cunningham and Ashley can be seen as uncertain prophets: circumspect and empirical as they grappled with a past that was quite uncertain enough.[44]

Both Cunningham and Ashley exemplify historians' ambivalence about the idea of the Industrial Revolution. While recognizing its crudeness, they returned to Toynbee's unifying concept when necessary, and it was used in marketing their work, even as they sought to question it in other ways. Both writers used the alternative phrase "industrial evolution" at different times and places, independently stressing the evolutionary nature of industrial change. The idea of evolution was omnipresent during this period, and so its appearance should come as no surprise; however, its use by Cunningham and Ashley as a countervailing metaphor in the discourse of economics can be added to the better-known instances of biological borrowing in the late nineteenth century.[45]

In a manner that perhaps ran contrary to his publisher's hopes, Cunningham opened the section of his magnum opus entitled "Laissez Faire" with a series of disclaimers. Even by the sixth edition of 1917, he claimed it was still too early to write the history of the Industrial Revolution, since it was "a new departure of which we do not even yet see the significance." He noted that "the time has not yet come to write the History of the Industrial Revolution in its broader aspects, for we only know the beginning of the story"; and, with relation to the changes themselves, he wrote, "We have no adequate means of gauging the rapidity and violence of the Industrial Revolution which occurred in England."[46] Despite this apparent historiographical nihilism, Cunningham endeavored to provide exactly such an account. As we have seen, his method was to stress the relatedness of the changes of 1800 to the state of contemporary affairs of 1900. "All the familiar features of our modern life, and all its most pressing problems, have come to the front within the last century and a quarter," and so the aim of his work was to do by honest historical graft that which prophecy pretended; that is, to make readers feel at home in the disorienting modern world.

Notes

1. Edward Slosson, ed. *Major Prophets of To-Day* (Boston, MA: Little, Brown, 1914); and *Six Major Prophets* (Boston, MA: Little, Brown, 1917). The "Major" in the title of this two-volume collection marked the editor's debt to, while also superseding, the "minor" prophets of the Bible.

2. Daniel C. S. Wilson, "Arnold Toynbee: The Science of History, Political Economy and the Machine Past," *History and Memory* 26, no. 2 (2014): 133–61.

3. Michael Bentley describes this period as the "long fin-de-siècle"; see "The Age of Prothero: British Historiography in the Long *Fin de Siècle*, 1870–1920," *Transactions of the Royal Historical Society* 20 (2010): 171–93.

4. For a British example, see Rosemary Elliott and Mark Freeman, "Economic History at Glasgow Before 1957," University of Glasgow, online at http://www.gla.ac.uk/media/media_213982_en.pdf, p3.

5. See William Cunningham, "The Perversion of Economic History," *Economic Journal* 2 (1892): 491–506.

6. The scientific status of economics was frequently debated at the British Association for the Advancement of Science, where Ashley and Cunningham made methodological interventions over several decades.

7. Works exploring such phenomena in Occidental settings include Bernhard Rieger, *Technology and the Culture of Modernity in Britain and Germany, 1890–1945* (Cambridge: Cambridge University Press, 2005); David E. Nye, *American Technological Sublime* (Cambridge, MA: MIT Press, 1994). For a similar project focusing on the colonial and developing worlds, see Michael Adas, *Machines As the Measure of Man: Science, Technology and the Ideologies of Western Dominance* (Ithaca, NY: Cornell University Press, 1990).

8. William Cunningham, *The Growth of English Industry and Commerce* (Cambridge: Cambridge University Press, 1882), 609.

9. Peter Mandler's chapter, "Looking Around the World," in this volume, outlines the uses of cross-cultural comparisons, the imperial historical imagination, and the role of such complex juxtapositions in the development of the human sciences.

10. Cunningham, *Growth of English Industry*, 613–14.

11. Cf. Karl Marx, *Capital: A Critique of Political Economy* (1867; London: Penguin, 1976), vol. 1, chap. 15.

12. W. J. Ashley, *The Economic Organisation of England: An Outline History* (London: Longmans, Green, 1914), 149.

13. Marx, *Capital*, 1: chap. 15. This question was addressed in the same terms in J. A. Hobson's popular book *The Evolution of Modern Capitalism*, which drew on both Ashley and Cunningham for its historical material; see Daniel C. S. Wilson, "J. A. Hobson and the Machinery Question," *Journal of British Studies* 54 (2015): 377–405.

14. Ashley here followed Cunningham's *Growth of English Industry* in using this description.

15. Max Weber is cited in the appendix as a source for this argument, but his work was not available in English translation at this time. Ashley was, therefore, an unheralded pioneer in bringing such insights to Britain.

16. Ashley, *Economic Organisation of England*, 160.

17. Ashley, *Economic Organisation of England*, 160.

18. Cunningham, *Growth of English Industry*, 295.

19. Cunningham, *Growth of English Industry*, 2.

20. See, most capaciously, Georges Canguilhem, "Machine and Organism," in *Incorporations (Zone 6)*, ed. Jonathan Crary and Stanford Kwinter (New York: Urzone, 1992).

21. Cunningham, *Growth of English Industry*, 3.

22. Cunningham, *Growth of English Industry*, 413–14.

23. This threat was most acutely associated with the British historian Thomas Henry Buckle, whose predictive histories drew on the work of French philosopher Auguste Comte.

24. *Report of the Meeting of the British Association for the Advancement of Science: Including Its Proceedings, Recommendations, and Transactions* (London: John Murray, 1891), 723 (henceforth *BAR*).

25. *BAR*, 1891, 724–25.

26. *BAR*, 1891, 724–25.

27. It was after the British Association meetings of 1889, 1890, and 1891 that methodological exchanges spilled acrimoniously into print; see, for instance, Cunningham, "The Perversion of Economic History," *Economic Journal* 2, no. 7 (1892): 491–506; and Marshall, "[The Perversion of Economic History]: A Reply," *Economic Journal* 2, no. 7 (1892): 507–19.

28. Editorial column, the *Times* (London), August 18, 1905, 7.

29. *BAR*, 1905, 468.

30. Cunningham returned in 1910 to the analogy between India and medieval England; see his "Presidential Address," *Transactions of the Royal Historical Society* 4 (1910): 5–6. The connection between temporality, timekeeping, and empire has been reprised by Simon Schaffer, who noted Henry Maine's claim that the colonial administrator was required to keep time using "two longitudes at once." The reason for this comment was Maine's belief that, whereas 250,000 Indians were living in the present, 250 million remained in the past. See Simon Schaffer, "An Antique Land," *Tarner Lectures*, Cambridge, February 23, 2010, online at http://upload.sms.cam.ac.uk/media/743664 (at 9'20").

31. *BAR*, 1905, 468.

32. *BAR*, 1905, 468.

33. *BAR*, 1905, 469.

34. This narrative in which the US is depicted as uninhabited, and native Americans as supine, has, of course, been contested widely. For one example, with special reference to the mythology of technology, see David E. Nye, *America As Second Creation: Technology and Narratives of New Beginnings* (Cambridge, MA: MIT Press, 2003).

35. See, for example, Cunningham, "Presidential Address," 18.

36. Donald Winch, "That Disputatious Pair: Economic History and the History of Economics," Discussion Paper: The Centre for History and Economics (Cambridge, 1997), http://www.histecon.magd.cam.ac.uk/docs/winch_disputatiouspair.pdf, accessed February 24, 2016)

37. Audrey Cunningham, *William Cunningham: Teacher and Priest* (London: SPCK Publishing, 1950), 86.

38. The label *catastrophist* was first used to describe geological theories that clung unempirically to theories of the Flood, and so the taint of irrationality is transferred to any historical interpretation to question the benefits of industrialization when the label is thus reprised.

39. Cunningham, *Growth of English Industry*, p8.

40. Maxine Berg, *The Machinery Question and the Making of Political Economy, 1815–1848* (Cambridge: Cambridge University Press, 1980).

41. Edward Cadbury, *Experiments in Industrial Organization* (London: Longmans, Green, 1912).

42. William Cunningham, *The Industrial Revolution* (Cambridge: Cambridge University Press, 1908).

43. Cunningham appears reluctant, however, to endorse wholeheartedly the abridgement, ending his prefatory note by saying that "it may sometimes be convenient to use this reprint, in classes or otherwise, along with the complete work."

44. The preceding chapter in this volume explores the work of William Henry Smith, who, writing earlier in the nineteenth century, exemplifies a predisciplinary version of the prophecy genre, which, in his case, was complex and idiosyncratic, and dramatizes his uncertainty about progress itself.

45. See, for example, John Burrow, *Evolution and Society: A Study in Victorian Social Theory* (Cambridge: Cambridge University Press, 1966).

46. Cunningham, *Growth of English Industry*, 6th ed. (Cambridge: Cambridge University Press, 1921), 609, 610, 613.

Acknowledgments

First and foremost, our thanks must go to Leverhulme Trust, whose generous grant funded the Cambridge Victorian Studies Group between 2007 and 2011. The grant was especially significant for the three PhD students and eight postdoctoral fellows among us—we were awarded time for research, interdisciplinary collaboration, and learning in ways that have shaped all our future work and lives.

Five members of the Cambridge Victorian Studies Group are not represented in this volume, though their contributions are felt on every page. Without James Secord, Anna Vaninskaya, and David McAllister as well as our incredible administrators, Denise Schreve and Hannah Spry, not only this book but the many monographs, collections, conferences, and essays that bear the names of members of the original project would look profoundly different. We owe each of them our thanks for their generosity, their feedback, and their friendship. In addition, we had an excellent advisory board committee: Crosbie Smith, Elizabeth Prettejohn, Sally Shuttleworth, and David Vincent all gave up time, ideas, and energy that shaped our work for the better, and we owe them special thanks.

During the five years we were funded by the Leverhulme, we were also privileged to be able to invite many speakers to workshops, symposia, and conferences. They are too numerous to mention, but for their critical contributions to our work, publications, and conversations we feel especially grateful to Anne Secord, Melanie Keene, Michael Wheeler, Ankhi Mukherjee, Vybarr Cregan-Reid, Sujit Sivasundaram, Elizabeth Kostova, Philip Hensher, Simon Schama, Gillian Beer, Matthew Kneale, Giles Waterfield, Leah Price, William St. Clair, Ruth Livesey, Alice Jenkins, Simon Dentith, Marcus Waithe, Timothy Larsen, Emma Francis, Caroline Arscott, Peter Galison, Ralph O'Connor, Robin Cormack, Indra Sengupta, Donna Yates, Melanie Hall, Edmund Richardson, Donald Malcolm Reid, Herbert Tucker, Mark Turner, and Josephine McDonagh.

Last but not least, our personal debts—also too many to mention. As editors, we wish to acknowledge (again) the many ways in which the research time, friendships, and intellectual collaborations afforded to us by the Leverhulme Trust and by the members of the Cambridge Victorian Studies Group were both career- and life-changing. But most important are the support and love of our parents, step-parents, and parents-in-law, our siblings, our partners, and our children: in particular ammi and abu ji, the late bhajii Sadiah, Sumayyah, Umar, Khalid, Sumera, Sidrah, and Sufyan Qureshi, Karrie and Cary Wood, Vince and Jo Garvin, Clive and Angela Buckland, Patrick, Isobel, and Orla Buckland, and magical Alice (whose turn it is to have a book dedicated to her, and who makes her mummy's heart sparkle).

Contributors

ADELENE BUCKLAND
>King's College London
>Department of English
>London

CLARE PETTITT
>King's College London
>Department of English
>London

DANIEL C. S. WILSON
>Alan Turing Institute
>British Library
>London

All contributors are based in the United Kingdom.

Contributors

DAVID GANGE
  Department of History
  University of Birmingham
  Birmingham

HELEN BROOKMAN
  King's College London
  Faculty of Arts and Humanities
  London

JOS BETTS
  Faculty of History
  University of Cambridge
  Cambridge

MARY BEARD
  Faculty of Classics
  University of Cambridge
  Cambridge

MICHAEL LEDGER-LOMAS
  King's College London
  Theology and Religious Studies
  Department of English
  London

PETER MANDLER
  Faculty of History
  University of Cambridge
  Cambridge

RACHEL BRYANT-DAVIES
  Department of Comparative Literature
  Queen Mary University of London
  London

SADIAH QURESHI
  Department of History
  University of Birmingham
  Birmingham

SIMON GOLDHILL
  Faculty of Classics
  University of Cambridge
  Cambridge

Illustrations

FW.1 "The New Zealander" x
1.1 Handbill, "Human Fossil" 5
1.2 Drawing of the "Human Fossil" 9
1.3 Aboriginal skull superimposed with a Neanderthal skull fragment 14
7.1 "Skelt's Favorite Horse Combats, No. 3." 127
7.2 "Hodgson's characters in the Giant Horse, Nos. 3 & 4" 132
7.3 "Further preparation of the Miller and his Men" 134
7.4 "Mr. Ducrow as the Roman Gladiator" 135
7.5 "Grand evening rehearsal of the Miller and his men" 137
7.6 "Madame Ducrow as Queen of the Amazons" 140
7.7 Playbill for *The Siege of Troy* 142
7.8 Plate 10 of Hodgson's characters in the *Giant Horse* 145
7.9 "Mr. Gomersal as Napoleon Buonaparte [Bonaparte]" 146
8.1 Unveiling the statue of John Bunyan, at Bedford 156
8.2 David Roberts, *Pilgrims Approaching Jerusalem* (1841) 162
8.3 William Holman Hunt, *The Miracle of the Holy Fire* (1892–99) 163
8.4 Bunhill Fields, London (2015) 169
10.1 Illustration of Poseidon, from "The Kings of Atlantis Become the Gods of the Greeks" 197
10.2 "At the Bottom of the Sea" 204
10.3 "A Walk Under the Waters" 206

10.4 "Life in the Primordial Sea" 208
10.5 "The Empire of Atlantis" 210
10.6 "The Diver in Search of the Atlantic Cable Gets into Hot Water" 214
10.7 Edward Burne Jones, *The Depths of the Sea* (1887) 216

Index

Abbeville, 7
Abbey Plain cross, 98–99
Aboriginal people, 13–16, 22n53, 24, 29, 34
Aborigines' Protection Society, 16
Academy, The (journal), 99, 186–87
Acton, Lord, 245
Adam (first man), 13, 69, 71, 75
ad fontes: abstract perfection and, 78; antiquity and, 72–73, 75, 79–80; Bible and, 68–76, 81; Christianity and, 69, 71–75, 78–79; Erasmus and, 68–69, 73, 75–76, 81; fossils and, 70; geology and, 69–72, 78, 80; Greeks and, 68–69, 72–73, 75; myths and, 68, 74; race and, 69; Renaissance and, 68–69, 71, 76; Romans and, 67–68, 72, 78; Tischendorf and, 73–74, 78, 81; Victorians and, 69, 77, 80–81; Wolf and, 78–80, 84n45

Aeneid (Virgil), 129, 137–39
Africa: civilization and, 24–25, 27, 29–30, 32, 34; East, 29; fakes and, 116; Marlow and, 218n43; origins and, 13, 18; slavery and, 25, 27, 29, 34; South, 18, 24, 27, 30; West, 29
afterlife, 176–82, 185, 188–89, 192, 230, 234
Agamemnon, 186
Age, The (newspaper), 139
agnostics, 162
Akenside, Mark, 53–54
Akhenaten, 189
Albert (prince), 180
Alfred the Great, 97
Also Sprach Zarathustra (Strauss), 192
Ancient Society (Morgan), 31
Andersen, Hans Christian, 213, 215
Anderson, Patricia, 138
Anglicans, 68, 72, 97, 155, 161, 168, 181, 190
Anglo-Saxons, 10; Bosanquet on, 93, 95–96; Cædmon

Anglo-Saxons (*continued*)
and, 87–100, 102n4, 102n10, 103n19, 104n26; Conybeare and, 92; Earle and, 97; epics and, 93–96; Freeman and, 88–89, 103n19; Gurteen and, 94; Oxford School of Literature and, 88–89; translation and, 93–96; use of term, 102n10
Annotations (Erasmus), 75
Antediluvian Child, 4–9, 11, 123n20
anthropology, 246; civilization and, 26, 28, 37–38; death and, 188; origins and, 3, 12, 16, 18, 21n34, 22n43, 23n56; pilgrimage and, 156, 158
Antiquary (Scott), 118
antiquation, 22n43
antiquity: *ad fontes* and, 72–73, 75, 79–80; Cædmon and, 97, 100; civilization and, 24, 30; English and, 94, 97, 100; fakes and, 108–9, 113, 115–16, 118, 121; geological, 3; Great Flood and, 3–4, 6, 13, 210, 260n38; interpreting, 11–15; origins and, 3–15, 18, 24; theaters and, 129–30, 136, 138–39, 144, 147–48; Victorians and, 24–25, 80, 94
Anti-Slavery Society, 16
Apostles, 74, 78, 159, 161, 171
archaeology: bones and, 6–11, 14, 18, 30, 33, 108, 116–17, 167, 179; civilization and, 30–33; death and, 185, 187; economic issues and, 253; fakes and, 108–12, 115–18; origins and, 3, 10, 12, 14–15, 18, 22n43; pilgrimage and, 156, 160–61, 172; Piltdown Man and, 111–12, 116; textual studies and, 87–88
arc of disillusionment, 28, 30, 34–36
Arnold, Thomas, 78
Arthur and George (Barnes), 176–77
Ascension Day, 159
Ashbee, C. R., 120
Asher, 186

Ashley, William, xxiv; assessment of, 254–57; catastrophists and, 254, 260n38; economic issues and, 245–50, 254–57; *The Economic Organisation of England*, 245, 248; Industrial Revolution and, 245–50, 254–57; machinery and, 245–50, 254–57; Marx and, 248–49; workers and, 245, 248–49
Assyria, xi
Astley's Amphitheatre, 126, 129–31, 136, 138–44
Athenaeum (newspaper), 138–39, 207, 234
Atlantis, 123n20, 197, 209–12
Atlantis (Donnelly), 197, 210–11
Austin, Alfred, 97
Aztecs, 34

Babylon, 27, 109
Bank of England, 115
Baptists, 167–68, 181, 190
barbarism, 11, 24, 26–27, 33, 36, 183, 247–48
Barkly, Henry, 187
Barnes, Julian, 176–77
Barnum, Phineas T., 115–16
Bartlett, William Henry, 160
Bastianini, Giovanni, 112, 114
Bathybius haeckelii, 205, 207–8
Battle of the Casts, 119
Battle of the Nile, 182
Battle of Waterloo, 146–47, 198–99
Baxter, Richard, 169
"Beachy Head" (Smith), 50
Becker, Carl Wilhelm, 115–16
Becket, Thomas, 164–65, 167
Bede, 86–87, 92, 100
Bedford, Duke of, 155
Ben Hur (Wallace), 79
Bennett, Alexander Morden, 171
Beowulf, 89, 106n84
Bergson, Henri, 212–13, 243
Bertram, Charles, 113
Bethany, 160
Bethlehem, 157, 163

Betts, Jocelyn, xxiii–xxiv, 118, 223–41
Bible, 59, 111, 192; Adam and, 13, 69, 71, 75; *ad fontes* and, 68–76, 81; Apostles and, 74, 78, 159, 161, 171; authenticity of, 108; Bodleian MS, 92; Book of the Dead and, 189; William Buckland and, 6, 20n14; calibration by, 3, 158; Catholics and, 164; chronology and, 6, 32, 71–73; Codex Bobbianus, 76; Codex Sinaiticus, 73; Codex Vaticanus, 76; Corinthians, 191; creation and, 110; as curriculum staple, xiii; David and, 97, 190; degeneration and, 31, 34; earliest manuscripts of, 76; Egyptian texts and, 189–90; Eichorn and, 79; Eve and, 13; fossil record and, 4; genealogies of, 71–72; Genesis, 86, 94, 210; Goldhill and, 123n20; Gospels and, 68, 73–76, 81, 160; Great Flood and, 3–4, 6, 13, 210, 260n38; Hebrews, 158; Hebrew vowel pointings and, 80; Humanist idealism and, 81; Inquisition and, 75–76; Jerusalem and, 243; Jesus Christ and, 78, 97, 157–59, 161, 166, 170, 172, 182, 223, 225, 232; John, 74, 159; King James, 76; Latin, 76; Mark, 76; Naville and, 185; Noah's Ark and, 3, 13, 210; original texts and, xix; Pentateuch, 80; pilgrimage and, 156, 158–61, 164, 172; prophecy and, 68, 223; Protestants and, 156, 164, 172; Psalms, 189–91; Resurrection and, 182; Revelation and, 25; Revised Version of, by Westcott and Hort, 76; Smith dictionary of, 161; Sodom and Gomorrah's destruction and, 108; Tischendorf and, 73–74, 78, 81; topography and, 158–60
Biographical History of Philosophy (Lewes), 234–35

Birch, Samuel, 185–86
Blackwood, John, 235
Blackwood's, 236–37
Blake, William, 170
Blavatsky, Madame, 211
Blindness and Insight (De Man), 60–61
Bodleian MS, 92
Boehm, Joseph Edgar, 155, 171
Boker, George Henry, 187
bones, 6–11, 14, 18, 30, 33, 108, 116–17, 167, 179
Book of the Dead, xxii, 180, 183–91
Book of the Dead, The (Boker), 187
"Book of the Dead and a Passage in the Psalms, The" (Cooke), 191
Book of the Dead (Budge), 188, 195n39
Bosanquet, William F. H., 93, 95–96
Botanic Garden, The (Darwin), 50, 56
Bournemouth, 167, 171
Bowker, Thomas H., 30
Boyle, Robert, 48
brain, 14, 33, 215
Brantlinger, Patrick, 29
Briefel, Aviva, 119
Bristol, Bishop of, 97
British Association for the Advancement of Science, 11, 252, 258n6
British Museum, xi, 31, 110–12, 116, 184, 186, 188, 190
Brixham Cave, 30
Brontë, Charlotte, 121
Bronze Age, 12, 33, 144
Brookman, Helen, xix–xx, 86–106, 109, 122
Brooks, Mary Abigail, 47
Bryant-Davies, Rachel, xx–xxi, 126–51
Buckland, Adelene, xiii–xxvii, 42–64, 107
Buckland, Francis, 8
Buckland, William, 6–8, 56–58, 82n11
Buddhists, 184, 188, 190–92
Budge, E. A. Wallace, 188, 195n39

Buffon, 48
Bulverhythe, 117
Bulwer-Lytton, Edward, 47
Bunhill Fields, 167–70
Bunyan, John, 155–56, 165–69, 171–72, 184
Burden of Nineveh, The (Rossetti), xi
Burke, Edmund, 26
burlesque, 122, 127, 129–30, 133, 145, 147, 148n2
Burnet, Thomas, 48
Bushmen, 24
Byron, Lord: *Cain*, 42; *Childe Harold's Pilgrimage*, 157, 198–201, 212–13, 216; Lucifer and, 42, 60; Neptunism and, 200–201; pilgrimage and, 161; Waterloo and, 198–99
Byzantium, 179

Cadbury, Edward, 255–56
Cædmon: Anglo-Saxon language and, 87–100, 102n4, 102n10, 103n19, 104n26; antiquity and, 97, 100; background of, 86; Bede on, 86–87, 92, 100; Bosanquet on, 93, 95–96; Brookman on, xix–xx, 86–101, 106n84; Christianity and, 96, 99–101; corpus of, 90–93; *The Dream of the Rood*, 93; English and, 86–101; epics and, 93–96; *Fall of Man*, 91; Gaskin on, 91, 96, 98–99; *Genesis*, 91–92, 94–95, 100–101; Gurteen on, 90, 92–96; Homer and, 87, 91–92, 94, 96; *Hymn*, 87, 91–92, 97, 101n1; lineages and, 96–100; local memorials and, 96–100; *Metrical Paraphrase*, 89, 91, 93, 95; Milton and, 87, 89, 92–96, 99, 101, 105n54, 105n59, 232; myths and, 86, 88, 91; Oxford School of Literature and, 88–89; poetry and, 86–101; race and, 91, 99; Thorpe on, 90–93, 101n1; translation and, 93–96; Victorians and, 89–91, 94, 100–101; Watson on, 90, 92–94, 97, 99
Cædmon cross, 97–99, 105n70, 157
Cædmon Manuscript, 92, 94
Cædmon: The First English Poet (Gaskin), 91, 98
Cædmon, The First English Poet (Watson), 97
Caesar, 10–11, 112–13
Cain (Byron), 42
Cambridge Victorian Studies Group, xi, xxv, 107
Cambyses, 53
Camden, 113
Canterbury Cathedral, 164–65, 167
capitalism, 88, 179, 223, 245, 249, 252
Caribbean, 29
Carlyle, Thomas, 165–66, 226–27, 245, 250
Carpenter, Edward, 236
Carpenter, W. B., 205
Carpenter, William, 203
Carthage, ix
Case, Thomas, 89
Castellani, Alexandro, 112
Castle of Otranto (Walpole), 118
catastrophists, 254, 260n38
Catholics, 75–76, 97, 114, 157–68, 171, 181, 187, 230, 232
Catlin, George, 4
Cat of Bubastes, The (Henty), 187–88
cave bears, 12, 41n37
Cave Committee, 31
cave-dwellers, 16, 34
Cave of Spleen, 46
Celestial City, 157, 166
Celts, 10–11, 13, 171
cemeteries, 98, 161, 179–80, 183
Chabas, 186
Chakrabarty, Dipesh, xiv
Chaldeans, 27
Chamberlain, Lord, 128
Chambers' Edinburgh Journal, 138
Champollion, 117
Chanson de Roland (French epic), 94
Chantrey, Francis, 8

Chapman, John, 225
Characteristics (Earl of Shaftesbury), 234
"Character of John Bunyan, Local, Ecclesiastical, and Universal, The" (Stanley), 155
Chatterton, Thomas, 114, 118
Chaucer, Geoffrey, 95–96, 99, 164–65
Chemical Essays (Watson), 48
Chesterton, G. K., 128, 133, 135–36, 243
Childe Harold's Pilgrimage (Byron), 157, 198–201, 212–13, 216
China, 24, 26, 108, 116, 121
Christians, xxii; *ad fontes* and, 69, 71–75, 78–79; Anglicans and, 68, 72, 97, 155, 161, 168, 181, 190; Apostles and, 74, 78, 159, 161, 171; Baptists and, 167–68, 181, 190; *Ben Hur* and, 79; Cædmon and, 96, 99–101; Catholics and, 75–76, 97, 114, 157–68, 171, 181, 187, 230, 232; Christian Socialism and, 228; civilization and, 26, 28–29, 32; creation and, 71; death and, 177, 182, 189–92; Dissenters and, 13, 155–56, 168, 170; forgeries and, 111; geology and, 69; German idealism and, 166; heaven and, 47, 49–50, 59–60, 101n1, 158, 179, 181; Hell and, 177, 181–82, 230; Jesus and, 78, 97, 157–59, 161, 166, 170, 172, 182, 223, 225, 232; Methodists and, 181; missionaries and, 15–16, 26–30, 38, 72; monogenesis and, 13, 21n34, 32, 69; pilgrimage and, 155–72; Presbyterians and, 40n27, 171, 181; Protestants and, 114, 156–58, 160–72, 177, 181, 187, 190, 193, 225, 249; Reformation and, 69, 72, 75, 158, 164, 166, 171, 225; Resurrection and, 182; William Henry Smith and, 228, 236; Stanley and, 155–60, 164–72; Tischendorf and, 73–74, 78, 81; Toleration Act and, 156; Unitarians and, 166, 170–71
Christmas, xxiv, 133
Church of the Holy Sepulcher, 77, 157, 161–63
Cirencester, Richard, 113
civilization: Aboriginal people and, 13–16, 22n53, 24, 29, 34; Africa and, 24–25, 27, 29–30, 32, 34; antiquity and, 24, 30; archaeology and, 30–33; arc of disillusionment and, 28, 30, 34–36; barbarism and, 11, 24, 26–27, 33, 36, 183, 247–48; Christianity and, 26, 28–29, 32; colonialism and, 28, 36–38; death and, 187; dehumanization and, 26, 34; Egypt and, 27; ethnography and, 31–32, 35, 37–38; fakes and, 107–8; fossils and, 24, 34–36; Greeks and, 27, 36; high cultures and, 24; Indians and, 4, 10, 15, 25, 27, 36–37, 38n4, 41n49, 259n30; Lubbock and, 30–31, 33, 35; Minoans and, 107; prehistoric times and, 24–25, 30–37, 40n24; race and, 28; Romans and, 36; salience of difference and, 37–38; savages and, 13–14, 16, 19, 24–26, 29–35, 41n37, 89; Sepoy Mutiny and, 27–28; slavery and, 16, 21n40, 25, 27, 29, 34, 53, 228; social evolution and, 12, 31, 33–37, 40n26, 188; stadial theories and, 25, 27, 33, 36; stages of, 25–30; technology and, 25, 33–35; time/space concepts and, 24–25; Victorians and, 24–25, 28, 32; Xhosa and, 27–29, 39n12
Clark, Kenneth, 112
Cleopatra's Needle, 190
Cockerell, Sydney, 107–8
Codex Bobbianus, 76
Codex Sinaiticus, 73
Codex Vaticanus, 76
Coleridge, Samuel Taylor, 225, 237
Colet, John, 165

Collier, John Payne, 116
Collins, John Churton, 89
colonialism, 259n30; Catholic Requiems and, 187; civilization and, 28, 36–38; extinction and, 35; origins and, 4, 17; postcolonial theory and, xviii; race and, 17–18, 22n43, 102n10; Survival of the Fittest and, 17
comparative method, 4, 14, 94, 117, 185–86, 188, 192, 246, 252–53
Comte, Auguste, 209, 225–26, 235, 259n23
Comte de Gabalis, Le (Montfaucon de Villars), 47
Conrad, Joseph, 201, 218n43
consequentialism, 228
conservationists, 16–17, 119
Consolations in Travel (Davy), 42
Contemporary Review (journal), 191
Conway, Moncure Daniel, 165–66, 168, 171
Conybeare, J. J., 90, 92, 96
Conybeare, William Daniel, 56–58
Cook, James, 9, 26
Cook, Thomas, 139, 161
Corinthians, Bible book of, 191
Corn Exchange, 155
Costa, Giovanni, 114
Council of Trent, 76
Cox, Jeffrey, 26
Cranmer, 171
Crimea, 114
Cromwell, Oliver, 169
Crystal Palace, 139
cult of death, xxii, 178–80
culturalists, 33–35
Cumming, Lucy Caroline, 233
Cunningham, William, xxiv; assessment of, 254–57; catastrophists and, 254, 260n38; economic issues and, 245–57, 259n30, 260n43; Great St. Mary's and, 254; *The Growth of English Industry*, 247, 250–51, 256; "Laissez Faire," 257; machinery and, 245–57; "Unconscious Assumptions in Economics," 252–53; workers and, 245, 248–49
Curiosities of Natural History (Buckland), 8
Curtin, Philip, 29
Cuvier, Georges, 4, 6–7, 117

Dale, Langham, 30
Darwin, Annie, 177
Darwin, Charles: *The Descent of Man*, 16–17, 116; evolution and, 11, 16–17, 25, 32, 36, 71, 80, 116, 177, 202, 205, 207, 213; extinction and, 16–17; fossil gaps and, 202; *On the Origin of Species*, 16; race and, 16–17; sea studies and, 202, 205, 207, 213; Survival of the Fittest, 17, 23n58
Darwin, Erasmus, 50–60, 62n18
David (Biblical king), 97, 190
Davies, Rachel Bryant, 126–51
Davy, Humphry, 42, 54, 58–59
Dawkins, William Boyd, 15–16, 22n51
Dawson, Charles, 116–17
Dead Sea, 108
death: afterlife and, 176–82, 185, 188–89, 192, 230, 234; archaeology and, 185, 187; Book of the Dead and, xxii, 180, 183–91; Buddhists and, 184, 188, 190–92; cemeteries and, 98, 161, 179–80, 183; Christianity and, 177, 182, 189–92; civilization and, 187; cult of, 178–80; Darwin and, 177; disease and, 15, 46, 177, 179, 223, 230, 251; Doyle and, 176–77; Egypt and, xxiii, 176–93; epics and, 192; Greeks and, 179, 183, 185, 192; Hell and, 177, 181–82, 230; myths and, 185; philosophy and, 191–93; photography and, 178–79; pilgrimage and, 184–87; Resurrection and, 182; Romans and, 179, 186; Sufi and, 188; tombs and, 97, 137, 157, 161, 167, 179, 184, 186,

189; Victorians and, 176–84, 188, 192; Zoroastrians and, 184
de Christol, Jules, 7
"Deep-Sea Cables, The" (Kipling), 203, 215, 217n23
degeneration, 31, 34, 223, 242
dehumanization, 26, 34
De Man, Paul, 60–61
Descent of Man, The (Darwin), 16–17, 116
diachronic theories, 16, 18
Dickens, Charles, 178
Dictionary of the Bible (Smith), 161
Dictionnaire Raisonné de l'Architecture (Viollet-le-Duc), 117
Diehgen, 7
Dirks, Nicholas, 28
disease, 15, 46, 177, 179, 223, 230, 251
Disraeli, Benjamin, 67–68, 75, 95
D'Israeli, Isaac, 95–96, 100
Dissenters, 13, 155–56, 168, 170, 172
Dome of the Rock, 161
Donati, Lucrezia, 114
Donation of Constantine (forgery), 109
Donnelly, Ignatius, 197, 210–11
Doré, Gustave, ix–xii
Doyle, Arthur Conan, 176–77
Dracula (Stoker), 178
Dream of the Rood, The (Cædmon), 93
Ducrow, Andrew, 134–36, 138–41, 144
Dumas, Alexandre, 114

Earle, J. A., 97
Early Man in Britain (Dawkins), 15
Earthward Pilgrimage, The (Conway), 165–66
Eaton, Charles, 121
Ecclesiologist, The, 78
economic issues: archaeology and, 253; Ashley on, 245–50, 254–57; capital and, 88, 179, 223, 245, 249, 252; catastrophists and, 254, 260n38; Cunningham on, 245–57, 259n30, 260n43; *fin-de-siècle* and, 242–44, 258n3; industry and, 242–57; "Laissez Faire" and, 257; land ownership and, 36, 245; machinery and, 242–57; Marshall and, 246, 249, 255; Marx and, 248–49; modernity and, 36–37, 70, 114, 212, 243; myths and, 249, 259n34; political economy and, 36, 41n49, 224, 228, 243–46, 255; self-reliance and, 224; slums and, 72, 242; technology and, 243, 246, 248–50; Victorians and, 25, 242, 244–46, 250, 255; workers and, 8, 230, 243, 245, 248–49
Economic Organisation of England, The (Ashley), 245, 248
Egypt: Akhenaten and, 189; art and, 111; Book of the Dead and, xxii, 180, 183–91; cemeteries and, 179; civilization and, 27; Cleopatra's Needle and, 190; Cooke and, 190–91; death and, xxii–xxiii, 176–93; deities of, 108, 185; fakes and, 108, 111; Horus and, 185; Middle Kingdom of, 184; Mill and, 27; mummies and, 144, 147; Napoleon and, 182–83; Naville and, 185–87; New Kingdom of, 184–85; Nile River and, 30, 182, 186; Old Kingdom of, 184; origins and, 4, 10; Queen Tetisheri and, 111; Rameses II and, 189; theaters and, 144, 147; Thothmes II and, 189
Egyptian Hall, 4
Eichorn, 79
Eliot, George, 237
Embankment Gardens, 169
Engels, Friedrich, 242, 255
England's Darling (Austin), 97
English: Anglo-Saxon language and, 87–100, 102n4, 102n10, 103n19, 104n26; antiquity and, 94, 97, 100; Bosanquet and, 93, 95–96;

English (*continued*)
  Cædmon and, 86–101; Collins and, 89; Conybeare and, 90, 92, 96; formation of, 87–90; Gurteen and, 90, 92–96; Middle, 89; Milton and, 87, 89, 92–96, 99, 101, 105n54, 105n59, 232; Old, 87, 89, 91–95, 99–100, 105n52, 105n59, 106n84; Oxford School of Literature and, 88–89; Thorpe and, 90–93, 101n1; translation and, 93–96; Watson and, 90, 92–94, 97, 99
Enlightenment, 25–27, 30–31, 33, 37, 58, 60, 199, 224
Enon Chapel, 179
Epic of Gilgamesh, 192
Epic of the Fall of Man, The (Gurteen), 94
epics: Anglo-Saxon language and, 93–96; Cædmon and, 93–96; death and, 192; fakes and, 108; Homer and, 78–80, 87, 91–92, 94, 96, 129, 137, 139, 144, 147; Milton and, 93–96; theaters and, 126, 129, 137–38, 141; Wolf and, 79
Epistle to Dr. Arbuthnot (Pope), 48
Erasmus, Desiderius, 68–69, 73, 75–76, 81, 165–66
Eskimos, 15
Essay on Sepulchres (Godwin), 157, 160
ethnography, 31–32, 35, 37–38
Ethnological Society of London, 12, 21n34
Etruscans, 52–53, 112–13
Evangelical Revival, 181
Evans, Arthur, 108
Evans, John, 115–18
Eve (first woman), 13
evolution, xvii; Darwin and, 11, 16–17, 25, 32, 36, 71, 80, 116, 177, 202, 205, 207, 213; fakes and, 108–9, 111–12, 116–17; fossil gaps and, 202; glaciation and, 11; Haeckel and, 205, 208, 218n31; Heidelberg Man and, 116; Java Man and, 116; Lyell and, 11, 25, 56, 71, 80; missing link and, 111–12, 116, 203; natural selection and, 17, 177; Neanderthals and, 11, 14, 30; origins of humans and, 3–4; Piltdown Man and, 111–12, 116; primeval protoplasm and, 205–9; sea studies and, 196, 205–9, 213; social, 12, 31, 33–37, 40n26, 188; Spencer and, 23n58, 32, 224, 235; Survival of the Fittest and, 17, 23n58
Exeter Book, 91
extinction, 202; Aboriginal people and, 16; bones and, 6–7, 11, 18; Carpenter and, 203; colonialism and, 35; Cuvier and, 6; Darwin and, 16–17; deists and, 6; exterminationists and, 16; fossils and, 4, 6–7, 11–12, 35; genocide and, 17–18, 23n61, 29; human causes of, 5; mammoths and, 6, 11–12; Mauritian dodo and, 5; as natural process, 5–6; origins of humans and, 3–7, 11, 15, 17–18; race and, 16–17; sea studies and, 203; theists and, 5–6

Fabian, Johannes, xiv
Faculty of Classics, xi
fakes: Africa and, 116; antiquity and, 115; archaeology and, 108–12, 115–18; authentic forgeries and, 118–22; Christianity and, 111; Cockerell and, 107–8; counterfeiters and, 108; Egypt and, 108, 111; epics and, 108; Arthur Evans and, 108; John Evans and, 115–18; evolution and, 108–9, 111–12, 116–17; forgeries in historical culture and, 111–15; Greece and, 107–12, 121; Moabite sculptures and, 111; multiplication of, 107–11; myths and, 113, 121; numismatics and, 115–18; prehistoric times and, 111–12, 116, 119; race and, 109, 111, 116–17; Rome

and, 112–13; scientific, 115–18; Victorians and, 107–14, 120–22
Fake? The Art of Deception (exhibition), 110
Fall of Man (Cædmon), 91
Fall of Man, The (Bosanquet), 95
Fergusson, James, 161
Figaro in London, The (newspaper), 145
Fiji, 29
fin-de-siècle, 242–44, 258n3
Fingal's Cave, 113
Fitzgerald, Edward, 191
Fitzwilliam Museum, 107
Flambard, Bishop, 171
flint, 7, 117
Folio of Shakespeare (forgery), 116
Forbes, Edward, 215, 217n14
Forster, E. M., 243
fossils: *ad fontes* and, 70; Antediluvian Child and, 4–9, 11, 123n20; civilization and, 24, 34–36; Cuvier and, 4, 6–7; Darwin and, 202; evolution and, 202; extinction and, 4, 6–7, 11–12, 35; gaps in record of, 202; geology and, 42–46, 49, 59; Kent's Hole and, 6; mammoths and, 6, 11–12; origins and, 4–14, 19, 24, 34–36, 42–46, 49, 59, 70, 202, 212; petrified, 6–11; pilgrimage and, 165; Red Lady of Paviland and, 6–7; sea studies and, 202, 212
Fossil Spirit, The (Mill), 42–43
Founder's Window, 171
Fox, Caroline, 237
Fox, Kate, 211
Fox, Robert Barclay, 237
Fraser, James, 192
Freeman, E. A., 77–78, 88–89, 103n19
French and Indian War, 25–26
Freud, Sigmund, 192
Frith, Francis, 159
Froude, 165

Galileo, 34
Gange, David, xxii, 40n24, 176–95
Garden of Eden, 69
Gaskell, Elizabeth, 121
Gaskin, Robert Tate, 91, 96, 98–99
Gasquet, J. R., 215
gender, 109, 111, 121, 127, 211
genealogy, 67–68, 71, 75, 80, 110
Genesis (Cædmon), 91–92, 94–95, 100–101
Genesis, Bible book of, 86, 94, 210
genocide, 17–18, 23n61, 29
Geoffrey of Monmouth, 113
Geological Evidences of the Antiquity of Man (Lyell), 11
Geological Society, 5, 8, 9, 11, 43, 58
geology: *ad fontes* and, 69–72, 78, 80; British popularity of, 43; Buckland and, 56–58; Christianity and, 69; civilization and, 42–46, 49, 59; Conybeare and, 56–58; Darwin and, 50–60, 62n18; De Man and, 60–61; Ethnological Society of London and, 12; gnomes and, 43–60, 63n25, 63n26; Great Flood and, 3–4, 6, 13, 210, 260n38; Greeks and, 43; Hutton and, 50–51, 58, 60; Lyell and, 44–47, 50–51, 56–61, 70–71, 201; myths and, 43; occult spirits and, 46–50; Paracelsus and, 47, 50–51; Pope and, 46–48, 54; Porden and, 54–59; race and, 61; Rosicrucians and, 47–48, 51, 54, 62n12, 62n18; Scafe and, 56–59; sea studies and, 196, 200–203, 205, 207, 209–12, 216n9; technology and, 47; Victorians and, 61; writings on, 42–43, 46–60
German idealism, 166, 229
Gethsemane, 162
Giant Horse (Hodgson), 129, 131–33, 136, 138–45
Gilgamesh, 192
glaciation, 11
gnomes, 43–60, 63n25, 63n26
Godwin, William, 47, 157, 160, 167

INDEX

Goethe, Johann Wolfgang von, 115
Golden Bough (Fraser), 192
Goldhill, Simon, xi, xix–xx, 67–85, 87, 91, 109, 122, 123n20, 157, 160, 165
Golgotha, 161, 168
Gollancz, Israel, 97
Gomorrah, 108
Gordon, Charles, 161
Gospels, 68, 73–76, 81, 160
Gothic style, 76–77, 99, 111, 114, 165, 171, 179
Grand Tour, 139
Great Exhibition, 139
Great Flood, 3–4, 6, 13, 210, 260n38
"Great Sea Serpent, The" (Andersen), 213, 215
Greeks: *ad fontes* and, 68–69, 72–73, 75; civilization and, 27, 36; death and, 179, 183, 185, 192; fakes and, 107–12, 121; geology and, 43; marble statues and, 107; origins and, 10; Plato, 209–11; sea studies and, 197, 210; Smith and, 235; theaters and, 130, 139–44, 147
Green, J. R., 245
Growth of English Industry, The (Cunningham), 247, 250–51, 256
Guadeloupe, 7
Gurteen, Stephen Humphreys, 90, 92–96
Gwilym, Dfydd ap, 113

Haeckel, Ernst, 205, 208, 218n31
Haggard, H. Rider, 187
Hall, Catherine, 27
Hamilton, William, 229–30, 235–36
Hammond, Barbara, 255
Hammond, Lawrence, 255
Hand-book for Travellers on the Continent, A (Murray), 139
Harbledown, 165
Hardy, Thomas, 70, 192
Hayley, William, 50
Hearne, 15

heaven, 47, 49–50, 59–60, 101n1, 158, 179, 181
Hebrews, Bible book of, 158
Hegelians, 165
Heidelberg Man, 116
Hell, 177, 181–82, 230
Henty, G. A., 187–88
Hieroglyphic Tales (Walpole), 118
Hilda, 97
Hilton, Boyd, 224
Hisarlik, 139
Historia Regum Brittaniae (Geoffrey of Monmouth), 113
Historical and Critical View of the Speculative Philosophy of Europe (Morell), 235
Historical Memorials of Canterbury Cathedral (Stanley), 164
historicism, xiv, 119–20, 215, 218n42, 249, 255
History of British India (Mill), 27
History of the Popes (von Ranke), x
HMS Challenger, 202–3, 207, 209–10, 216
HMS Cyclops, 205
Hodgson, Orlando, 129, 131–33, 136, 138–45
Hogath, William, 129
Holden, William, 28
Holy House, 163–64
Holy Land, 139, 157–62
Homer, 78–80, 87, 91–92, 94, 96, 129, 137–39, 144, 147
Horta, Victor, 120
Horus, 185
Hottentots, 24
Howard, John, 172
Hugo, Victor, 213
Hull, George, 115–16
Hume, David, 44–45, 52, 58, 60
Hunt, William Holman, 162–63
Hunt Cooke, John, 190–91
Hutton, James, 44, 50–51, 58, 60, 70–71, 200
Huxley, T. H., 14, 202, 205, 207, 218n36
Hyde Park, 139

Hymn (Cædmon), 87, 91–92, 97, 101n1
Hymn to the Aten, 189

Iliad (Homer), 129, 137–39
India, 27–28, 35–37, 38n4, 41n49, 253–54, 259n30
Indians, American, 4, 9–10, 15, 25–26
Industrial Revolution, 244–50, 253–57
industry: Cadbury on, 255–56; economic issues and, 242–57; entertainment, 144; historicism and, 119; machinery and, 242–57; moral progress and, 230; new classes and, 108; Toynbee and, 245, 253–55, 257; United States and, 253–54; workers and, 8, 230, 243, 245, 248–49
In Memoriam (Tennyson), 187
Inquisition, 75–76
Iron Age, 12, 33
Isenberg, Carl von (prince), 116
Isle of Wight, 97
Iuefankh papyrus, 185
Ivimey, Joseph, 167–70

Jameson, Robert, 200
Java Man, 116
Jenkin, Fleeming, 202
Jerusalem, 111, 120, 157, 159, 161–62, 167, 243
Jesus Christ, 78, 97, 157–59, 161, 166, 170, 172, 182, 223, 225, 232
Jevons, Stanley, 244
Jews, 68, 119, 159, 167
Johannine Comma, 75
John, Bible book of, 74, 159
John Bunyan Museum, 172
Jones, Chris, 88, 94, 99
Jones, Mark, 117
Julius Caesar, 11, 112
Junius, Franciscus, 92, 94
Juvenile Drama, 131, 133, 136, 144, 149n5, 150n16

Keats, John, 114, 157
Kenrick, Timothy, 170
Kent's Hole, 6
Khaib, 191
Kidderminster, 169
King Coal's Levee (Scafe), 56–58
King James Bible, 76
King Priam's Treasure, 139
King Saitaphernes, 114
Kingsley, Charles, 211–13
Kipling, Rudyard, 203, 215, 217n23
Kitchener, Lord, 111
Klein, Lawrence, 223

Lachmann, 73, 75, 79
"Laissez Faire" (Cunningham), 257
Lambert, David, 39n11
land ownership, 36, 245
Languedoc, 6
Lanney, William, 16
Latimer, 171
Latin, 76, 79, 86, 139, 164
Leader (newspaper), 225
League of Nations, 120
Lebour, G. A., 97
Ledger-Lomas, Michael, xxii, 85n49, 155–75
Leech, John, 133, 136, 141
Lees, James Cameron, 169
Lepsius, Karl Richard, 185–86
Lester, Alan, 39n11
Leverhulme Trust, xi–xii, 85n49
Lewes, George Henry, 225, 234–35
Lewis, C. S., 189
Life of Jesus (Renan), 78
Life of Mr. John Bunyan, Minister of the Gospel at Bedford (Ivimey), 167–68
Light, Alfred, 170
Linnaeus, Karl, 49
Literary and Philosophical Society, 99
Literary Gazette, 143
Livingstone, David, 29
Livy, 68, 78
Locke, John, 52
London Bridge, ix, xi
London Review, 236

Longfellow, Henry Wadsworth, 91, 101
Loti, Pierre, 161–63
Louvre, 114
Lubbock, John, 11–15, 30–31, 33, 35
Lyell, Charles: evolution and, 11, 25, 71, 80; *Geological Evidences of the Antiquity of Man*, 11; geology and, xvii, 44–47, 50–51, 56–61, 70–71, 80, 201; origins and, 7, 11–12

Macaulay, Thomas Babington, x–xi, 100, 168, 224
machinery: Ashley and, 245–50, 254–57; Cunningham and, 245–57; economic issues and, 242–57
Macleod, Norman, 160
Macpherson, James, 113, 118
Maine, Henry, 35–37, 41n46, 224, 259n30
Malory, Via, 113
mammoths, 6, 11–12
Mandler, Peter, xi, xviii, xxiii, 13, 15, 24–41, 127, 224–25
Man's Place in Nature (Huxley), 14
Mantell, Gideon, 42
Mantena, Karuna, 28, 36, 41n47
Maori, 108
Marble Faun, The, 119
Marie Antoinette, 109
Mark, Bible book of, 76
Marshall, Alfred, 246, 249, 255
Marshall, Henrietta Elizabeth, 113
Martin, Luther H., 218n42
martyrs, 81, 158, 170–71, 187
Marx, Karl, 248–49
Masorah, 80
Matabeleland, 17
materialism, 207–8, 211, 215, 228–30
Maurice, Frederick Denison, 225–26, 228, 235–36
Mauritius, 5, 187
Maya, 115
Mayor, J. E. B., 113
McLennan, J. F., 33–35
McNeil, Maureen, 63n26

medieval era, 77; Cædmon and, 88, 91, 94, 99, 102n10; death and, 179–80, 183; economic issues and, 245, 253, 259n10; fakes and, 108, 117–21; pilgrimage and, 164, 171; Smith and, 226; theaters and, 144
Mediterranean, 198, 202
Mehta, Uday, 26
Melanesian peoples, 29
Mendelssohn, Felix, 113
Mesopotamia, 108, 179, 184, 192
Methodists, 181
Metrical Paraphrase (Cædmon), 89, 91, 93, 95
Middle Ages, 91, 164
Middle East, 31, 38n4, 40n24, 73
Middlemarch (Eliot), 237
Mill, James, 27–28, 38n4
Mill, John, 42–43
Mill, John Stuart, 36–37, 224–29, 235, 237
Miller and His Men, The (production), 133, 141
Millingen, James, 112
Milton, John, 48, 50, 87, 89, 92–96, 99, 101, 105n54, 105n59, 232
Mine, The (Sargent), 47–48, 50
Miracle of the Holy Fire, The (Hunt), 162–63
Mirror of the Sea, The (Conrad), 201
missing link, 111–12, 116, 203
missionaries, 15–16, 26–30, 38, 72
Moabites, 111
Mobius, 207
modernity, 36–37, 70, 114, 212, 243
monism, 196, 208
monogenesis, 13, 21n34, 32, 69
Montfaucon de Villars, Abbé N. de, 47
Moral and Metaphysical Philosophy (Maurice), 235–36
Morant Bay uprising, 27
Morell, John Daniel, 235
Morgan, Lewis, 31–32
Morganwg, Iolo, 113
Morning Chronicle (newspaper), 140–41, 143

Morris, William, 119–20
Morris dancing, xv, xxiv
Morte d'Arthur (Malory), 113
Moulin Quignon, 116
Mount Ida, 137
Mount Posilippo, 137
mummies, 8, 108, 132, 144, 147, 165
Munch, Peter Andreas, 10
Murray, John, 139
Muthesius, Hermann, 120
myths: *ad fontes* and, 68, 74; Cædmon and, 86, 88, 91; death and, 185; economic issues and, 249, 259n34; fakes and, 113, 121; geology and, 43; origins and, 13; sea studies and, 196, 209–10; theaters and, 126–30, 139, 143–44, 147

Napier, A. S., 97
Napoleon, 145–47, 182–83, 199, 230
Napoleon III, 111
National Cast Museum, 120
National Museum of Denmark, 10
National Museums of Scotland, 20n23
Native Americans, 9
natural selection, 17, 177
Naville, Edouard, 185–87
Ndebele wars, 17
Neanderthals, 11, 14, 30
Nelson, Horatio, 182
Neolithic Age, 12, 15, 33, 117, 210
Neptunism, 200–201
New Hebrides, 29
Newman, John Henry, 164
Newton, Isaac, 34, 251
"New Zealander, The" (Doré), ix–xii
Niebuhr, Reinhold, 67–68, 71, 78
Nietzsche, Friedrich, 192
Nile River, 30, 182, 186
Noah's Ark, 3, 13, 210
nomads, 24
Normans, 70, 77, 117
Norse skalds, 91
North American Indian Gallery, 4
Novilos, M. de, 114
numismatics, 115–18

oceanography, 196, 201–2, 216
O'Connell, Daniel, 178
Ojibwa, 34
Old Curiosity Shop, The (Dickens), 178
On the Origin of Species (Darwin), 16
"On the Physical Basis of Life" (Huxley), 205, 207
Ordnance Survey, 113
Origin of Civilisation and the Primitive Condition of Man, The (Lubbock), 31
origins: Aboriginal people and, 13–16, 22n53, 24, 29, 34; Adam and, 13, 69, 71, 75; Africa and, 13, 18; antiquity and, 3–15, 18, 22n43, 24; archaeology and, 3, 10, 12, 14–15, 18, 22n43; colonialism and, 4, 17; debate over, 3–4; diachronic theories and, 16, 18; Egypt and, 4, 10; evolution and, 3–4 (see also evolution); extinction and, 3–7, 11, 15, 17–18; fossils and, 4–14, 19, 24, 34–36, 42–46, 49, 59, 70, 202, 212; Great Flood and, 3–4, 6, 13, 210, 260n38; Greeks and, 10; Huxley and, 14, 202, 205, 207; identity and, 4; Indians and, 4, 10, 15, 25, 27, 36–37, 38n4, 41n49, 259n30; Lubbock and, 11–15, 30–31, 33, 35; Lyell and, 7, 11–12; mammoths and, 6, 11–12; myths and, 13; Noah's Ark and, 3; petrified bones and, 6–11; polygenists and, 13, 21n34; prehistoric times and, 4, 11, 14–15, 22n51; race and, 4, 11–12, 15, 17, 19; savages and, 13–14, 16, 19, 24–26, 29–35, 41n37; skulls and, 7, 10–11, 14, 30, 33, 116; synchronicity and, 14, 16, 215; Three Age System and, 10–12, 33; tools and, 7, 11, 13, 18, 30, 33–35, 48, 111, 248; vanishing peoples and, 15–18; Victorians and, 17

Ossian, 91, 113
Our Island Story (Marshall), 113
Oxford School of Literature, 88–89

Palace of Westminster, 169
Paleolithic Age, 12, 15, 33–34
Palestine Exploration Fund, 111
Palgrave, Francis, 13, 91–92
Paracelsus, 47, 50–51
Paradise Lost (Milton), 48, 89, 94
Paris Salon medal, 114–15
parodies, 47, 54, 118–19, 127, 129
Past and Future of the Kaffir Races, The (Holden), 28
"Past vs Present in Victorian Britain" (Cambridge Victorian Studies Group), xi
Pattison, Mark, 226
Paviland Cave, 6
Peacock, Thomas Love, 47
Penelli, Enrico, 112
Penelli, Pietro, 112
Pengelly, William, 11
Pennines, 113
Penny Magazine, The, 138
penny publications, 131, 136–38, 141
Penny Satirist (magazine), 141
Pentateuch, 80
Pentland, 8
Perthes, Boucher de, 7, 30
Peru, 53, 108
Petrie, William Matthew Flinders, 189
Pettitt, Clare, xi, xxiii, 157, 196–219
Philp, John, 11
physiology, 32, 228, 230
pilgrimage, xvi; archaeology and, 156, 160–61, 172; Bible and, 158–61, 164, 172; Bunyan and, 155–56, 165–69, 171–72, 184; Byron and, 161, 198–201; Christianity and, 155–72; death and, 184–87; fossils and, 165; Holy Land and, 139, 157–62; Ivimey and, 167–70; Ledger-Lomas and, xxii, 85n49, 155–75; relics and, 155, 160, 165, 172; Romans and, 161; Stanley and, 155–60, 164–72; steamships and, xxii–xxiii; tourism and, 157–62; Victorians and, 156–62, 165, 170–72
Pilgrims Approaching Jerusalem (Roberts), 162
Pilgrims of the Nineteenth Century (Ivimey), 168
Pilgrim's Progress, The (Bunyan), 155–56, 172, 184
Pillars of Hercules, 209
Piltdown Man, 111–12, 116
Pisistratid recension, 79
Plagiaulax dawsoni, 116–17
Plato, 209–11
playbills, 128, 139–44
Playfair, James, 70–71
Playfair, John, 44
political economy, 36, 41n49, 224, 228, 243–46, 255
polygenists, 13, 21n34
Polynesians, 24, 26
Pope, Alexander, 46–48, 54
Popular Science Review, 210
Porden, Eleanor Anne, 54–59
pottery, 6, 30, 53, 121, 170
Prehistoric Time (Lubbock), 11–12, 14, 31
prehistoric times: civilization and, 24–25, 30–37, 40n24; fakes and, 111–12, 116, 119; origins and, 4, 11–12, 14–15, 22n51; sea studies and, 203
Presbyterians, 40n27, 171, 181
Price, Richard, 30
Primitive Culture (Tylor), 31
Primrose Hill, 183
Principles of Geology (Lyell), 44, 58, 70, 71, 201
Prolegomena to Homer (Wolf), 79–80, 84n45
prophecy, xxiv, 17, 56, 60, 68, 156, 223, 226, 243–44, 257–58, 260n44
Protestants, 114, 156–72, 177, 181, 187, 190, 193, 225, 249
Psalms, Bible book of, 189–91

Punch (magazine), xi, 146–47, 214
Puritans, 95, 167, 237
Pyramid General Cemetery Company, 183

Queen's College, 120
Qureshi, Sadiah, xvii–xviii, xxiii, 3–23, 30–31, 35, 107, 123n20, 225

race: Adam and, 13, 69, 71, 75; *ad fontes* and, 69; Cædmon and, 91, 99; chauvinistic narratives and, 88; civilization and, 28; colonialism and, 17–18, 22n43, 102n10; critical race theory and, 102n10; Darwin and, 16–17; dehumanization and, 26, 34; extinction and, 16–17; fakes and, 109, 111, 116–17; Gaskin and, 91; genocide and, 17–18, 23n61, 29; geology and, 61; origins and, 4, 11–12, 15, 17, 19; Survival of the Fittest and, 17, 23n58; theories of innate differences and, 21n40
Ramaswamy, Sumathi, 212
Rameses II, 189
Rape of the Lock, The (Pope), 46–47
Ratione Studii ac legendi interpretandique auctores, De (Erasmus), 68
Ravenna, 200
Rawnsley, Canon Hardwicke, 96–97
Red Lady of Paviland, 6–7
Red Sea, 202
Reform Act, 34, 168
Reformation, 69, 72, 75, 158, 164, 166, 171, 225
Rehbok, Philip, 207, 218n31
relics, 10, 18, 45, 155, 160, 165, 172
Religious Tract Society, 190
Renaissance, 47, 68–69, 71, 76, 89, 108, 112, 114, 155
Renan, Ernest, 78, 166, 232
Renouf, Peter le Page, 188
Researches into the Early History of Mankind and the Development of Civilization (Tylor), 31

restorationists, 77
Resurrection, 182
Revised Version of the Bible (Westcott and Hort), 76
Ridley, 171
Rilke, 192
Robert Elsmere (Ward), 78–79
Roberts, David, 162
Robertson, William, 26
Rodin, Auguste, 120
Romans, ix–x, 171; *ad fontes* and, 67–68, 72, 78; Caesars and, 10–11, 112–13; civilization and, 36; death and, 179, 186; fakes and, 112–13; Niebuhr and, 67; pilgrimage and, 161; Red Lady of Paviland and, 6–7; theaters and, 135, 138–39, 144, 147; travel guides and, 139
Rose Cross, 47
Rosicrucians, 47–48, 51, 54, 62n12, 62n18
Rossetti, Dante Gabriel, xi
Rouchomovsky, Israel, 114–15
Royal Institution Building, 9
Royal Society, 9, 62n18
Royal Surrey Zoological Gardens, 4
Rubaiyyat of Omar Khayyam (Fitzgerald), 191
Ruskin, John, 71, 114, 119–20
Russell Square, 183
Ruth (Gaskell), 121
Ruthwell Cross, 97

Sargent, John, 47–51, 54–55, 57–59
Sartor Resartus (Carlyle), 227
savages, 13–14, 16, 19, 24–26, 29–35, 41n37, 89
Savonarola, 114
Say, Jean-Baptiste, 230
Scafe, John, 56–59
Scaliger, Joseph, 71
Schaffer, Simon, 259n30
Schliemann, 139
Schmerling, Philippe-Charles, 7, 20n17, 30
Scott, 118

sea studies: Atlantis and, 123n20, 197, 209–12; Byron and, 198–201, 212–13, 216; chronology and, 196–97; Darwin and, 202, 205, 207, 213; electrical cables and, 201–5, 213–15; evolution and, 196, 205–9, 213; extinction and, 203; fossils and, 202, 212; geology and, 196, 200–203, 205, 207, 209–12, 216n9; Greeks and, 197, 210; Haeckel and, 205, 208, 218n31; *HMS Challenger* and, 202–3, 207, 209–10, 216; *HMS Cyclops* and, 205; Huxley and, 202, 205, 207, 218n36; marine zoology and, 196, 203, 207, 215–16; Mediterranean and, 198, 202; monism and, 196, 208; myths and, 196, 209–10; Neptunism and, 200–201; ocean bottom and, 201–4; oceanography and, 196, 201–2, 216; prehistoric times and, 203; primeval protoplasm and, 205–9; Red Sea and, 202; technology and, 202; undersea cities and, 209–13; Verne and, 205, 206, 209, 212, 214; Victorians and, 196; Wyville Thomson and, 202–3, 207

Secord, Jim, xi

Secret Doctrine, The (Blavatsky), 211

Selous, Frederick Courteney, 17, 23n58

Sepoy Mutiny, 27–28

Seven Lamps of Architecture (Ruskin), 119

Seven Years' War, 25–26

Seward, Anna, 50

Shaftesbury, Earl of, 234

Shakespeare, William, 95–96, 99, 108, 115–16, 157

Shakespeare Society, 116

Shapira, Moses, 111

Shaw, George Bernard, 192

Shelley, Mary, 167, 171

Shelley, Percy Bysshe, 47, 88, 167

Siege of Troy, The (exhibition): Astley's Amphitheatre and, 126, 129–31, 136, 138–44; combat scenes and, 126–27, 129, 138, 141, 144; revival of, 141; theaters and, 126–47

Simpson, Edward "Flint Jack," 117

Sinai and Palestine (Stanley), 159–60

Skeat, Walter, 97, 114

Skelt, Martin, 131, 135–36, 141, 143–44

skulls, 7, 10–11, 14, 30, 33, 116

slavery, 16, 21n40, 25–29, 34, 53, 228

slums, 72, 242

Smith, Charlotte, 50

Smith, William, 121, 161

Smith, William Henry, 260n44; background of, 225; Christianity and, 228, 236; didacticism and, 227, 232–33, 237; Earl of Shaftesbury and, 234; Greeks and, 235; Hamilton and, 229–30; hypocrisy and, 231–32; Lewes and, 225; Maurice and, 225–26, 228, 235–36; Mill and, 224–29, 235, 237; psychological viewpoint of, 228; reviews of, 236–37; Sterling and, 226, 237; *Thorndale*, xxiv, 118, 225–36

Smith Woodward, 117

social evolution, 12, 31, 33–37, 40n26, 188

Social Statics (Spencer), 224

Society for Promoting Christian Knowledge, 99, 190

Society of Antiquaries, 9, 90, 93, 102n4, 113, 116

Sodom, 108

Solinus, 113

Sommières caves, 7

South Place Chapel, 165

Southwark Fair (Hogath), 129

souvenirs, 126–30, 133, 136–38, 143, 145, 148

Speaight, George, 133

Spears, Robert, 170

Spencer, Herbert, 23n58, 32, 224, 235
"Spirit of the Age" (Mill), 226
spiritualism, 177, 211
stadial theories, 12, 19, 25, 27, 33, 36
Standard (newspaper), 143
Stanley, Arthur Penrhyn, 155–60, 164–72, 173n18
Stanley, Augusta, 155
Stephen, Leslie, 226
Sterling, John, 226, 237
Stevenson, Robert Louis, 131, 133–35
Stewart, Dugald, 26
St. George and the Dragon (performance), 133, 141
St. Giles' Cathedral, 169
St. Januarius, 164
Stoker, Bram, 178
Stone Age, 12, 14–15, 33–34
Stone Period, 10
St. Pancras, 167
St. Paul's Cathedral, ix, xi
St. Peter's, 167, 171
Strauss, David Friedrich, 78, 209
Strauss, Richard, 192
Street, G. E., 171
Stukeley, William, 113
Sufi, 188
Supreme Intelligence, 32
Survival of the Fittest, 17, 23n58
Swenson, Astrid, xx, 107–25, 160
Sybil (Disraeli), 67, 75
Sydenham, 139
synchronicity, 16, 215
syncretism, 192, 196, 208–12, 218n42
Synopsis of the Museum of the Society of Antiquaries of Scotland (Wilson), 9–10
System of Logic (Mill), 226

Table Mountains, 116
Tacitus, 113
Tahiti, 29
Talbot, William Henry Fox, 186
Tasmania, 13, 16, 22n53, 29
Tawney, R. H., 255
technology, 70; agency and, 248; civilization and, 25, 33–35; communication, 211; economic issues and, 243, 246, 248–50; geology and, 47; immortality and, 178–79; Industrial Revolution and, 244–50, 253–57; measured confidence and, 224; restorationists and, 77; sea studies and, 202; theaters and, 139; workers and, 243, 245, 248–49; writing, 79
Temple Church, 237
Ten Commandments, 189
Tennyson, Alfred, 97, 187
Tertiary species, 203
Thames, ix, 30, 121
theaters: antiquity and, 129–30, 136, 138–39, 144, 147–48; Astley's Amphitheatre, 126, 129–31, 136, 138–44; as brilliant jumble, 143–48; burlesque, 122, 127, 129–30, 133, 145, 147, 148n2; cheap knowledge and, 136–43; Ducrow and, 134–36, 138–41, 144; Egypt and, 144, 147; epics and, 126, 129, 137–38, 141; extraordinary feats and, 130–36; *Giant Horse* and, 129, 131–33, 136, 138–45; Greeks and, 130, 139, 141, 143–44, 147; Juvenile Drama and, 131, 133, 136, 144, 149n5, 150n16; *The Miller and His Men* and, 133, 141; myths and, 126–30, 139, 143–44, 147; parodies and, 127, 129; playbills and, 128, 139–44; Romans and, 135, 138–39, 144, 147; *The Siege of Troy* and, 126–47; Skelt and, 131, 135–36, 141, 143–44; souvenirs and, 126–30, 133, 136–38, 143, 145, 148; *St. George and the Dragon* and, 133, 141; technology and, 139; toy, 127–31, 133–36, 141, 143, 147, 149n5, 150n16; *Wallace, the Hero of Scotland* and, 130–31

Theatres Act, 128
Theory of the Earth (Hutton), 200
Theory of the Earth (Playfair), 44
Theosophical Society, 211
Thomas, Nicholas, 28
Thomsen, C. J., 10, 33
Thorndale (Smith), xxiv, 118, 225–36
Thorpe, Benjamin, 90–93, 101n1
Thothmes II, 189
Three Age system, 10–12, 33
Tiara of Saitaphernes, 114–15
Times (newspaper), 96–97, 99, 107, 138–39, 141, 143, 252
Tischendorf, Constantine, 73–74, 78, 81
Toilers of the Sea (Hugo), 213
Toleration Act, 156
Tolkien, J. R. R., 106n84
tombs, 97, 137, 157, 161, 167, 179, 184, 186, 189
tools, 7, 11, 13, 18, 30, 33–35, 48, 111, 248
tourism, 99, 137, 139, 157–62, 183
Tourist, The (penny paper), 137
Tournal, Paul, 6–7
Toynbee, Arnold, 245, 253–55, 257
toy theaters, 127–31, 133–36, 141, 143, 147, 149n5, 150n16
Tractarians, 165, 171
Transactions of the Ethnological Society of London, 12
Treatise of Human Nature (Hume), 44
Tremendous Trifles (Chesterton), 135–36
Trilby, 119
Troad, 139, 147
Trojan War, 126–27, 129–30, 137, 186, 198
Trouillebert, Paul Desire, 114
Trukanini, 16
Tulloch, John, 227, 236
Twenty Thousand Leagues under the Sea (Verne), 205, 206, 209
Tylor, Edward, 31, 33–35, 37–38
Tyndale, William, 169

"Unconscious Assumptions in Economics" (Cunningham), 252–53
Unitarians, 165, 166, 170–71
Ussher, James, 70
utopianism, 27, 35, 167, 230–31, 243

Valla, Lorenzo, 109
van Flekwyk, Herman, 75–76
Veils, The (Porden), 54
Vercelli Book, 91
Verne, Jules, 205, 206, 209, 212, 214
Vernon, Matthew X., 90
Victoria (queen), 158, 180
Victoria and Albert Museum, 97, 114
Victorians: *ad fontes* and, 69, 77, 80–81; antiquity and, 24–25, 80, 94; arc of disillusionment and, 28, 30, 34–36; Cædmon and, 89–91, 94, 100–101; civilization and, 24–25, 28, 32; death and, xxii, 176–84, 188, 192; economic issues and, 25, 242, 244–46, 250, 255; Enlightenment and, 25–27, 30–31, 33, 37, 58, 60, 199, 224; fakes and, 107–14, 120–22; fascination with past of, xiii–xviii; geology and, 61; origins and, 17; pilgrimage and, 156–62, 165, 170–72; progress and, 223–37; sea studies and, 196; Smith and, 223–26, 237
Vie de Jésus (Renan), 166
Village-Communities (Maine), 36–37
Villette (Brontë), 121
Viollet-le-Duc, Eugène Emmanuel, 77–78, 81, 117–18, 120
Virgil, 129, 137–39, 144
Voltaire, 167, 251
von Ranke, x

Wales, 6, 56–57, 113
Wallace, the Hero of Scotland (production), 130–31
Wallis, Henry, 114, 187
Walpole, Horace, 47, 118
Ward, Mrs. Humphry, 78–79
Water-Babies, The (Kingsley), 211–13

Waterhouse, Alfred, 97, 187
Watson, Richard, 48
Watson, Robert Spence, 90, 92–94, 97, 99
Weber, Max, 254, 259n15
Wedgwood, Josiah, 52–53
Weiner, J. S., 116
Wells, H. G., 243, 257
Wernerian theory, 200
Westcott and Hort, 76
Westminster Abbey, 155, 165, 168–70
Westminster Review, 207
Wex, Karl, 113
Whitby, 86, 96–99, 101, 157
Wightman, Edward, 170
Wilberforce, Bishop, 78
Wilde, Oscar, 178, 184
William III (king), 168
Williams, Edward, 113
Williams, John, 29
Willson, Thomas, 183
Wilson, A. E., 131
Wilson, Daniel, 9–10
Wilson, Daniel C. S., xxiii, xxiv, 242–60

Wilson, John, 237
Wolf, Friedrich August, 78–80, 84n45
Wollstonecraft, Mary, 167
Wonders of Geology (Mantell), 42
Woodward, B. R., 113
Woolf, Virginia, 42
Woolford, Louisa, 136
Wordsworth, William, 114, 199, 237
workers, 8, 230, 243, 245, 248–49
World War I, 180, 245, 255
Worsaae, J. J. A., 10
Wyville Thomson, Charles, 202–3, 207

Xhosa, 27–29, 39n12

Young England, 67
Young Troublesome (comic picture-book), 133, 136
York, Archbishop of, 97

zoology, 196, 203, 207, 215–16
Zoroastrians, 184
Zulu, 34

Printed in Great Britain
by Amazon